197
Topics in Current Chemistry

AF769727

Springer

Berlin
Heidelberg
New York
Barcelona
Budapest
Hong Kong
London
Milan
Paris
Singapore
Tokyo

Dendrimers

Volume Editor: F. Vögtle

With contributions by
V. Balzani, T. Butz, A.-M. Caminade, S. Campagna,
N. Feuerbacher, G. Greiveldinger, A. Juris,
J.-P. Majoral, V.V. Narayanan, G.R. Newkome,
P.B. Rheiner, A.-D. Schlüter, D. Seebach, H. Sellner,
S. Serroni, M. Venturi, F. Vögtle

Springer

This series presents critical reviews of the present position and future trends in modern chemical research. It is addressed to all research and industrial chemists who wish to keep abreast of advances in the topics covered.

As a rule, contributions are specially commissioned. The editors and publishers will, however, always be pleased to receive suggestions and supplementary information. Papers are accepted for "Topics in Current Chemistry" in English.

In references Topics in Current Chemistry is abbreviated Top. Curr. Chem. and is cited as a journal.

Springer WWW home page: http://www.springer.de
Visit the TCC home page at http://www.springer.de/

ISSN 0340-1022
ISBN 3-540-64412-1
Springer-Verlag Berlin Heidelberg New York

Library of Congress Catalog Card Number 74-644622

Printed in Germany

Cover design: Friedhelm Steinen-Broo, Barcelona; MEDIO, Berlin
Typesetting: Fotosatz-Service Köhler OHG, 97084 Würzburg

SPIN: 10649254 66/3020 – 5 4 3 2 1 0 – Printed on acid-free paper

Volume Editors

Prof. Armin de Meijere

Institut für Organische Chemie
Georg-August-Universität
Tammannstraße 2
D-37077 Göttingen, Germany
E-mail: ameijer1@uni-goettingen.de

Editorial Board

Preface

Undoubtedly, the chemistry of dendritic molecules has become more and more interesting and gained in importance as can be observed from an ever increasing number of publications on this topic in the chemical literature exceeding now 2000.

This volume of "Topics in Current Chemistry" aims to highlight major developments in the field and it shows the efforts of chemists from various subdisciplines of chemistry to design new dendritic molecules focusing on novel properties, functions, and potential applications. Hyperbranched/dendritic molecules containing silicon, phosphorus, and other elements are reported apart from hydrocarbon, carbohydrate and nucleic acid cascade molecules.

The present activity in the field reflects the unique capabilities available with dendrimers: Basically the whole world of chemistry can be conceived as a dendritic molecule in the shape of almost any molecular building blocks of organic, inorganic or biochemical origin. Practically every chemist should be able to enter the field of dendrimer chemistry or take advantage of it for his specific needs.

The development of a number of dendritic systems (Polyamines, PAMAMs, carbosilane dendrimers and others) in the early years of dendrimer chemistry has made available dendritic skeletons that can be used for further functionalizations. Building up from e.g. dendritic polyamine cores or from dendritic polyarylether "wedges" multiple functional groups can be introduced and thus *functional dendrimers* can be obtained. Many functions have been reported up to now targeting future applications and including supramolecular recognition/assembly processes, ion transport, guest enclosure, adhesion, catalysis, microstructuring, coatings, and diagnostics.

This Topics volume also shows that in the development of dendrimer chemistry there is still a need for efficient synthetic methods ensuring *multiple high-yield* conversion and leading to more or less pure (monodisperse, structurally perfect) dendritic molecules. Large "libraries" of imperfect substances are often formed at higher generations, only differing in small structural details.

Dendrimers are taking their space in the "tool box" of the modern synthetic chemist. *Dendritization* might offer solutions to problems yet unsolved. Dendritic wedges, i.e. "*dendryl" substituents* of well-chosen size and generation allow us to tune molecular properties like solubility, steric accessability of reactive sites, redox behaviour, and other features. Easy-to-make dendrimers and dendrons will thus become extremely helpful for any chemist in the covalent as well as in the supramolecular "world".

This Topics volume starts with a general survey on the development of repetitive/iterative synthesis in general (F. Vögtle et al.) and of dendritic molecules in particular during the last few years whereby reactions or inclusions in the inner side of dendritic molecules are emphasized (G. Newkome et al). The following contributions by J.-P. Majoral and D. Seebach et al. reflect further synthetic, stereochemical, and analytical aspects in the field of rigid, inorganic, and chiral building blocks, proceed to the higher-molecular-weight dendrimers ("dendripolymers") described by A.-D. Schlüter and and pass over to photophysical and electrochemical properties of designed dendrimers (V. Balzani et al.).

Chiral and switching dendrimers, holography and imaging aspects show that dendritic molecules are interesting for *applications in the material and biosciences,* being commercially available as synthetic building blocks, but also as contrasting agents. In the future, we expect more knowledge about *intra- and inter-dendritic interactions* like "dendritic crowding", antenna effects, self assembly to high molecular mass aggregates, surface modification, switching devices and catalysis.

The present volume gives a general and at the same time rather detailed review on main research developments in the field of dendrimers (*oligomer and polymer*) during the past several years, but also offers *views and visions of the future* – of what could soon be achieved in this area at the interface between small organic molecules and macromolecules (polymers). We are sure that the rapid development of fractal-shaped molecules will continue in academic institutes as well as in industry – there is still more to come.

This Topic issue might therefore be considered as a timely *update report* of the Newkome/Moorefield/Vögtle book on Dendritic Molecules published in 1996.

F. Vögtle, A. Archut
Kekulé-Institut für Organische Chemie
und Biochemie der Universität Bonn

May 1998

Contents

Contents of Volume 195

Biosynthesis
Polyketides and Vitamins

Volume Editors: F. J. Leeper, J. C. Vederas

Contents of Volume 196

Carbon Rich Compounds I

Volume Editor: A. de Meijere

Iterative Synthesis in Organic Chemistry

Nina Feuerbacher · Fritz Vögtle*

Kekulé-Institut für Organische Chemie und Biochemie der Universität Bonn,
Gerhard-Domagk-Str. 1, D-53121 Bonn, Germany. *E-mail: voegtle@uni-bonn.de*

Iterative synthesis is a useful method in organic chemistry, allowing automation of reaction sequences to synthesize nano-scale molecules. This methodology has recently gained importance as it is the prevailing synthetic strategy to build up dendrimers and other supramolecular systems. Iterative pathways open up an approach to the synthesis of monodisperse oligomers or polymeric precursors. Moreover, linear aliphatic molecules constructed of identical building blocks can be assembled. This contribution gives a brief overview of iterative syntheses with emphasis on organic preparative chemistry. In order to elucidate this principle some biological and biochemical examples are given. The scope and limits of selected iterative applications are discussed and compared to other methods.

Keywords: Dendrimers, linear aliphatic molecules, oligophenylenes, repetitive synthesis, supramolecular chemistry.

---

* Corresponding author.

Dedicated to Professor Achim Müller on the occasion of his 60th birthday.

List of Symbols and Abbreviations

ACP	acyl carrier protein
Boc	*tert*-butyloxycarbonyl group
CoA	Coenzyme A
Et	ethyl group
Iter.	iteration
S_N	nucleophilic substitution
Tos	tosyl group

1 Introduction

The repeated succession of similar reaction sequences for the stepwise synthesis of molecules which consist of repeated building blocks has been termed in literature as repetitive or iterative synthesis. To a molecule with a defined framework having one or more functional groups A (Fig. 1) an identical molecule unit is attached (Iter. 1). Subsequently the functional groups are then reobtained (Iter. 2) so that the reaction sequences can be repeated over and over with the molecule growing by one unit in each reaction cycle (Iter. 1 + 2). This growth

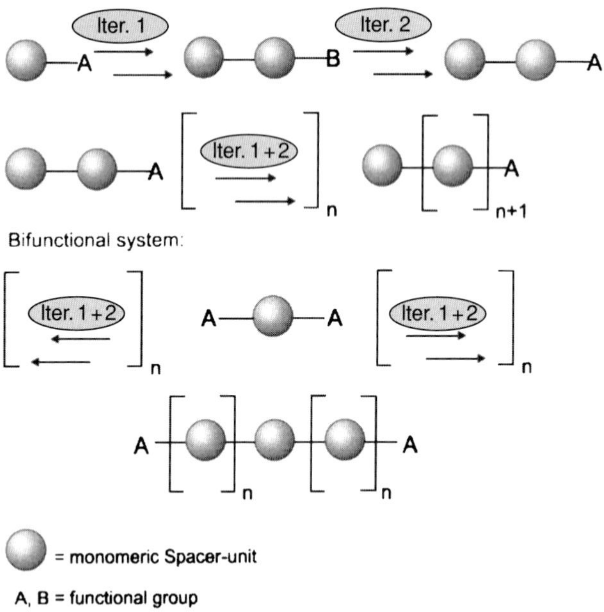

Fig. 1. Principle of iterative synthesis. In the first reaction step (Iter. 1) a monomeric unit, which may possess another functional group, is attached to the starting material. In the second step of the reaction sequence (Iter. 2) the functional group A is reobtained, so that Iter. 1 and 2 can be repeated

process can also be achieved exponentially in several directions depending on the number of functionalities at the initial molecule [1]. The present paper reviews this synthetic strategy, its application and compares it to other synthetic methodologies.

In mathematics the repeated application of an algorithm is called iteration (from Latin *iterare*, *iterum* again, i.e. "going the same way") where the solutions of one iteration step are the arguments for the following iteration [2]. A repetitive operation on the other hand describes the repetition of a calculation process with fixed initial conditions (from Latin *repetere* to seek again, i.e. "retrieving") [3]. Taking this into account the term "iterative synthesis" seems to be the proper term [4].

2
Biochemical Processes and Solid Phase Synthesis

Nature gives us some illustrative examples of iterative methodologies in its biochemical mechanisms. The fatty acid-polyketide biosynthesis is one of them. The assembly of acyl units by sequential *Claisen*-type condensations to form a polyketide or fatty acid takes place at a multi-enzyme complex, at which the initial molecule is lengthened by one C_2-unit per pass of a reaction cycle (Fig. 2).

The active centre of the enzyme consists of a cystein group and the sulfhydryl group of the ACP-phosphopanthein [5]. The enzyme is supposed to associate in a head-to-tail dimer, so the two active groups are placed side by side [6]. The cystein group is charged with an acetyl unit, the ACP-group with a malonyl unit. Subsequently an ester condensation takes place liberating the cystein group. In the three following reaction steps the β-ketone is reduced to the aliphatic chain and finally the molecule is transferred back to the cystein group. After this complete C_2-elongation the fatty acid can undergo the reaction cycle again. Sevenfold iteration leads to palmitic acid. This process was carried out biomimetically with tetramethylglycoluril as a template in 1994 [7].

Such biosyntheses were models for the *Merrifield*-synthesis [8] (Fig. 3), which culminated in the development of fully automated peptide synthesizers [9]. In a repeated reaction cycle a N-terminal protected amino acid, which is attached with its C-terminal end to an insoluble solid support, is deprotected, activated and lengthened by a second protected amino acid unit. The deprotecting and coupling steps can be repeated until the entire peptide is assembled.

The advantage of this method compared to classical peptide syntheses in solution rests basically in the the fact that purification becomes more simple. The starting reactants can be used in excess and can be removed simply by filtrating and washing, which means a substantial saving of time and increases the overall yield of the peptide. Today peptide syntheses up to a molecular mass of 10 kDa are possible. Nevertheless, oligomers with default amino acid sequences are difficult to separate from the desired polypeptides, which is still a problem. In the future, improved separation methods will allow the synthesis of larger peptides, so purification criteria of peptide-active substances can be satisfied [10]. The same repeated solid-phase methodology was used for the synthesis of oligosaccharides [11] and oligonucleotides [12].

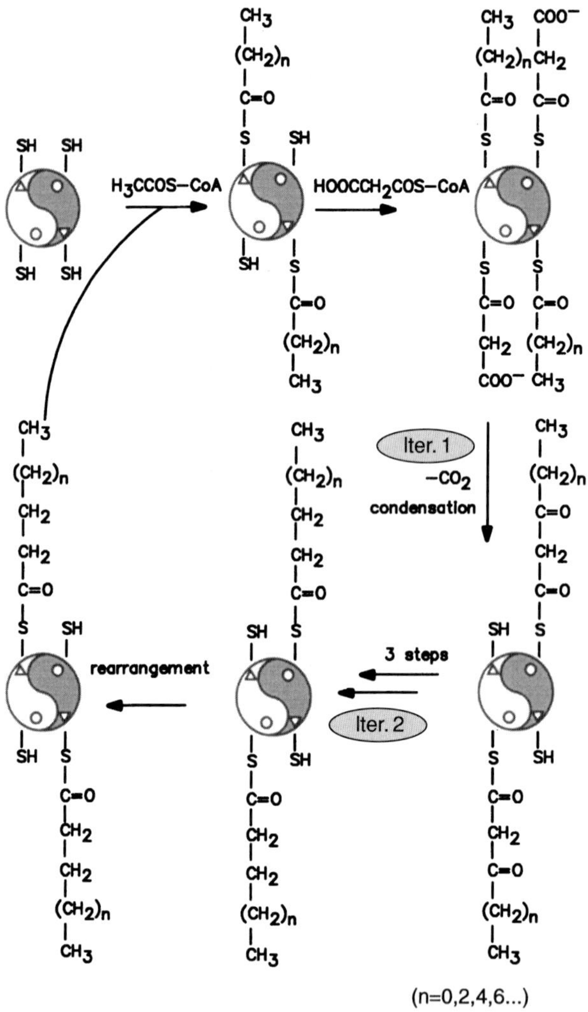

Fig. 2. Fatty acid biosynthesis (schematically)

Recently, a solid-phase synthesis was used iteratively for the synthesis of organic substances like oligocarbamates [13] and oligoureas [14] by repeated coupling to amino-functionalized supports. In this way substance libraries [15] have been developed showing that iterative methods can also be employed in combinatorial chemistry [16].

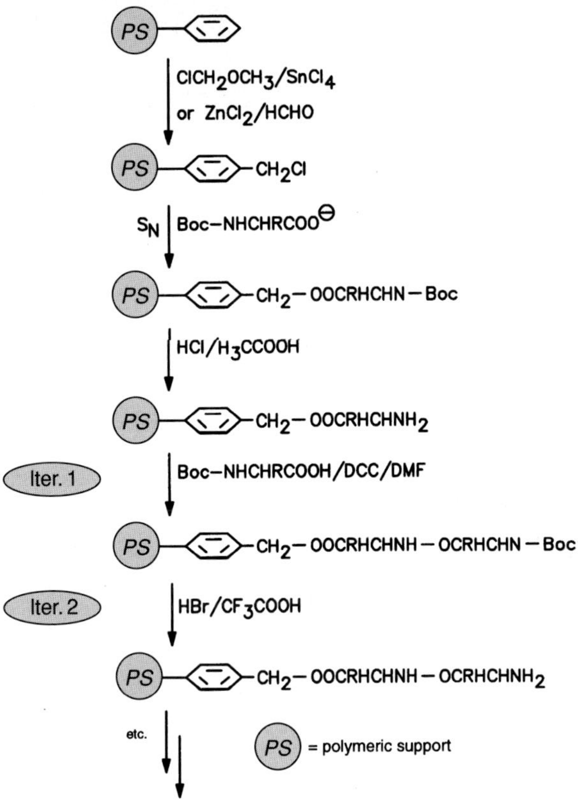

Fig. 3. Merrifield solid-phase synthesis (schematically)

3 Organic Synthesis

In 1978 we described a synthetic methodology as a "repeating-step principle", which led us to the first "cascade molecules", today known as dendritic molecules [17]. We recognized then that a synthetic pathway, which allows consecutive repetition, implies the advantage of likewise reactants and reaction conditions and is suited for the building of more or less structure perfect highly branched molecules, particularly of polyamines (Fig. 4).

Treatment of mono- and diamines with acrylonitrile (Iter. 1) yields the nitriles, which are then reduced (Iter. 2), recovering the initial functional groups and thus a repetition of the acrylonitrile addition is possible.

In the same manner we succeeded in the formation of multicyclic nonskid-chain-like polyaza compounds with an increasing number of macrocyclic chains per iteration step [17] (Fig. 5).

Monocyclic compounds with two amino functionalities like **1** were converted to the dinitriles **2** and subsequently reduced to the amines **3**. An intramolecular

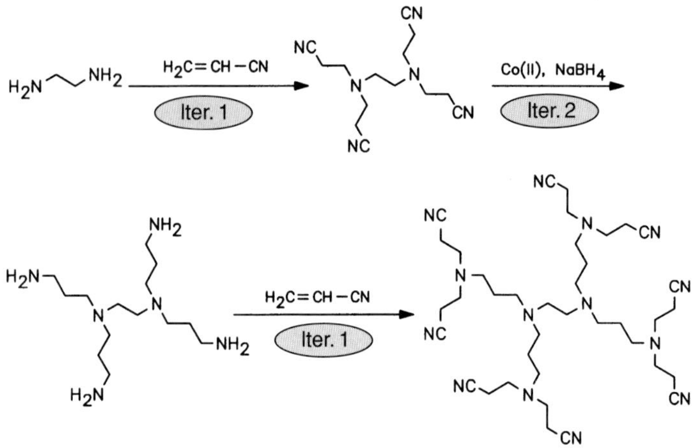

Fig. 4. Preparation of cascade molecules by a repeated addition of acrylonitrile to amines (according to Vögtle et al.)

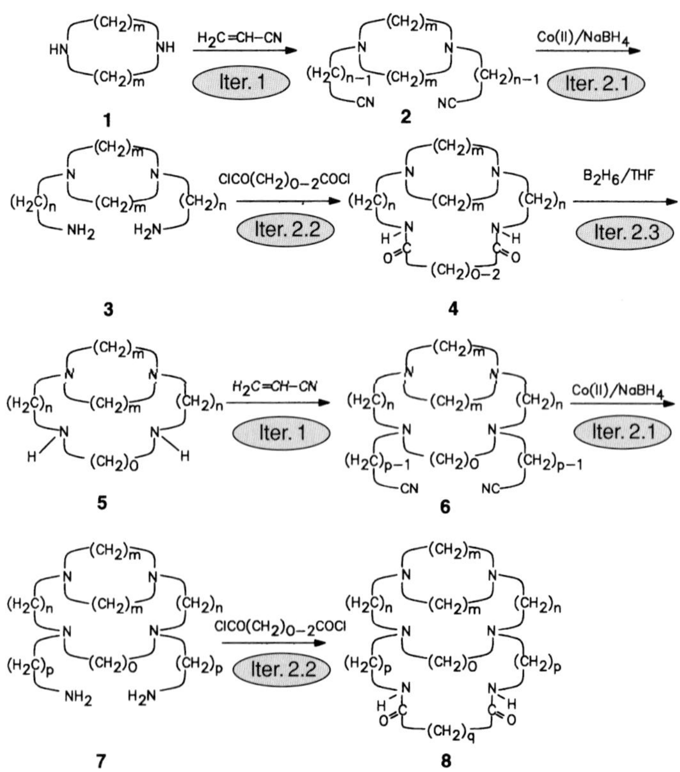

Fig. 5. Iterative construction of polycyclic molecules in a nonskid-chain-like manner

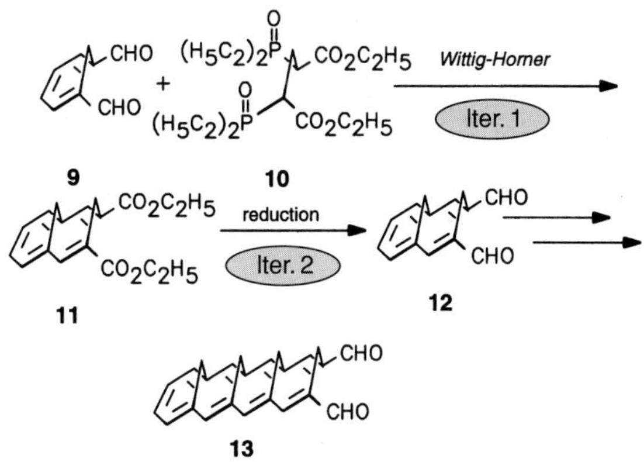

Fig. 6. Preparation of [4n + 2] annulenes according to a modular-constructive system

cyclization reaction of the two amino arms led, under dilution conditions, to the bicycle **4**. The final reduction afforded compound **5**, carrying again two amino functionalities, which can be subjected to the reaction sequence again, so that in only seven steps the macrotricycle **8** could be obtained.

Vogel et al. [18] reported the stepwise synthesis of [4n + 2]-annulenes according to a modular-constructive system (Fig. 6). Starting from the cycloheptatriene dicarbaldehyde **9** the authors prepared the higher homologues up to the dicarbaldehyde **13**, which already consists of four annelated cycloheptatriene rings. The dialdehydes were first reacted with the bifunctional *Wittig-Horner*-reagent **10** to afford the diester **11** (Iter. 1), and in two following steps the corresponding homologous dialdehyde **12** was obtained (Iter. 2).

4 Dendrimers

Our 1978 publication laid the foundation of the preparation of dendritic molecules, which have attracted considerable attention in the last decade in the field of supramolecular chemistry, but also in theoretical, physical, polymer, and inorganic chemistry due to their material properties as well as in biotechnology [1]. Such branched or even hyperbranched molecules called arboroles [19], cascade molecules [17], dendritic molecules or starburst-dendrimers [20] are constructed from identical monomeric building blocks carrying branching sites which are located in a spherical way around a core. The shells of monomers are called generations. At the periphery dendrimers can carry numerous functional groups that can finally lead to a surface congestion due to their sterical interactions (dense-packed stage or "starburst") [21] and thus render the molecules towards further reactions.

The synthesis of uniform dendritic molecules can proceed in two major iterative ways: the divergent-iterative pathway (Fig. 7), which was used in the early work in 1978, starts from an initial core with one or more functional

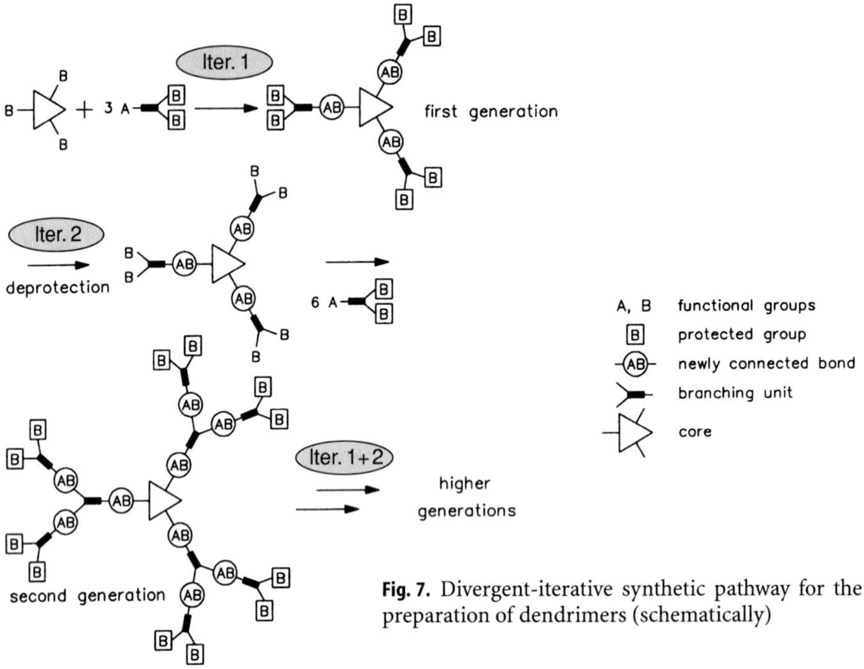

Fig. 7. Divergent-iterative synthetic pathway for the preparation of dendrimers (schematically)

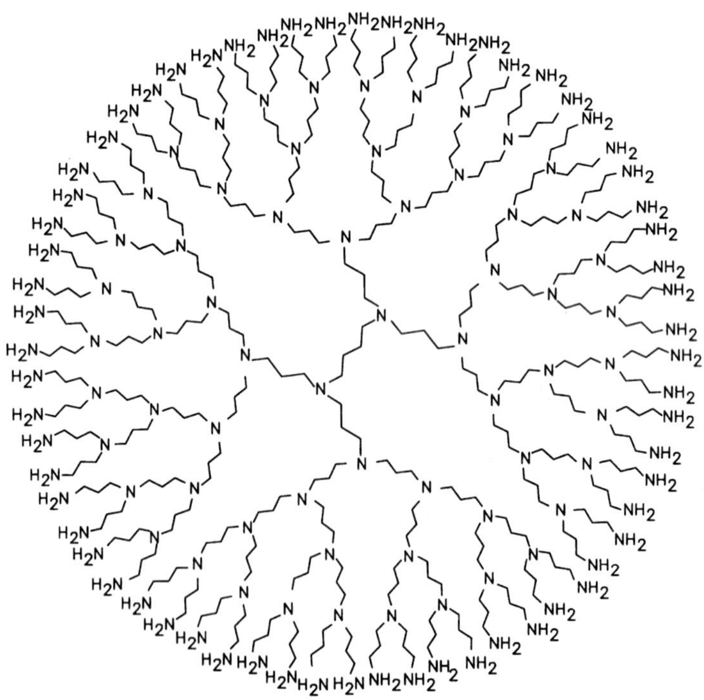

14

Fig. 8. Polyamine dendrimer of the fifth generation obtained on a kilogram scale groups. These are converted using monomers with protected reactive sites. The removal of the protecting groups and the repeated reaction with monomer units leads to an exponential increase of functional groups on the surface of the spherical molecule.

With this method new dendrimers were prepared in the following years by Denkewalter et al. [22], Tomalia et al. [23], Newkome et al. [24], and ourselves [25]. Following a reaction pathway similar to the one used in 1978, Meijer et al. successfully synthesized a polynitrile dendrimer **14** up to the fifth generation on a large scale [26] (Fig. 8).

A potential source of structural imperfection is the rapid increase of reactive groups as growth is pursued. Their incomplete conversion leads to defects inside the molecule [27]. In convergent-iterative syntheses these problems are avoided by directing the dendritic growth from the surface inwards to a focal point. In a final step several dendrons are connected with a multifunctional core to yield the desired dendrimer (Fig. 9).

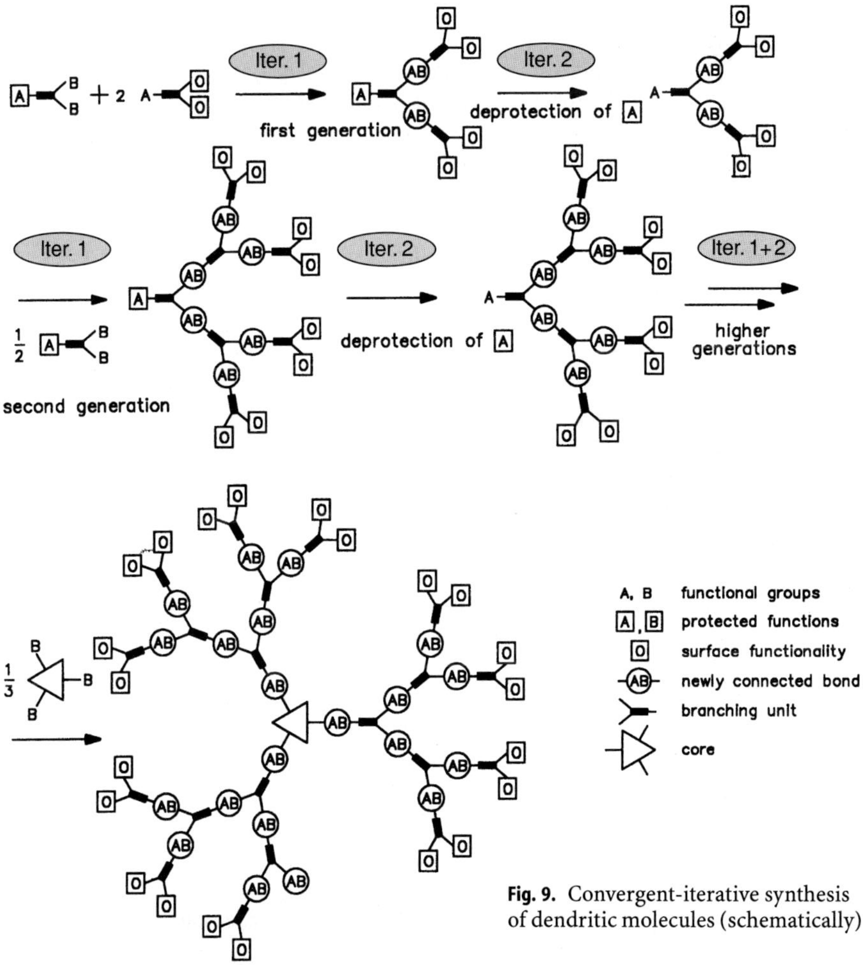

Fig. 9. Convergent-iterative synthesis of dendritic molecules (schematically)

A large family of new dendrimers has been synthesized following this divergent method. Hawker and Fréchet developed polyaryl(-benzyl)ether dendrimers [27], Miller and Neenan [28], and also Moore and Xu [29] prepared hydrocarbon dendrimers. The latter have reported the largest monodisperse organic hydrocarbon dendrimer with a molecular mass of 18 kDa and a diameter of 12.5 nm [30].

5
Ribbon and Belt Shaped Molecules

In supramolecular chemistry [31] iterative synthesis is frequently used for the construction of ribbon and belt shaped molecules [32]. The repeated Diels-Alder reaction is suitable for the preparation of beltenes, collarenes and cyclacenes containing partially or fully unsaturated annelated six-membered rings. They are potential molecular receptors for appropriate substrates. Stoddart et al. synthesized the cyclacene **18** by Diels-Alder cyclization of the bisdiene **15** and the bisdienophil **16** [33] (Fig. 10).

The obtained 2:1-adduct **17** can react with two molar equivalents of the bisdienophil **16**. The reaction sequence can be terminated by converting the 2:1-adduct **17** with one molar equivalent of the bisdienophil **16** resulting in the macrocyclus **18**, a considerable stereogenic uniform product with 16 stereo and 4 pseudostereo centres. The high stereoselectivity of this synthesis can be explained by the stereospecification, the regioselectivity, and the *endo* stereoselectivity of the Diels-Alder reaction, by the authors referred to as a "molecular lego set" [34]. By using this high-pressure reaction an acyclic nonadecacene derivative with 19 laterally fused 6-membered rings was isolated [35]. A similar strategy was followed by Klärner et al. [36] for the construction of a *Kohnken*-analogous molecule, bearing methylene instead of oxygen bridges, as well as for

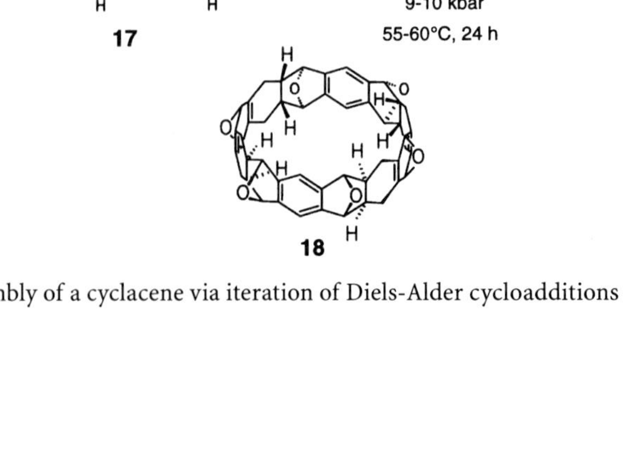

Fig. 10. Assembly of a cyclacene via iteration of Diels-Alder cycloadditions

the preparation of molecular tweezers [37]. Iterative Diels-Alder methods also provide an economical route to several linear precursors for [*n*]beltenes [38].

In the field of polymer chemistry the regio- and stereoselectivity of the Diels-Alder reaction is used for the concerted synthesis of structurally homogeneous double-stranded ladder polymers [39], which are useful materials with non-linear optical properties and high electrical conductivity. It has turned out that the repeated Diels-Alder method is superior to an alternative two-step process, in which first an open chain precursor is formed followed by polymer ring closure as structural defects can occur [40].

Schlüter et al. have isolated the beltene derivative **20**, which was developed as byproduct when synthesizing the ladder-type polymer **21** by a repeated Diels-Alder reaction (Fig. 11). The monomer **19** can act both as diene and as dienophil [41]. The iterative Diels-Alder method is however limited by the decreasing solubility of the products of higher molecular masses.

A different iterative method for the construction of ribbon and belt type molecules using tetrafunctionalized arenes and cyclophane monomers was developed by us [42]. The newly designed key compound **23** allows a repeated elongation of a ribbon type molecule by two benzene units (Fig. 12) [43].

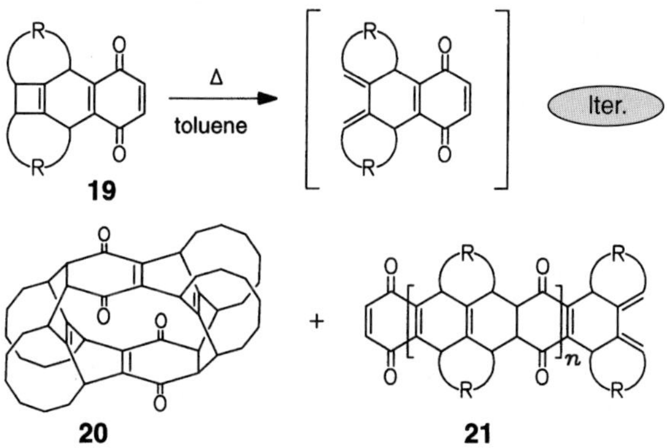

Fig. 11. Synthesis of the beltene **20**

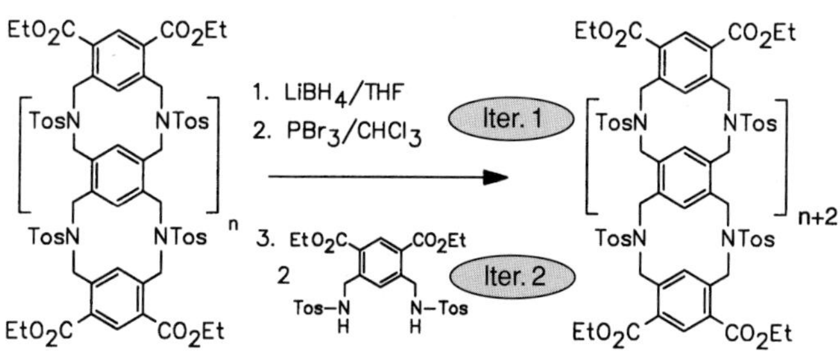

Fig. 12. Successive construction of molecular ribbons according to Vögtle et al.

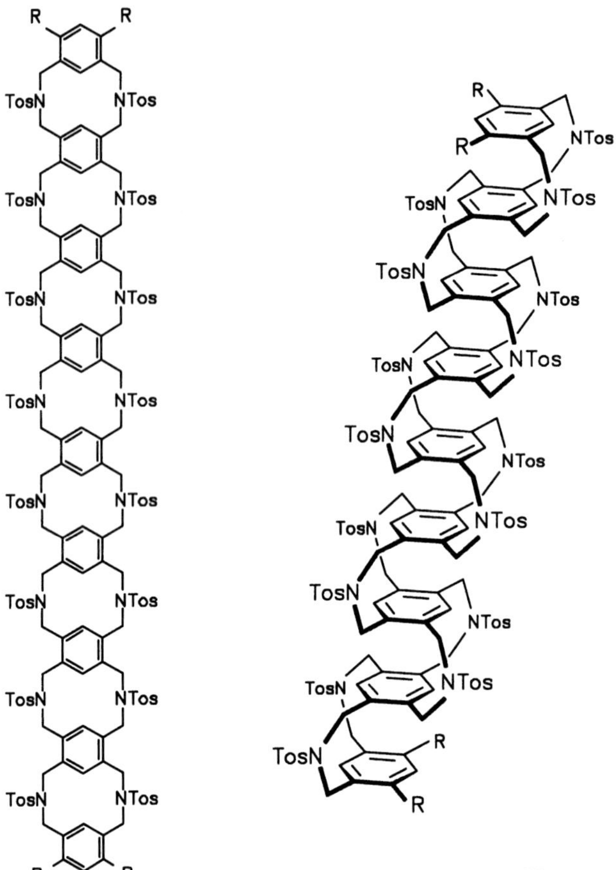

25 $R = CO_2Et$

Fig. 13. Nine-layered ribbon-type molecule synthesized by an iterative methodology. On the *right hand side* the stacked arrangement of the benzene units is shown

The reaction sequence starts with the reduction of the four ester groups of the cyclophane **22** yielding the corresponding alcohols. Subsequent bromination affords the tetrakis (bromomethyl) compounds (Iter. 1). Cyclization with two molar equivalents **23** provides the elongated cyclophane **24** with the initial four ester functionalities (Iter. 2). In twelve iterative steps we succeeded in preparing a structurally perfect benzenophane **25** containing nine benzene units forming a stacked structure of nanometer size as shown in Fig. 13 [44]. The preparation of even longer cyclo- [45] and pyridinophanes [46] might be possible as their solubility is good and appears to be independent of their length.

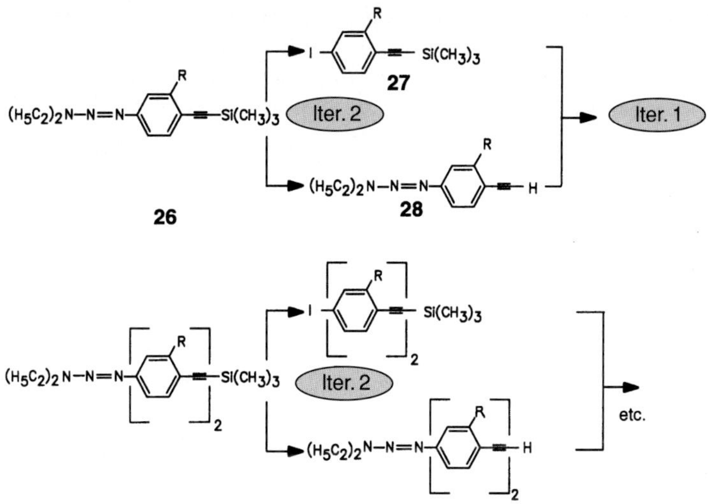

Fig. 14. Iterative convergent-divergent preparation of oligophenylenes according to Tour et al.

6 Oligophenylenes

In order to enhance the understanding of the properties in polymers, iterative pathways have been chosen for the synthesis of structurally perfect molecules. Data obtained from the analysis of precisely defined oligomers and polymers may relate chain length and conformation to physical, electronic and optical properties. Statistical polymerization processes are not suitable as they yield polydisperse material.

For instance, the repeated cross coupling of terminal acetylenes **28** with aryl iodides **27** using an appropriate protecting technique [47] leads to nanostructured molecular wires with 16 monomeric units (Fig. 14) [48, 49]. The efficiency of this iterative divergent-convergent methodology results from the complementarity of the trimethylsilyl and the 3,3-triazene protecting groups which can be removed selectively. By this means a chain growth of 2^n monomeric units is achieved after n-fold repetition of the reaction sequence [50]. Exponential growth of oligophenylenes by repeated *Suzuki* coupling of monomers is also possible with trimethylsilyl and bromo protective groups [51, 52]. Recently the stepwise synthesis of oligophenylene vinylenes forming mesophases has been published [53].

7 Linear Aliphatic Molecules

Synthetic pathways with repeating reaction sequences are also attractive for the preparation of linear aliphatic molecules constructed of identical building blocks. One example is polyspiro linkages with four-membered rings [54], which have been recently used as rigid building blocks in hydrophobic chains of tensides, leading to restricted micelle formation [55] (Fig. 15).

In three reaction steps the ditosylate **29** leads to the bistosylate of the monospiro[3.3]heptane **32**, which can be converted in successive iteration steps to polyspiranes of any length. Starting from the tetratosylate of the cyclobutane compound **33** the application of two iterative reaction sequences at the same time is possible. Nonaspiro compound **34** was thus synthesized in only twelve reaction steps [56].

Based on a known synthesis of spiro compounds with six-membered rings [63] we succeeded in the stepwise assembly of terminal substituted mono-[64] and dispiranes [65]. Hereby the cyclisation of the dibromide **35** with TOSMIC (Iter. 1) is the keystep of the reaction sequence as the initial ketone functional group (**36**) is recovered. Four subsequent reactions led to the spirodibromide **40** (Iter. 2). Final spirocyclisation afforded the dispirane **41**, representing a precursor for new calamitic liquid crystals (Fig. 16) [66].

Furthermore, iterative approaches are useful methods to construct polyhydroxy chains with 1,2- or 1,3-diol units of any length as chiral precursors for the synthesis of complex natural products [57] because automated synthesis becomes feasible. A preparation of *trans*-fused polytetrahydropyranes as structural unit for polycyclic ether biotoxines by repeated reaction sequences was recently named reiterative synthesis [58].

The iterative synthesis of bi- and tercyclohexyl derivatives as liquid crystals shows that this method is widely applicable in various fields of chemistry [59].

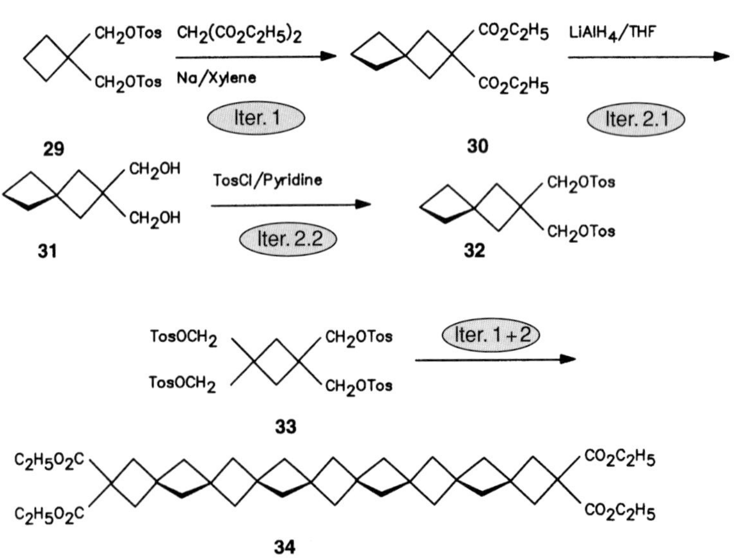

Fig. 15. Iterative pathway for polyspiro linkage of four-membered rings

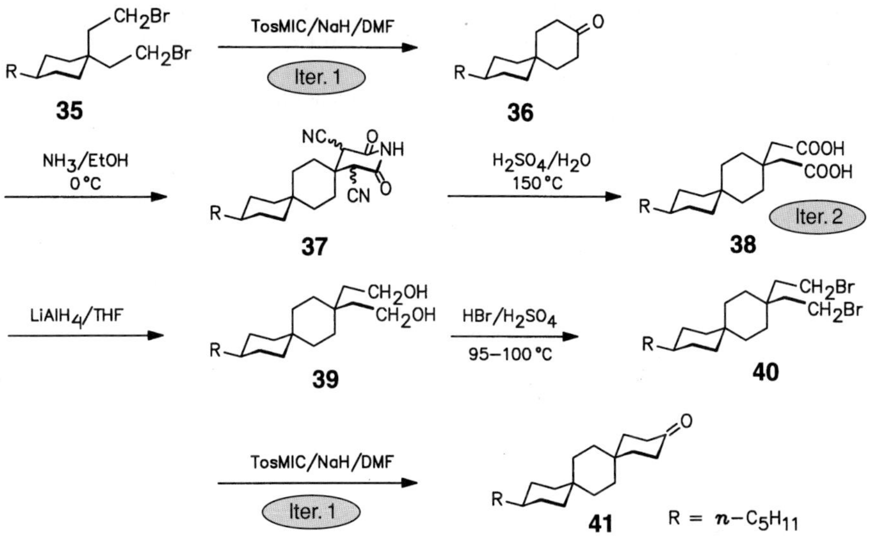

Fig. 16. Stepwise synthesis of terminal substituted dispiranes as precursors for new liquid crystals

8 Conclusions

Due to the rising interest in supramolecular structures and nanoarchitectures, as well as their broad applicability, including the field of combinatorial chemistry, iterative strategies will gain more and more importance in preparative organic chemistry [60].

The advantage of iterative strategies is based on the specific preparation of well defined structures and structurally perfect spacers of nanometer scale. This stepwise approach yields monodisperse material in contrast to other statistical routes. The use of the same reactants and the conversion of the same functional groups facilitates the synthetic effort compared with non-iterative methods.

Iterative procedures hold the potential of automated production as employed reliably and frequently in the case of the *Merrifield* synthesis. Furthermore, the specific isolation of intermediates and the precise insertion of substituents compensates the enhanced preparative effort compared to polymer syntheses.

Rapid growth can be achieved by improved iterative methods, especially following a convergent-divergent pathway. In future this will be the method of choice as it facilitates the separation of unreacted material and minimizes the total number of reaction steps needed for the synthsesis of large molecules [50, 61].

There are surely many more examples in preparative organic chemistry that follow the iterative synthesis principle for successive construction of larger molecules [62]. Yet the few examples described here illustrate its versatile applications.

9
References

1. Newkome GR, Moorefield CN, Vögtle F (1996) Dendritic Molecules, 1st edn. VCH, Weinheim
2. Schneider HJ (1991) Lexikon der Informatik und Datenverarbeitung, 3rd edn. Oldenbourg München, Wien
3. IBM (1985) Fachausdrücke der Informationsverarbeitung. C H Beck, Nördlingen
4. For this reason in the rest of the text the term "iterative" is used exclusively
5. Voet D, Voet JG, Maelicke A (ed), Müller-Esterl W (ed) (1992) Biochemie. VCH, Weinheim
6. Wahil SJ, Stoops JK, Joshi VC (1983) Ann Rev Biochem 52:568
7. Sun S, Harrison P (1994) J Chem Soc, Chem Commun 2235
8. Merrifield RB (1963) J Am Chem Soc 85:2149
9. Merrifield RB (1969) Adv Enzymol 32:221
10. (a) Bayer E (1991) Angew Chem 103:117; (b) Bayer E (1991) Angew Chem Int Ed Engl 30:113
11. (a) Douglas SP, Whitfield DM, Kerpinsky JJ (1991) J Am Soc 113:5095 (b) Danishefsky SJ, McClure KF, Randolph JT, Ruggeri RB (1993) Science 260:1307
12. (a) Gassen HG, Lang A (eds) (1982) Chemical and enzymatic synthesis of gene fragments. VCH, Weinheim (b) Beaucage SL, Iyer RP (1992) Tetrahedron 48:2223
13. Cho CY, Moran EJ, Cherry SR, Stephans JC, Fodor SPA, Adams CL, Sundaram A, Jacobs JW, Schultz PG (1991) Science 251:767
14. (a) Burgess K, Limthicum DS, Shin H (1995) Angew Chem 107:975; (b) Burgess K, Limthicum DS, Shin H (1995) Angew Chem Int Ed Engl 34:907
15. Fodor SPA (1993) Science 261:1303
16. (a) Früchtel JS, Jung G (1996) Angew Chem 108:19; (b) Früchtel JS, Jung G (1996) Angew Chem Int Ed Engl 35:17; (c) Jung G (ed) (1996) Peptide and nonpeptide libraries. VCH, Weinheim
17. Buhleier E, Wehner W, Vögtle F (1978) Synthesis 155
18. (a) Vogel E, Deger HM, Sombroek J, Palm J, Wagner A, Lex J (1980) Angew Chem 1:43; (b) Vogel E, Deger HM, Sombroek J, Palm J, Wagner A, Lex J (1980) Angew Chem Int Ed Engl 19:41
19. Newkome GR, Yao Z-Q, Baker GR, Gupta VK (1985) J Org Chem 50:2003
20. Tomalia DA, Baker H, Dewald J, Hall M, Kallos G, Martin S, Roeck J, Ryder J, Smith P (1985) Polym J 17:117
21. de Gennes PG, Hervet HJ (1983) Phys Lett (Paris) 44:51
22. (a) Denkewalter RG, Kolc JF, Lukasavage WJ (1981) US Pat 4289872; (b) Denkewalter RG, Kolc JF, Lukasavage WJ (1985) Chem Abstract 102:79,324q
23. (a) Tomalia DA, Naylor AM, Goddard III WA (1990) Angew Chem 102:119; (b) Tomalia DA, Naylor AM, Goddard III WA (1990) Angew Chem Int Ed Engl 29:138
24. (a) Newkome GR, Moorefield CN, Baker GR, Johnson AL, Behera RK (1991) Angew Chem 103:1205; (b) Newkome GR, Moorefield CN, Baker GR, Johnson AL, Behera RK (1991) Angew Chem Int Ed Engl 30:1176
25. (a) Mekelburger H-B, Jaworek W, Vögtle F (1992) Angew Chem 104:1609; (b) Mekelburger H-B, Jaworek W, Vögtle F (1992) Angew Chem Int Ed Engl 31:1571 (c) Review: Issberner J, Moors R, Vögtle F (1994) Angew Chem 106:2507; (d) Issberner J, Moors R, Vögtle F (1994) Angew Chem Int Ed Engl 33:2413, (e) Moors R, Vögtle F (1995) Cascade molecules. In: Newkome GR (ed) Advances in dendritic macromolecules. JAI Press Inc., 2:41 (f) Issberner J, Böhme M, Grimme S, Nieger M, Paulus W, Vögtle F (1996) Tetrahedron Assym 7:2223 (g) Issberner J, Vögtle F, De Cola L, Balzani V (1997) Chem Eur J 3:706
26. (a) Brabander-van den Berg EMM, Meijer EW (1993) Angew Chem 105:1370; (b) Brabander-van den Berg EMM, Meijer EW (1993) Angew Chem Int Ed Engl 32:1308 (c) Wörner C, Mülhaupt R (1993) Angew Chem 105:1367; (d) Wörner C, Mülhaupt R (1993) Angew Chem Int Ed Engl 32:1306; (e) BASF has prepared similar polyamine dendrimers in technical scale, Paulus W (1997) BASF Research Laboratory. Personal communica-

tion; (f) Issberner J, Vögtle F, Paulus W (1997) German Patent 196 21 510.2-O.Z. 0050/46935
27. Hawker CJ, Frechet JMJ (1990) J Chem Soc, Chem Commun 1010
28. Miller TM, Neenan TX (1990) Chem Mater 2:346
29. Moore JS, Xu Z (1991) Macromolecules 24:5893
30. (a) Xu Z, Moore JS (1993) Angew Chem 105:1394; (b) Xu Z, Moore JS (1993) Angew Chem Int Ed Engl 32:1354
31. (a) Vögtle F (1992) Supramolekulare Chemie, 2nd edn Teubner, Stuttgart; (b) Vögtle, F (1993) Supramolecular chemistry. Wiley, Chichester; translated into Japanese and Chinese. (c) Lehn J-M (1995) Supramolecular chemistry. VCH, Weinheim
32. Schröder A, Mekelburger H-B, Vögtle F (1994) Top Curr Chem 172:180
33. (a) Kohnke FH, Slawin AMZ, Stoddart JF, Williams DJ (1987) Angew Chem 99:941; (b) Kohnke FH, Slawin AMZ, Stoddart JF, Williams DJ (1987) Angew Chem Int Ed Engl 26:892
34. Kohnke FH, Mathias JP, Stoddart JF (1989) Angew Chem Adv Mater 101:1129
35. (a) Ashton PR, Isaacs NS, Kohnke FH, Mathias JP, Stoddart JF (1989) Angew Chem 101:1266; (b) Ashton PR, Isaacs NS, Kohnke FH, Mathias JP, Stoddart JF (1989) Angew Chem Int Ed Engl 28:1258
36. Benkhoff J, Boese R, Klärner F-G, Wigger AE (1994) Tetrahedron Lett 35:73
37. (a) Klärner F-G, Benkhoff J, Boese R, Burkert U, Kamieth M, Naatz U (1996) Angew Chem 108:1195; (b) Klärner F-G, Benkhoff J, Boese R, Burkert U, Kamieth M, Naatz U (1996) Angew Chem Int Ed Engl 35:1130
38. (a) Graham RJ, Paquette LA (1995) J Org Chem 60:5770; (b) Alder RW, Allen PR, Edwards LS, Fray GI, Fuller KE, Gore PM, Hext NM, Perry MH, Thomas AR, Turner KS (1994) J Chem Soc Perkin Trans 1:3071
39. Scherf U, Müllen K (1992) Synthesis 23
40. Schlüter AD (1991) Adv Mater 3:282
41. (a) Godt A, Enkelmann V, Schlüter AD (1989) Angew Chem 101:1704; (b) Godt A, Enkelmann V, Schlüter AD (1989) Angew Chem Int Ed Engl 28:1680
42. Josten W, Karbach D, Nieger M, Vögtle F, Hägele K, Svoboda M, Przybylski M (1994) Chem Ber 127:767
43. Breidenbach S, Ohren S, Vögtle F (1995) J Chem Soc Chem Commun 1237
44. Breidenbach S, Ohren S, Vögtle F (1996) Chem Eur J 2:832
45. (a) Breidenbach S (1995) PhD thesis, University of Bonn; (b) Ohren S (1996) PhD thesis, University of Bonn
46. Breidenbach S, Ohren S, Herbst-Irmer R, Kotila S, Nieger M, Vögtle F (1996), Liebigs Ann:2115
47. Igner E, Paynter OI, Simmonds DJ, Whiting MC (1987) J Chem Soc Perkin Trans 1:2247
48. Zhang JS, Moore JS, Xu Z, Aguirre RA (1992) J Am Chem Soc 114:2273
49. (a) Schumm JS, Pearson DL, Tour JM (1994) Angew Chem 106:1145; (b) Schumm JS, Pearson DL, Tour JM (1994) Angew Chem Int Ed Engl 33:1360
50. Review: Tour JM (1996) Chem Rev 96:537
51. Liess P, Hensel V, Schlüter A-D (1996) Liebigs Ann 1037
52. Hensel V, Schlüter A-D (1997) Liebigs Ann/Recueil 303
53. Maddux T, Li W, Yu L (1997) J Am Chem Soc 119:844
54. (a) Buchta E, Geibel K (1961) Liebigs Ann 648:36; (b) Rice LM, Grogan CH (1961) J Org Chem 26:54
55. (a) Menger F, Ding J (1996) Angew Chem 108:2266; (b) Menger F, Ding J (1996) Angew Chem Int Ed Engl 35:2137
56. Buchta E, Merck W (1966) Liebigs Ann 694:1
57. (a) Stork G, Rychnovsky SD (1987) J Am Chem Soc 109:1564 (b) Danishefsky J, Halcomb RL (1989) J Am Chem Soc 111:6661 (c) Dondoni A, Merino P (1993) Synthesis 903
58. (a) Mori Y (1997) Chem Eur J 3:849; (b) Alvarez E, Candenas M-L, Pérez R, Ravelo JL, Martin JD (1995) Chem Rev 95:1953
59. Sücrow W, Rädecker G (1988) Chem Ber 121:219

60. For other strategies see: (a) Bindra JS, Bindra R (1975) Creativity in organic synthesis. Academic Press, New York, London; (b) Lindberg T (ed) (1984) Strategies and tactics in organic synthesis. Academic Press, Orlando, London; (c) Mulzer J, Altenbach H-J, Braun M, Krohn K, Reissig, H-U (1991) Organic synthesis highlights. VCH, Weinheim, New York, Basel, Cambridge
61. (a) Wooley KL, Fréchet JMJ, Hawker CJ (1994), Angew Chem 106:123; (b) Wooley KL, Fréchet JMJ, Hawker CJ (1994) Angew Chem Int Ed Engl 33:82; (c) Zeng F, Zimmerman SC (1996) J Am Chem Soc 118:5327
62. For other concepts in nanometer size architectures see: Fuhrhop J, Penzlin G (1994) Organic synthesis: concepts, methods, starting materials, 2nd rev ed. VCH, Weinheim, New York, Basel, Cambridge, Tokyo
63. a) Rice LM, Freed ME, Grogan CH (1964) J Org Chem 29:2637; b) Rice LM, Scott KR (1968) J Med Soc 11:378
64. Schmidt W, Vögtle F, Poetsch E (1995) Liebigs Ann:1319
65. Schmidt W (1995), PhD thesis University of Bonn
66. Feuerbacher N, Schmidt W, Vögtle F, Poetsch E (1997) German patent office Patent application: 97129

Supramolecular Chemistry within Dendritic Structures

Venkatraj V. Narayanan · George R. Newkome

Center for Molecular Design and Recognition, Department of Chemistry, University of South Florida, Tampa, FL, 33613, USA. *E-mail: gnewkome@research.usf.edu*

This chapter describes reports on supramolecular interactions, i.e., molecular recognition, between dendritic hosts and guest molecules, and non-covalent self assembly of branched monomers. The emphases are associated with the chemistry *within* dendrimers rather than on the periphery. A brief introduction to the field of supramolecular chemistry is accompanied by a short overview of dendrimers as well as a brief analysis of the syntheses of these macromolecules. Initially, accounts of random molecular inclusion or preliminary non-specific, hydrophilic/hydrophobic-based binding interactions between small guests and dendritic molecules are described, followed by examples of physical encapsulation of guest(s) molecules and their shape selective release. Molecular recognition processes via site-specific interactions including *H*-bonding host-guest complexations are discussed. Reports dealing with electro- and photo-active dendrimers will be considered. The self-assembly processes involving dendrimers or dendritic building-blocks assisted by forces such as those associated with hydrophilic and lipophilic behavior of amphiphilic species, π-π interactions, and liquid crystalline behavior are reviewed. This is supplemented by a section based on self-assembly processes of dendrons, mediated by intermolecular *H*-bonding, and mono- and multi-layer aggregation processes of dendritic structures. Lastly the construction and supramolecular chemistry of metallo-dendrimers are considered.

Keywords: Dendrimers, supramolecular chemistry, molecular recognition, self-assembly, macromolecular.

Topics in Current Chemistry, Vol. 197

1 Introduction

Supramolecular chemistry [1–7] is a rapidly growing field and has been succinctly defined by Lehn as "the chemistry beyond the molecule." The principles behind this concept have long been recognized to play an important role in molecular events such as those that comprise the evolution of life and in the biosynthesis of contemporary biological systems [8]. Molecules, made up of atoms and bound by strong covalent bonds, define molecular chemistry. On the other hand, supramolecular chemistry is based on molecular recognition, i.e., a study of an ensemble of polymolecular species and their assemblies in which non-covalent intermolecular forces – aromatic stacking, H-bonds, polar and van der Waal's forces – bring molecules together as structurally interrelated species. The chance discovery of crown ethers by Pederson [9] in 1967 followed by the pioneering efforts of the Cram [10] and Lehn [11] research groups formed the basis of supramolecular chemistry for which they were awarded the Nobel prize in 1987. The work of the above researchers resulted in artificial chemical host systems such as crown ethers, cyclophanes, and cryptands, which are the synthetic counterparts to the naturally occurring materials, e.g., cyclodextrins. Since then a multitude of reports describing molecular recognition and self-assembly aspects of new chemical systems are constantly gracing the literature. From Stupp et al.'s self-assembling, miniaturized, triblock copolymers [12] to Moore et al.'s networks [13], originating from molecular building blocks, a wide spectrum of new materials with potentially useful properties have been generated. Interested readers are referred to excellent accounts by Whitesides et al. [14], Lehn et al. [15], Philp and Stoddart [16], and Percec et al. [17] for creative efforts in this field. Rebek et al.'s pH sensitive host [18] and Ghadiri et al.'s self assembling channels [19] are also recent examples of remarkable advances in this field.

Since the experiments of Friedrich Wöhler in 1828 in which a clearly inorganic compound was transformed to an obvious organic material, urea, one of the most significant aspirations of the chemical community has been to invent novel materials with unusual yet predetermined properties. Dendrimers, a relatively new family of oligomeric macromolecules offer potential for such materials [20–25]. These macromolecules of nanometric dimensions also known, in part, as arborols [26], starburst [27], cascade [28] or cauliflower [29] polymers are now bridging the gap between traditional unnatural molecules (MW < 2000 amu) and "classical" polymers.

Reports of molecular recognition and self-assembly aspects of supramolecular chemistry associated with dendrimers and related structures will be the chief focus of this review.

2
Building Block Concept in Dendrimer Design

At the outset, it would be worthwhile to review the factors that affect dendrimer design and architecture. These highly branched macromolecules are synthesized by a step-wise approach with either *linear* or *branched building blocks* (Fig. 1). Unlike most traditional polymers, one can take precise structural control over the molecular weight as well as chemical and physical properties of dendritic macromolecules. Typically, an *n*-directional core molecule with *n*-number of surface functional groups is treated with either a *linear* or a *branched monomer* to yield what is called a first tier or first generation dendrimer. Deprotection of the surface terminal groups of the first tier molecule followed by treatment with an adequate amount of monomer provides the next generation. This iterative approach of synthesizing dendrimers provides an opportunity to control the architecture and the molecular weight of resulting generations of macromolecules. These fractal-like constructs generally possess excellent solubilities, when compared to "classical" polymers, which therefore facilitate their purification and characterization. It should be noted that the unique three-dimensional, branched structure and the nature of the peripheral functional groups are important factors that determine the chemico-physical

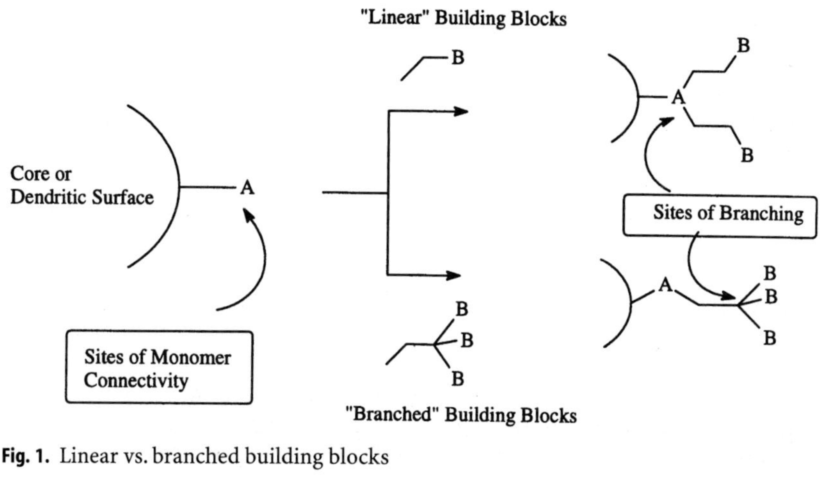

Fig. 1. Linear vs. branched building blocks

properties of these cascade molecules. Additionally, dendritic polymers in general possess low intrinsic viscosities when compared to their linear counterparts.

The effects of using either a *linear* or a *branched monomer* in the construction procedure are critical to the instilled architectural features of the dendrimer. The terms divergent and convergent ("outward" vs "inward" tier growth, respectively) are used to describe dendritic synthetic strategies. Typically *linear building blocks* afford branching only by the direct attachment of two or more units at a particular site on the substrate (Fig. 1). Hence only divergent strategies have been used with these monomers. Advantages include ready availability and decreased costs of monomers owing to simple structures.

On the other hand, *branched building blocks* [30] possess inherent branching points remote from the site of connection. These monomers can be simple or complex in design and have been used in both divergent and convergent synthetic strategies. In addition, *branched monomers* can also be used to install utilitarian functionality within cascade molecules.

As will be demonstrated, surface branching has afforded a series of dendrimers lacking site-specific [31–33] characteristics; whereas, the use of *branched monomers* has the unique ability to instill tailored loci capable of site-specific molecular recognition and encapsulation.

2.1
Linear Building Blocks in Dendrimer Construction

The history of dendrimer chemistry can be traced to the foundations laid down by Flory [34] over fifty years ago, particularly his studies concerning macromolecular networks and branched polymers. More than two decades after Flory's initial groundwork (1978) Vögtle et al. [28] reported the synthesis and characterization of the first example of a cascade molecule. Michael-type addition of a primary amine to acrylonitrile (the *linear monomer*) afforded a tertiary amine with two arms. Subsequent reduction of the nitriles afforded a new diamine, which, upon repetition of this simple synthetic sequence, provided the desired tetraamine (**1**, Fig. 2); thus the advent of the "iterative synthetic process" and the construction of branched macromolecular architectures was at hand. Further growth of Vögtle's original dendrimer was impeded due to difficulties associated with nitrile reduction, which was later circumvented [35, 36]. This procedure eventually led to DSM's commercially available poly(propylene imine) dendrimers.

The addition of ammonia to excess methyl acrylate (a *linear monomer*), followed by amidation with excess ethylenediamine afforded the resultant cascade molecule, and thus Tomalia [37] created the commercially available PAMAM starburst series of dendrimers (**2**, Fig. 2). Related core molecules such as ethylenediamine and aminoalcohols and other functionalizable groups such as thiol moieties were used to prepare similar dendrimers [38]. This methodology is applicable to most primary amines, resulting in a $1 \rightarrow 2$ branching pattern. Recently, examples of related Si-, [39] P-, [40] and metallo systems [41], which follow this *linear monomer* "protocol" have been reported.

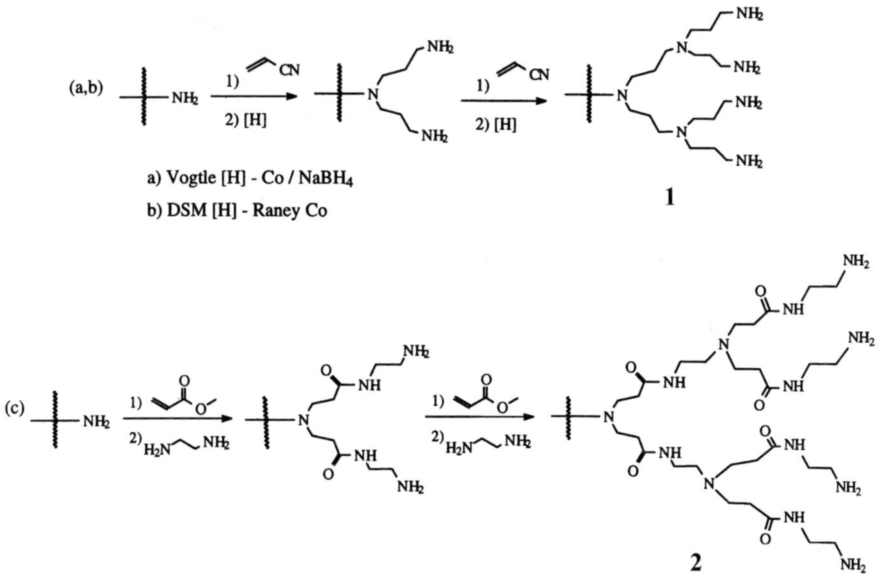

Fig. 2a–c. Dendrimer synthesis using linear building blocks: **a** Vögtle's original procedure; **b** DSM's commercial process; **c** Tomalia's synthesis of PAMAM dendrimers

2.2
Branched Building Blocks in Dendrimer Construction

The second approach to dendrimer construction, as first suggested in a patent by Denkewalter et al. [42], employed a protected amino acid, *N,N′*-bis(*tert*-butoxycarbonyl)-L-lysine, as the monomer. The two-directional, asymmetric core (**3**) was constructed from L-lysine and benzhydrylamine. Coupling of the building blocks was accomplished by the use of an activated *p*-nitrophenyl ester (**4**) followed by removal of the *tert*-butoxycarbonyl (*t*-BOC) protecting groups. The new free polyamine moieties (**5**) were now available for the construction of the next generation (Fig. 3). Although Denkewalter et al.'s work was studied [43, 44], their patents did not afford the critical proof of structure. In 1985, Newkome et al. devised a workable 1→3 building block procedure which generated (e.g., **6**) the structurally proven "arborols" [26]. This synthesis is also divergent in that it was constructed from the inside out. Use of the related 1→3 building block (Behera's amine) under DCC coupling conditions with carboxylic acids terminated cores followed by treatment with formic acid (to hydrolyze terminal *t*-butyl ester moieties) and repetition of this sequence has afforded a variety of polyamide dendrimers (e.g., **7**, Fig. 3) [45]. The beauty of *the branched monomer* approaches is that convergent procedures are also possible as demonstrated by Fréchet et al. [46, 47], where large branched monomers called "dendrons" were synthesized from the periphery to the core using 3,5-dihydroxybenzyl alcohol as the key monomer. Two main synthetic transformations, the selective alkylation of phenolic hydroxyl groups (**8**), and conversion of a benzylic alcohol (**9**) to a

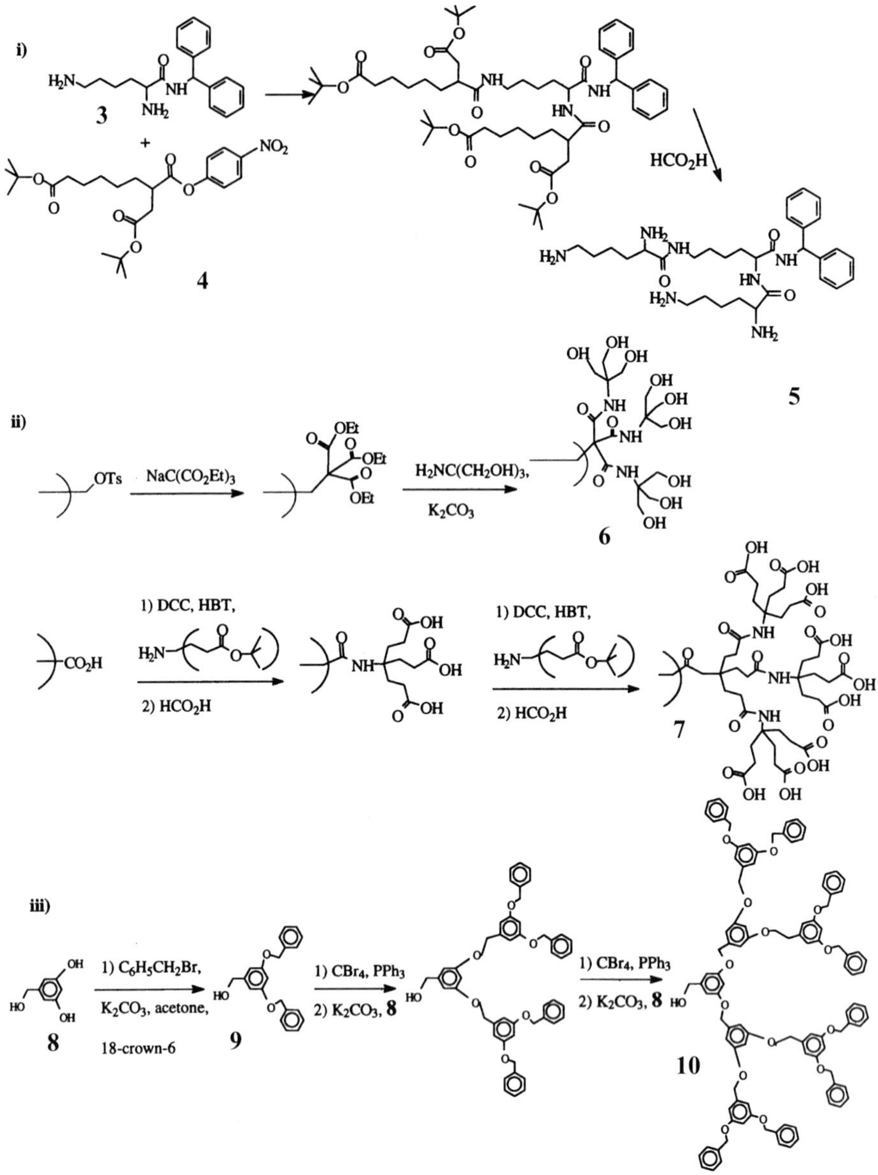

Fig. 3i–iii. Dendrimer synthesis using branched building blocks: **i** Denkewalter's polylysine dendrimers; **ii** Newkome's polyamide dendrimers; **iii** Fréchet's aryl ether dendrimers

benzylic bromide thus affording a reactive focal moiety, were responsible for the construction of these "dendritic wedges" or dendrons (**10**, Fig. 3). Other examples of complex branched monomers can be seen in the related work of Miller and Neenan, [48] Moore and Xu [49, 50] and L'abbé et al. [51]. More recently Zimmerman and Zeng [52] reported a rapid synthesis of dendrimers via an "*orthogonal coupling*" strategy. Two different *branched building blocks* (of the type AB_2) containing two complementary coupling functionalities were coupled by the Mitsunobu esterification reaction or by the Sonogashira reaction of a terminal acetylene with an aryl iodide. Most notable features of this strategy include the absence of a normally required deprotection or activation step.

Dendrimers synthesized via a divergent approach using either *linear* or simple *branched monomers* could produce some side products in the form of partially substituted molecules, due to the large number of reactions that need to occur at multiple reaction sites of the previous generation dendrimer. Thus careful purification of dendrimers obtained by this approach is necessary due to the small difference in molar masses of fully and partially substituted generations. However, this particular problem is less apparent with the convergent approach followed for complex *branched building blocks* (e.g., **10**, Fig. 3) because of the large differences in the molar masses of the products and by-products. Both assembly approaches have their advantages and disadvantages; for example higher generation dendrimers are difficult to obtain by the convergent approach since the focal point of the larger dendrons (due to steric reasons) cannot get close enough to the core's reactive site. Despite these architectural defects, very low monodispersities can be achieved and have been reported for dendrimers constructed following both approaches.

Alternatively, the one-step polymerization of *branched monomers* results in what is called a hyperbranched polymer [53] possessing a higher degree of polydispersity and lower degree of branching compared to the analogous dendrimer.

3
Molecular Inclusion in Dendritic Systems

The concept of trapping guest molecule(s) by a host molecule with a spatial, egg shell-like structure was proposed for the first time by Maciejewski in 1982 [54]. Unlike linear polymers which possess random-coil structures the three dimensional motif of dendrimers impart to them unique structural features. Compared to the relatively open structures of lower generation dendrimers, at higher generations, they tend to adopt a spherical surface with pockets of spaces in the interior, thus acting as "unimolecular" micelles [26] capable of guest inclusion [55]. These empty spaces within the dendritic infrastructures may be envisioned as "*dendritic voids*". Thus synthetic models that can mimic functions of natural proteins such as molecular recognition of substrates either in the interior or on the surface are now available in the form of dendrimers. The terms *endo*-receptors and *exo*-receptors can also be used to describe binding sites on the inside and periphery of dendrimers, respectively [27].

3.1
Hydrophobic Non-Specific Interior Complexations

Tomalia and coworkers [55] reported the *random* complexation of host molecules in dendritic interiors by monitoring the change in guest ^{13}C spin-lattice relaxation times (T_1). Theoretical studies on PAMAM dendrimers (**11 a** and **11 b**, Fig. 4) predicted that above the fourth generation they adopt a globular structure with hollow cavities on the inside, and that at a certain stage of dendritic growth the surface becomes increasingly congested, thus eventually restricting further uniform growth [56]. PAMAM dendrimers with methyl ester termini (**12 a**, **12 b**) were used as the dendritic host; whereas, aspirin and 2,4-dichlorophenoxyacetic acid were chosen as guest molecules. The T_1 values of these guests in $CDCl_3$ change in the presence of dendrimer and these are in turn dependent on the generation number of the specific dendrimer used. The values for T_1 decreased as the generation number increased from 0.5 to 3.5 (the ester terminated dendrimer was considered to be a half-generation), but remained constant for the higher generations. Thus, good agreement between experimental and theoretical data led to the inference that interior binding occurs in the larger dendrimers.

Chemistry of micelles is an important area of dendrimer research. A micellar structure depends on numerous factors, such as temperature, concentration, and mainly the molecular framework of the given amphiphiles. Revolutionary research in micellar chemistry is exhibited in the work of Menger et al. [57] and by Shinkai et al. [58].

We reported the first synthesis of a symmetrical, four-directional, saturated hydrocarbon dendrimer containing 36 carboxylic acid moieties equidistant

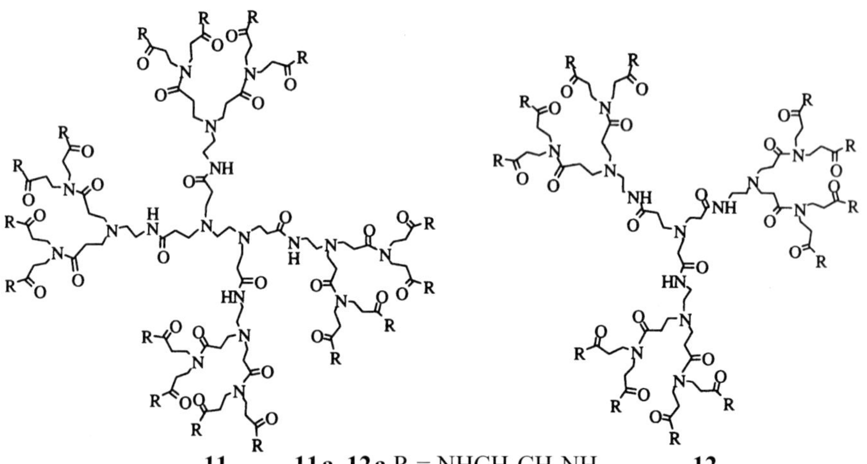

11 **11 a, 12 a** R = $NHCH_2CH_2NH_2$ **12**

11 b, 12 b R = R = $-OCH_3$

Fig. 4. Early generation PAMAM dendrimers with amine and ester/acid termini from a central quaternary carbon-based core [59–61]; the term "unimolecular micelle" was coined for this novel dendrimer. Interestingly, in our first 1985 paper [26] we suggested that dendritic structures were examples of a unimolecular micelle, three-dimensional macromolecules possessing architectural similarities to that of polymolecular surfactant aggregates (known as "classical micelles" [62]) and, under certain conditions capable of molecular inclusion. The acid groups were converted to their corresponding ammonium (**13**) and tetramethylammonium (**14**) carboxylate derivatives to facilitate the study of micellar properties (Fig. 5). The color of an aqueous solution of pinacyanol chloride (PC) changes from pink to blue in the presence of these anionic micelles [63, 64]. An aqueous solution of PC (3.9×10^{-6} mol/l) has absorptions at 554 and 600 (λ_{max}) nm in the absence of amphiphiles. In solutions of PC (3.9×10^{-6} mol/l) and of sodium dodecylsulfate (SDS) at concentrations above the critical micelle concentration (cmc, $> 10^{-2}$ mol/l), these absorptions shift to 564 and 610 nm, although SDS solutions at concentrations below 10^{-2} mol/l show an absorption at 478 nm. Upon addition of Micellanoate™ **13** or **14** (6.8×10^{-4} and 3.3×10^{-4} mol/l, respectively) to an aqueous solution of PC (3.9×10^{-6} mol/l, pH 10.0), λ_{max} shifts to 610 nm and the absorption becomes more intense and narrowed. In both cases, this absorbance ($\lambda_{max} = 610$ nm)

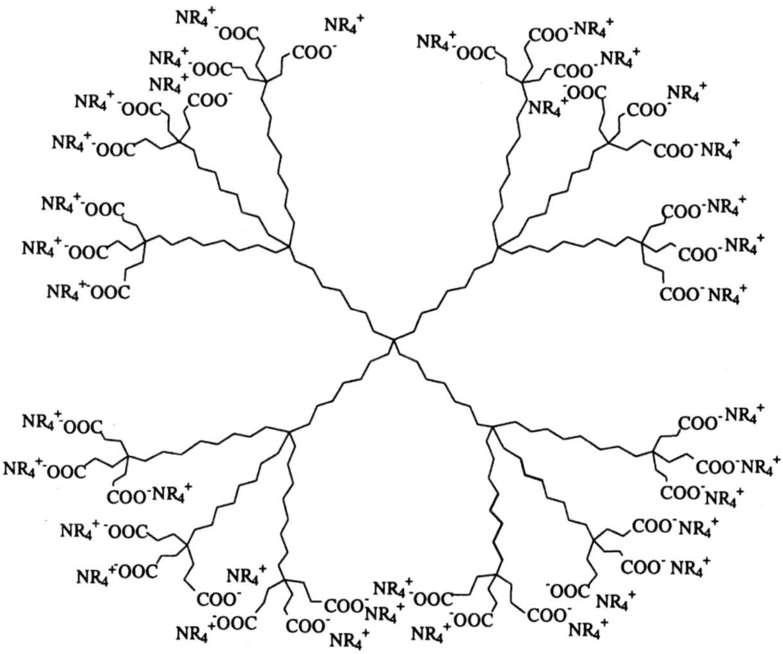

13 R = H
14 R = CH_3

Fig. 5. All-hydrocarbon micelle reported by Newkome et al.

gradually shifts (toward $\lambda_{max} = 600$ nm) as the concentration of either decreases to 1.6×10^{-6} mol/l; absorptions appear at 478 nm when concentrations fall below 3.9×10^{-7} mol/l. These results were consistent with those of others [57, 58] or non-unimolecular micelles and if a cmc existed it must be $< 3.9 \times 10^{-7}$ mol/l. Micellanoate™ **14**, has also been shown to be monomeric in aqueous media over a broad range of concentrations as indicated by dynamic light scattering studies. Additional evidence for molecular inclusion was provided by the UV spectrum of ($\lambda = 214, 276$ nm) of an aqueous solution of Micellanoate™ **13** (2.3×10^{-4} mol/l) with naphthalene ($\lambda = 214, 276$ nm; 3:1 H_2O/EtOH). Also, the absorption intensity for **13** and naphthalene at 276 nm is approximately four times greater (3.35) than that observed for naphthalene in a 2:1 mixture of water and ethanol (0.844).

Fréchet et al. reported the synthesis of a unimolecular micelle (Fig. 6) generated by a convergent approach starting with 3,5-dihydroxybenzyl alcohol, as the key branched monomer unit [65]. Fourth generation dendrimer **15** with 32 carboxylic acid moieties on the periphery was synthesized and its correspond-

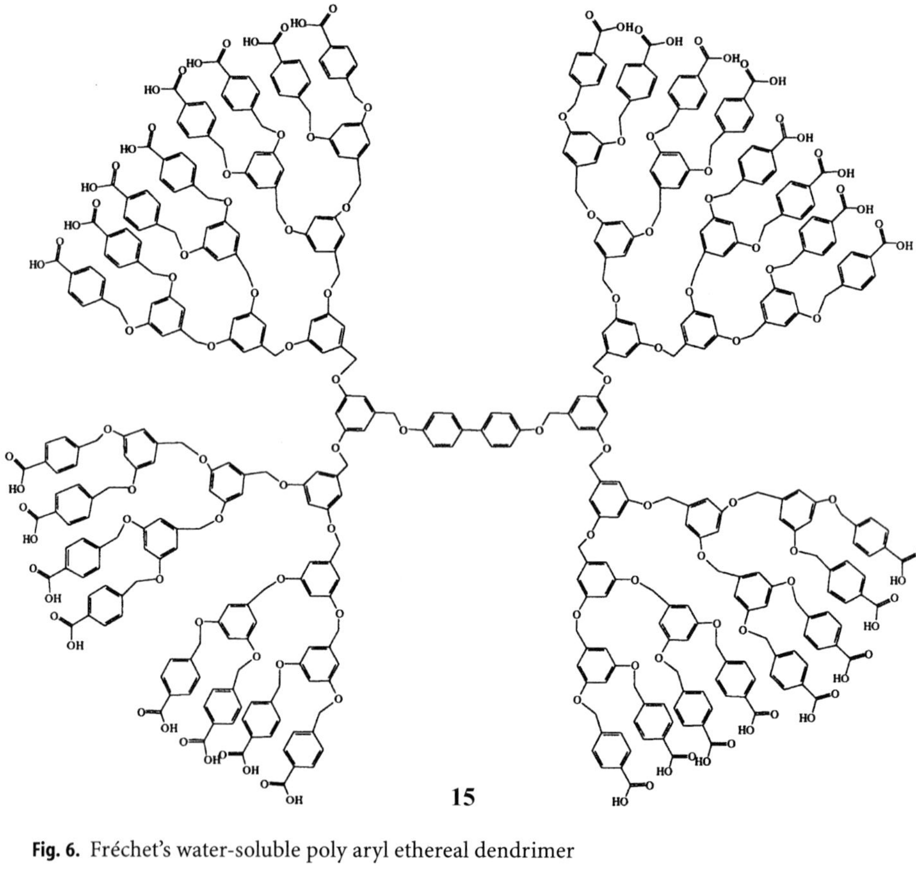

15

Fig. 6. Fréchet's water-soluble poly aryl ethereal dendrimer ing water soluble potassium salt was tested for micellar characteristics. Solubilization of a commonly used apolar organic molecule (pyrene) in doubly distilled water was tested with sonication and under high temperature conditions. A 120-fold increase in the solubilization of pyrene in water was observed in the presence of dendrimer **15**. Also, a linear relationship was observed for the amount of solubilized pyrene and the dendrimer concentration. On comparison with SDS (CMC: 8.1×10^{-3} mol dm^{-3}) **15** was found to solubilize pyrene at concentrations as low as 5×10^{-7} mol dm^{-3}. Further, addition of NaCl increased the number of pyrene molecules solubilized to 1.9 for each dendrimer compared to 0.45 for each dendrimer in the absence of NaCl. This was attributed to the displacement of water molecules from the dendritic interior thus making the inner-environment less polar. The high solubilization efficiency of these polyethereal dendrimers was attributed to possible π-π interactions with aromatic guest molecules. Other guests were investigated to retest this behavior. Approximately, twice as much of 2,3,6,7-tetranitrofluorenone was solubilized compared to pyrene while 1,4-diaminoanthraquinone and anthracene were solubilized to the degree of about one half the amount of pyrene. The pyrene-dendrimer complex was precipitated from solution with acetic acid. Almost all the pyrene was recovered by dissolving this precipitate in THF and extracting it with an aqueous base. The recovered dendrimer was found to solubilize pyrene with the same efficiency as above, thus, demonstrating its potential as a novel recyclable solubilizing agent.

Synthesis and supramolecular chemistry of novel dendrimers with hydrophilic interiors and hydrophobic shells (inverse micelles) were reported by Meijer and coworkers [66]. These dendrimers were synthesized by reacting amine terminated poly(propylene imine) cascade molecules with long-chain, aliphatic acid chlorides. Hydrophilic guest molecules, such as Bengal Rose, were shown to bind within the dendritic interior in an ethanol solution. Hexane solution of the dendrimer-dye complex(es), released the dye molecules by addition of toluene but not by water. Evidence for non-aggregation and unimolecular micellar behavior (hydrodynamic diameter: 2–3 nm in dichloromethane) of the higher generation dendrimers ($n \geq 4$) was obtained from ^{1}H NMR and dynamic light scattering studies.

3.2 Non-Specific Guest(s) Encapsulation and Release

Examples of guest binding into dendritic interiors have been demonstrated but are based on dynamic processes where the guests are free to diffuse in and out of the dendrimer depending on the conditions applied. Meijer et al. went one step further where they demonstrated physical encapsulation of small guest molecules into the dendritic cavities [67–70]. This was achieved by reacting (or capping) the 64 terminal amino groups of a fifth generation poly(propylene-imine) dendrimer **16** with tBOC or FMOC protected amino acids in the presence of guest molecules, for example Bengal Rose and tetracyanoquinodimethane (TCNQ). The surface groups are close-packed and probably hydrogen bonded as suggested by their low mobility, as indicated by NMR relaxation measurements.

Followed by exhaustive dialysis, EPR, fluorescence, and UV spectroscopic methods supplied evidence for the number and type of guest molecules trapped in the dendritic cavities. This method could have potential applications ranging from drug delivery systems to fluorescent markers. In a later report the above researchers further demonstrated in a remarkable way the shape-selective liberation of guest molecules trapped in the "dendritic box." In one example, a dendritic box containing Bengal Rose *and* 2,2,3,4,5,5-hexamethyl-3-imidazolinium-1-yloxy methyl sulfate (a small EPR probe) was deprotected with formic acid to remove the tBOC groups on the surface. This, followed by dialysis, afforded a perforated dendritic box (Fig. 7) in which all of the smaller guests were removed. Further hydrolysis of the surface amide groups by refluxing 12 N HCl for 2 h followed by dialysis with 100% water resulted in release of Bengal Rose molecules and about 50–70% of the intact, original poly(propylene imine) dendrimer was recovered. Thus, systems like these hold potential promise towards development of drug delivery devices.

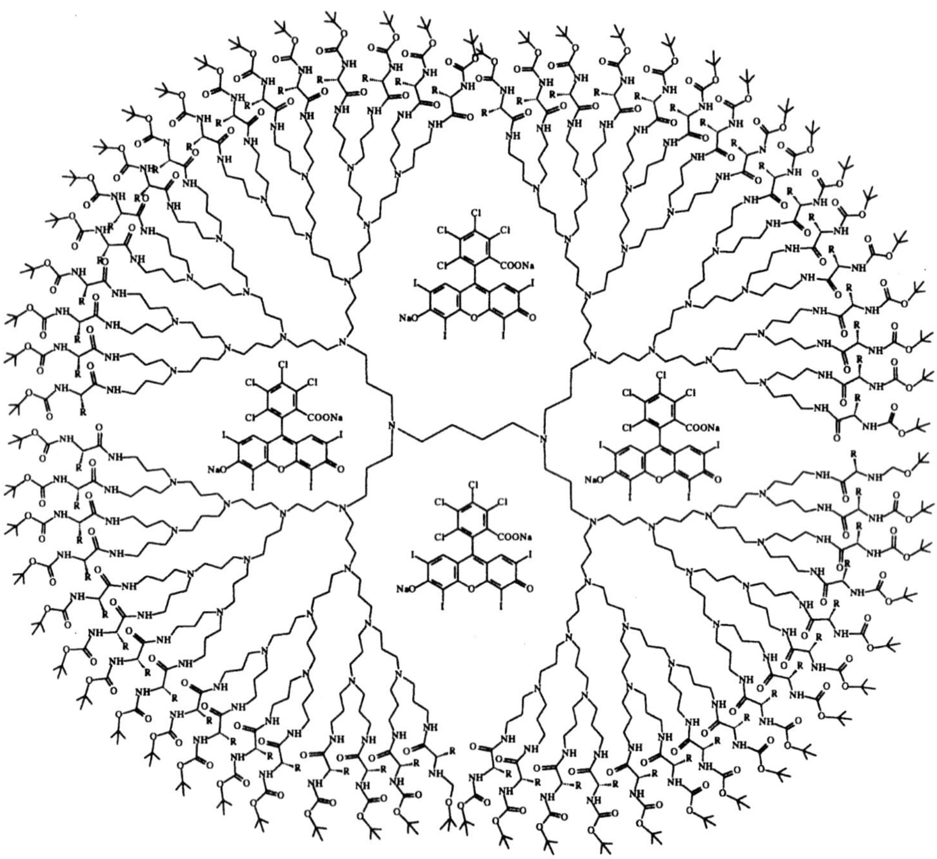

16

Fig. 7. Meijer et al.'s "dendritic box" with encapsulated Bengal Rose molecules

3.3 Site-Specific Interior Complexations

Precise placement of metal complexing sites within the infrastructure of a cascade molecule is of importance from a variety of perspectives. In the construction of the above noted Micellane™ family (cf. Sect. 3.1), we reported the construction of dendrimers with four alkyne moieties at sites equidistant from each other in the interior (**17**, Fig. 8) [60]. These were treated with decaborane ($B_{10}H_{14}$) to afford 1,2-dicarba-closo-dodecaboranes (*o*-carboranes) [71]. Rendering boron clusters soluble in water is of interest because of their use in cancer treatment by Boron Neutron Cancer Therapy. First and second generation water-soluble dendrimers containing four and twelve precisely located boron cluster sites, respectively, were synthesized (e.g., **18**). These water soluble dendrimers and their precursors were characterized by ^{1}H-, ^{13}C-, and ^{11}B-NMR spectroscopy (Fig. 8).

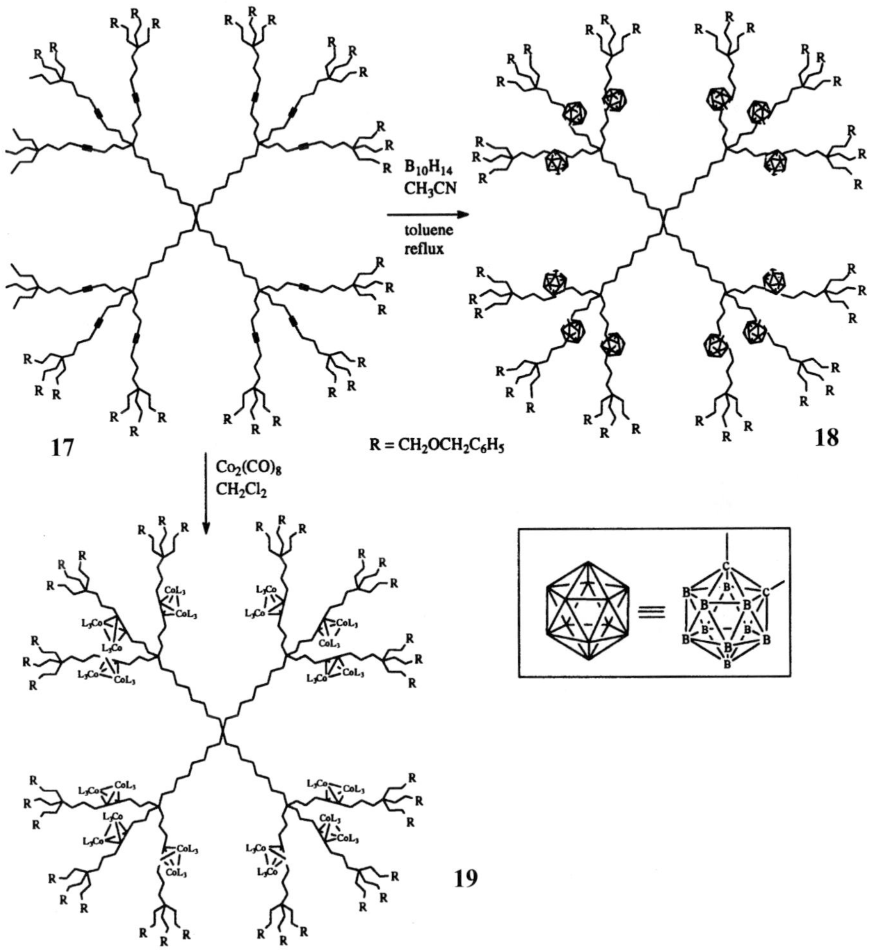

Fig. 8. "Site-specific" metal complexing dendrimers reported by Newkome et al.

Dicobaltoctacarbonyl is known to form addition complexes to acetylenes as well as act as a protecting group [72]. The resultant dicobalt complexed alkynes also act catalytically to promote ene-yne cyclopentenone cyclizations [73–76]. Thus, treatment of **17** with dicobaltoctacarbonyl afforded deep red, viscous dicobalthexacarbonyl adduct **19**, which was demonstrated from ^{13}C NMR spectroscopy. Thus, metal-binding sites at precise locations were designed and the binding of metals at these specific positions were shown; these demonstrated the broader concept of metallo-superclusters [77–80].

Shinkai and coworkers reported the synthesis and metal complexation chemistry of a novel series of crown ether based cascade molecules, with benzene-1,3,5-tricarbonylchloride, as the core [81]. The 1→2 monomer used was constructed from *N*-benzyloxycarbonyl-1,4,10,13-tetraoxa-7,16-diazacyclooctadecane and 3,5-bis(ethoxycarbonylmethoxy)benzoyl chloride. A convergent technique was used to build three generations of dendrimers (e.g., **20**). Reducing the amide groups in the arborols increased the ion-binding affinity of the crown ether moieties (Fig. 9). It was noted that the dendrimers extracted metal ions in a generation independent manner and formed a 1:1 complex with Cs^+ unlike the sandwich complexes formed with Cs^+ by polymeric crown ethers [82]. Thus,

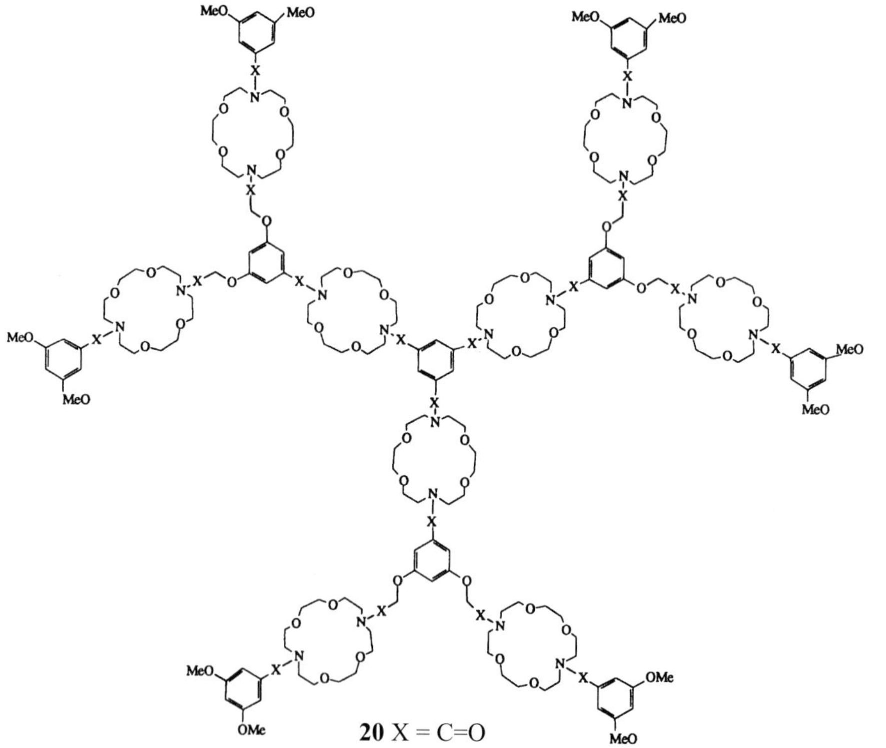

20 X = C=O
21 X = CH_3

Fig. 9. Crown ether arborols reported by Shinkai et al.

the above results indicated that the crown-ether moieties in the dendrimers do not necessarily function in a cooperative manner. The ability of the dendrimers to solubilize myoglobin in organic solvents via peptide-NH_3^+-crown ether associations was also reported. Interestingly, a DMF solution of the first generation reduced dendrimer (**21**) was the only one which increased the solubility of myoglobin. This result was attributed to the fact that the first generation of the reduced dendrimer could easily interact with the NH_3^+ and CO_2^-–M^+ moieties on the surface of myoglobin, while in the higher generation dendrimers the crown ethers are buried within the arborols, resulting in poor interactions with the charged moieties on guest peptide surface.

In designing models for globular proteins, Diederich et al. reported a novel series of dendritic host molecules which he termed "dendrophanes" (Fig. 10). These dendrimers (e.g., **22**), containing a core made up of a cyclophane derived from diphenylmethane and dendrimerized with a branched 1→3 monomer [83], have been studied [84]. These water-soluble dendrophanes (e.g., **22a**) were shown to bind with small aromatic guests, and in another study dendrophanes with a larger cyclophanic core were found to host a variety of steroidal molecules [85]. The complexation chemistry was monitored by ^{1}H NMR or spectrofluorimetric titrations of the guest molecules. Binding inside the cyclophanic cavity was suggested by the upfield shifts of aromatic moieties on both the host

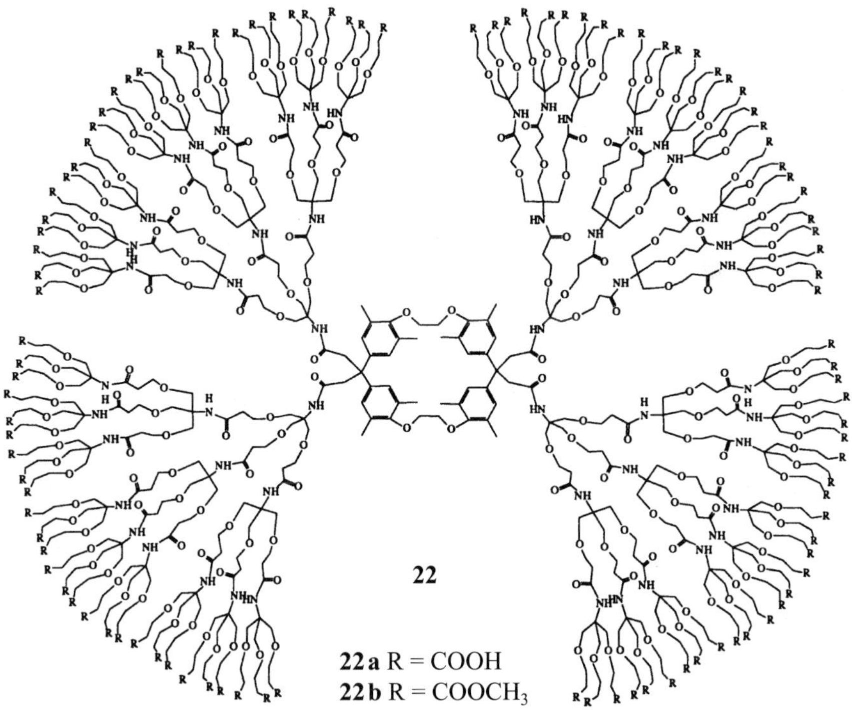

22

22a R = COOH
22b R = $COOCH_3$

Fig. 10. Diederich's guest hosting dendrophanes

and guest species. They noted that the binding constants were not affected to any large extent with change in generation number suggesting a relatively open structure even at the third generation. Thus, these examples demonstrated specific hydrophobic binding of guest molecules, thus exhibiting their potential to act as models for drug delivery systems.

Aida et al. reported a spectroscopic investigation on the interior properties of dendritic zinc porphyrins and their interpenetrating action with dendritic imidazoles [86]. A series of Fréchet dendrons were attached to a tetraphenyl zinc porphyrin core (LnPZn: n = 1 – 5) and to one nitrogen atom of imidazole (LnIm: n = 1, 2 and 4; **23**). The effect of solvent on the microenvironment of the porphyrin core for this dendritic LnPZn series was monitored by UV spectroscopy. In particular, the Soret band wavelength was used as an indicator of the changes within the Zn porphyrin microenvironment. Dichloromethane and 1,3-dimethoxybenzene were used as the solvents for this particular purpose. It was observed that the Soret band wavelength of L5PZn in dichloromethane was very close to that of 1,3-dimethoxybenzene ($\Delta\lambda = 0.7$ nm). This suggested that at higher generations the solvent effect on the Zn-porphyrin center is totally nullified due to the dendritic shell.

The interactions of dendritic imidazoles (**23**) with the dendritic porphyrins (LnPZn: n = 1 – 5; see **24, 25**) were also reported. Binding between LnPZn and LnIm was found to follow a 1 : 1 stoichiometry and the binding affinity decreased with increasing generations of both the guest and host systems (Fig. 11). For instance, there was a notable difference in the binding constants between L3PZn and L4PZn. They also noted that L4Im could weakly bind to L5PZn, indicating a considerable amount of interpenetration that must be occurring between the branches of both dendrimers.

Aida and Jiang recently reported an iron(II) porphyrin 1-methylimidazole complex covalently encapsulated within a large aryl ethereal dendrimer cage, as

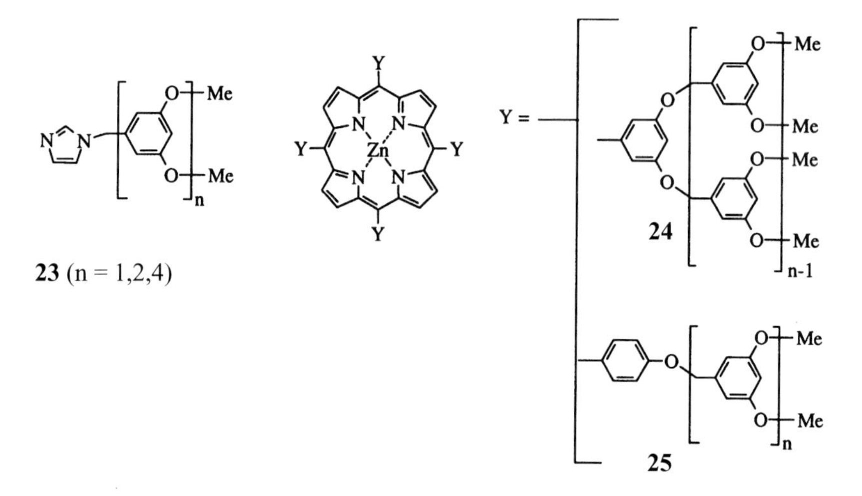

Fig. 11. Aida's dendritic Zn-porphyrins and dendritic imidazoles

a monomolecular model of dioxygen carrying hemeproteins [87]. This complex of the Fe-porphyrin dendrimer and 1-methylimidazole in anhydrous or water saturated toluene was shown to form long-lived reversible adducts with dioxygen.

Recently, Collman and coworkers reported [88] remarkable observations from a gas equilibrium binding study with first (**26**) and second (**27**) generation iron(II) porphyrin dendrimers (Fig. 12). They noted that these dendrimers exhibited reversible O_2 and CO binding activities. The O_2 binding affinities were measured to be about 1500 times *greater* than those of a hemoglobin and "picket fence" porphyrin. These authors suggested a possible *H*-bonding interaction between the terminal O of the bound O_2 molecule and an amide N-H group.

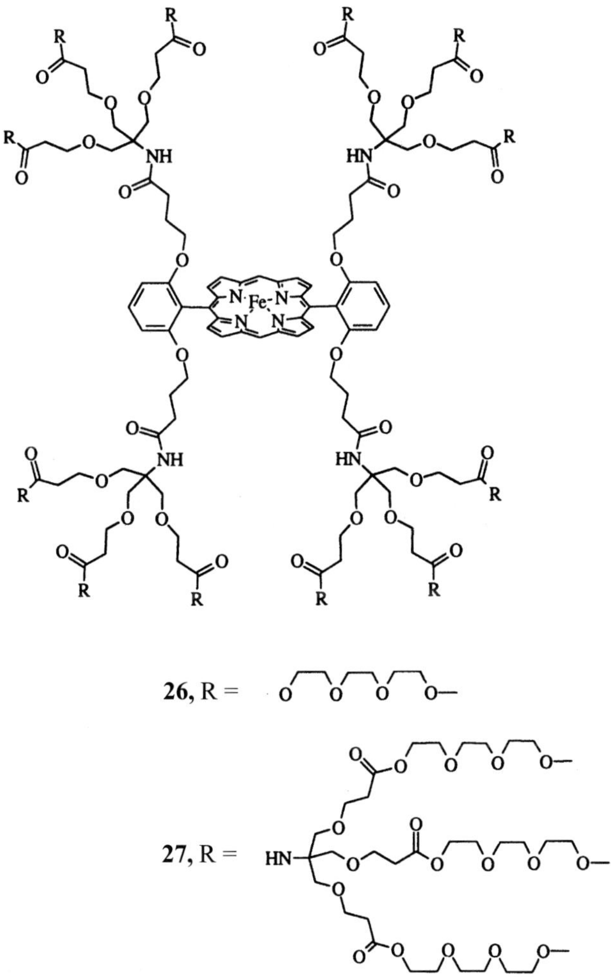

Fig. 12. O_2 and CO_2 binding dendritic porphyrins

3.4 Molecular Recognition via Site-Specific *H*-Bonding

One of the elegant features in biological recognition observed in natural systems is the formation of *H*-bonds. For example in DNA, the two complementary strands are held in place by *H*-bonding interactions between the bases adenine and thymine, as well as cytosine and guanine. Also, proteins and substrates are held together at complementary sites by *H*-bonds. Such non-covalent interactions [89] have attracted synthetic chemists to construct receptors capable of binding to biologically-important as well as interesting unnatural molecules.

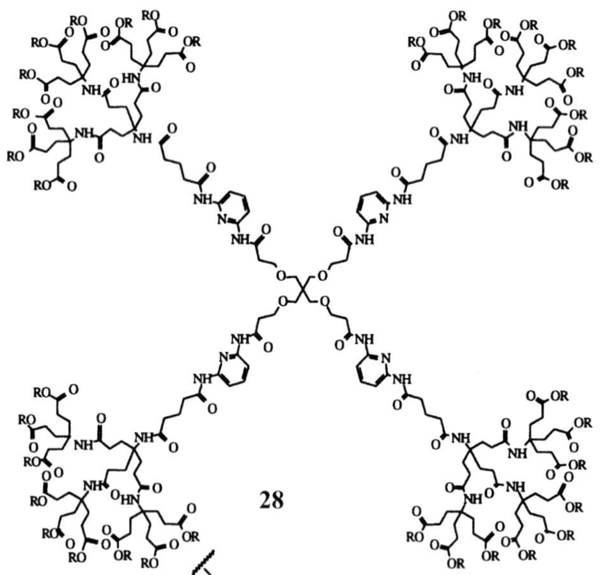

28

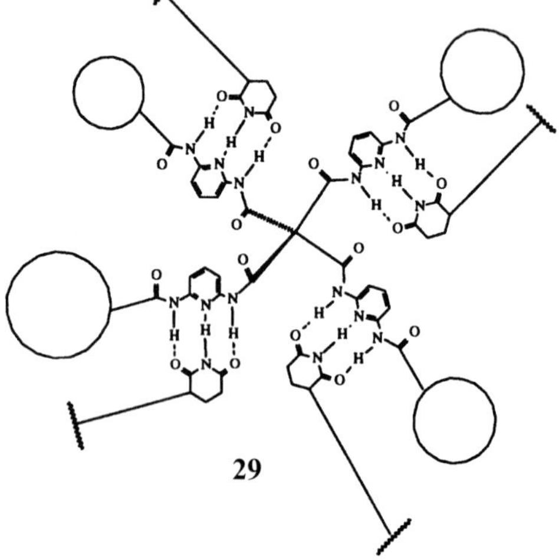

29

Fig. 13. Newkome's dendritic host with internal *H*-bonding sites

We recently described [90] a variety of dendritic hosts (e.g., **28**) containing 2,6-diacylaminopyridine *H*-bonding units (Fig. 13). ^{1}H-NMR titration was used to evaluate host-guest interactions with complementary guests such as glutarimide, barbituric acids as well as the drug AZT. The findings supported discrete host-guest interactions, although, for the higher generation dendritic hosts, it was complicated by the fact that guest molecules could bind at alternate, albeit weaker *H*-bonding sites within the host's infrastructure [91]. It is envisioned that molecular recognition mediated self-assembly between **28** and complementary binding moieties should lead to novel tetrahedral dendritic arrays, such as **29**.

4
Electro- and Photo-Active Dendrimers

Dendrimers with a covalently linked porphyrin at the core have been prepared and studied as synthetic models of globular heme proteins. For example, Diederich et al. reported [92] the divergent synthesis of zinc porphyrin dendrimers such as **30** employing a simple $1 \rightarrow 3$ aminotriester building-block [83]. These dendrimers (Fig. 14) were designed to act as models for electron transfer proteins such as cytochrome c, in which the hydrophobic peptide shell affects the interior metal-porphyrin redox potential. Cyclic voltammetry data for these dendrimers revealed first oxidation potentials up to 0.3 V less positive than the corresponding values obtained for unshielded Zn-porphyrin tetraester. The results were rationalized on the basis that, with increasing generation numbers for the dendrimers, addition of electrons destabilized the molecule due to the increasing electron rich microenvironment of the dendritic shell.

These researchers also described [93] the design and synthesis of iron(II) porphyrin dendrimers with triethylene glycol monomethyl, ether surface groups (e.g., **31**) which render them soluble in a wide range of organic solvents and water. The potential difference between the first ($1 \cdot FeCl$) and second generation ($2 \cdot FeCl$) Fe-porphryin dendrimers was found to increase more in water than in dichloromethane (0.42 vs 0.08 V). This remarkable potential difference between $2 \cdot FeCl$ and $1 \cdot FeCl$ in water was comparable with that found between cytochrome c and a similarly ligated, more solvent-exposed cytochrome c heme model compound.

Vögtle et al. reported the first example of a photoswitching dendrimer [94] with six azobenzene moieties attached to a derivative of 1,3,5-trisubstituted benzene as the central core. Irradiation of the all (*E*)-isomer at 313 nm led to a photostationary equilibrium where most of the azobenzene units were switched to the (*Z*)-configuration. Conversely, irradiation of this species again at a lower energy frequency (436 nm) led to a second photostationary equilibrium where the (*E*)-form was dominant; however, it was not proven as to how many azobenzene units isomerized after irradiation.

Delivery of physically encapsulated small molecules at targeted sites by dendrimers has been superbly envisioned in a recent report by McGrath and Junge [95]. Second generation Fréchet type dendrons linked to a central azobenzene-derivative (**32**, Fig. 15) underwent reversible *cis*/*trans* isomerization of

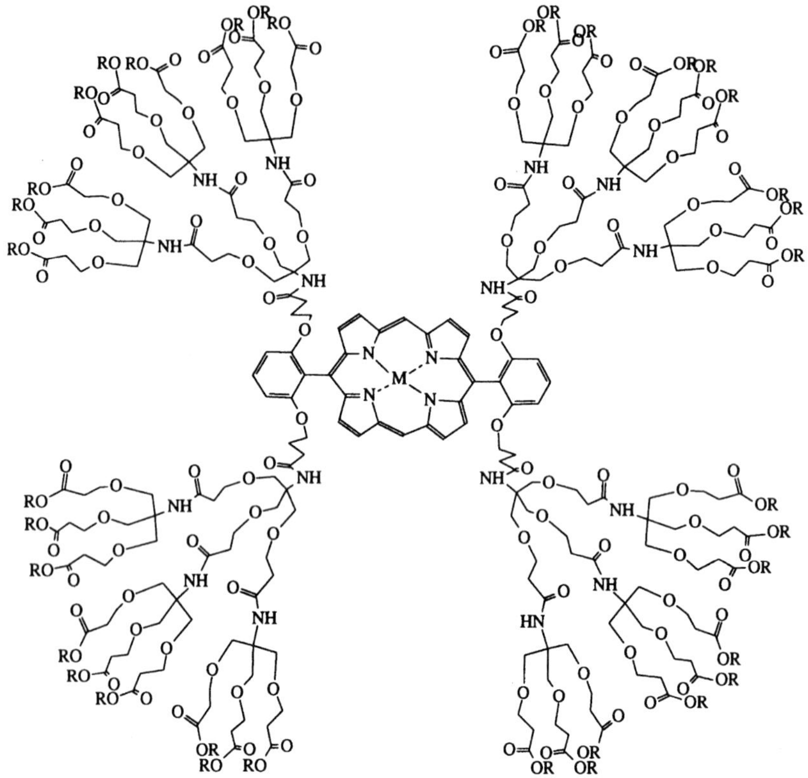

30 M = Zn (II), R = CH_3
31 M = Fe (III), R = $(OCH_2CH_2)_3CH_3$

Fig. 14. Diederich's dendritic metallo porphyrins

the azobenzene moiety upon exposure to ultraviolet light. The authors hope that the higher generation dendrimers could be used to encapsulate and eject selected molecules by photoisomerization, thus acting as novel dendritic switches.

Recently, Aida and Jiang [96] synthesized a series of dendrimers similar to those reported by McGrath and coworkers and reported photoisomerization of their core azobenzene units using infrared radiation. This new strategy holds immense potential for using dendrimers as light-harvesting matrices.

Molecular antennas based on directional transduction of energy along a convergent path have a variety of applications ranging from enhancing rates of photochemical reactions to nanosensors to use as materials in efficient solar energy conversion. In a seminal paper [97], Moore and Xu reported the synthesis of a phenylacetylene monodendron **33** with a perylene luminophore at the focal point (Fig. 16). Dendron **33** which was designed and synthesized by a convergent procedure consists of a highly conjugated system with multiple, peripheral chromophoric units which can collect light energy and channel it to

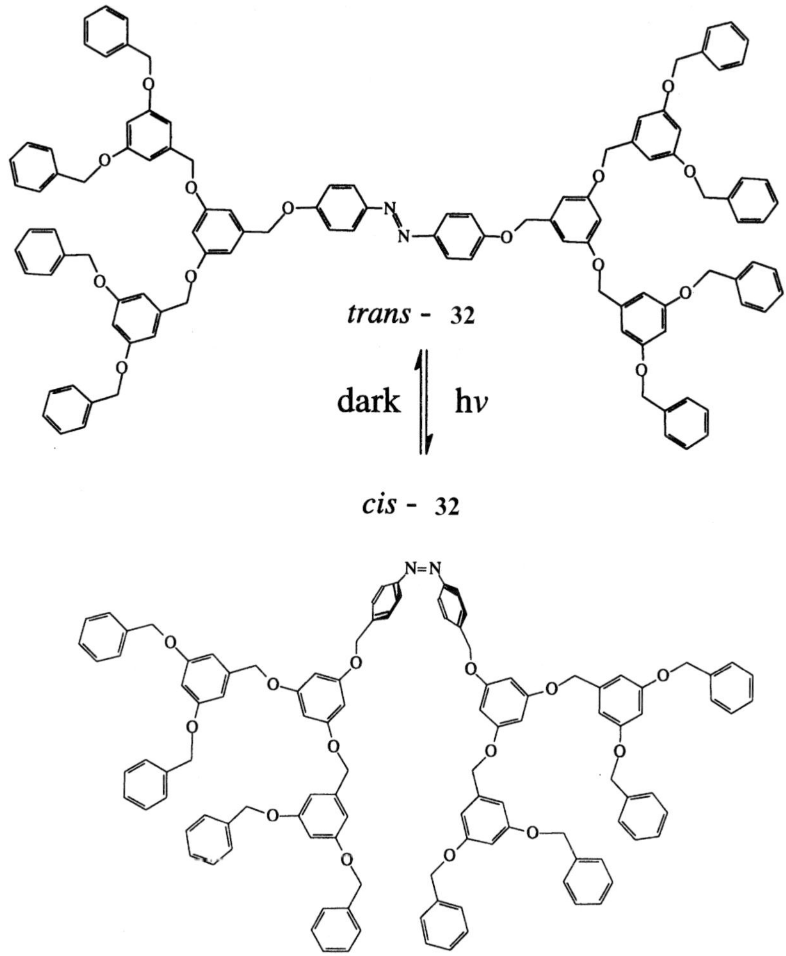

Fig. 15. Photoactive dendrimer reported by McGrath et al.

the focal point which is an efficient fluorescence emitter. Dendron **33** was characterized by NMR and UV spectroscopy, SEC and combustion analysis. Preliminary photophysical experiments indicated doubling of light energy enhancement for **33** as compared with 3-ethynylperylene.

Moore et al. later reported [98] the design and synthesis of triarylamine based dendrimers. These fluorescent macromolecules exhibited reversible redox processes and their potential use in electro-optic film applications was envisioned.

Solvatochromic probes have been used for a variety of applications like the study polarity of pure and mixed solvents [99], and the retention behavior in reverse-phase liquid chromatography [100] among other applications. Fréchet et al. used 4-(*N*-methylamino)-1-nitrobenzene (*p*-MANB), as the chromophore, to probe the microenvironment of polyaromatic ether based dendrimers [101].

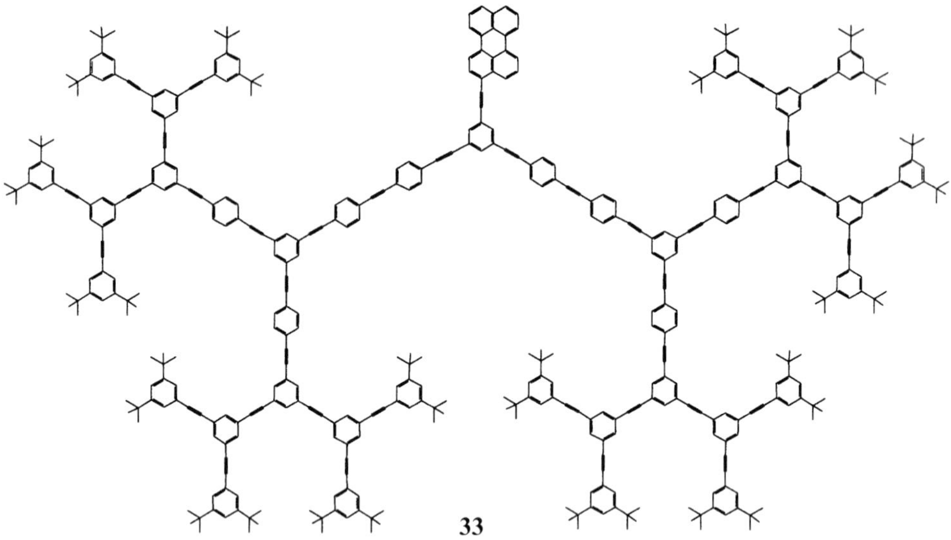

33

Fig. 16. Moore's dendritic molecular antenna

p-MANB, which has a wide solvatochromic absorption range, was attached to a series of polyethereal dendrimers of increasing generation numbers. They noted that, for low polarity solvents, there was a drastic change in the absorption maximum going from generation 3 to 4, which coincides with the proposed shape change for cascade molecules from an extended to a more globular shape. Hanson et al. obtained similar results [102] with comparable dendrimers having pyrene, as the chromophoric unit, attached to the focal point of Fréchet type dendrons.

Fox and Stewart reported [103] the first photophysical study of electron donor-acceptor pairs covalently attached to a dendritic structure. A series of second generation aryl ethereal dendrons (**34**, Fig. 17) attached to an electron donor moiety at the focal point [C_1 = 3-(dimethylamino)phenoxy), -OCH_2CH_2N-$(CH_2CH_3)_2$], and electron acceptor moieties at the periphery (C_2 = 2-naphthyl, 1-pyrenyl) were synthesized. Photoexcitation of the acceptor moiety (C_2) should result in electron transfer from the donor (C_1) through the dendritic backbone to the outer array of C_2 moieties. Detailed fluorescence experiments suggested that there was extensive electronic coupling between appended chromophores and quenchers across the dendritic framework, indicating the potential viability of these novel dendrons for use as molecular light harvesters.

Dendritic molecules with electroactive units at either the focal point or core have been reported [92, 97]. There are, however, only a few examples of such moieties specifically pinned within cascade infrastructure. Our recent efforts in this direction [104–106] involve the incorporation of chromophoric 1,4-diaminoanthraquinone (**35**) within the cascade infrastructure. Dendrimers based on a four-directional pentaerythritol core were synthesized using the extended $1 \rightarrow 3$ building block **36**. A high dilution technique was applied to synthesize **36**

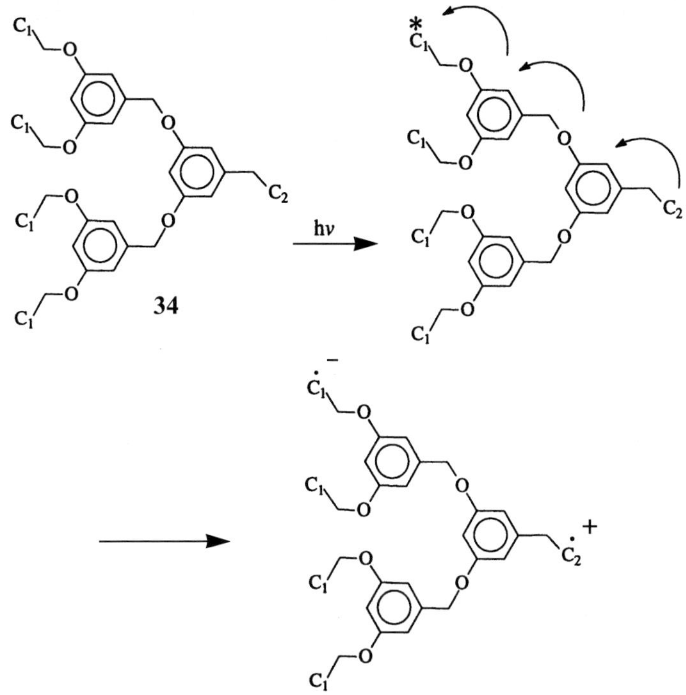

Fig. 17. Light harvesting dendrons reported by Fox et al.

when commercially available **35** was reacted with glutaryl chloride and 1→3 monomer **37**, "Behera's Amine" [107]. The previously developed divergent methodology was employed in synthesizing the first generation dendrimer (**38**) from **35**; the subsequent generations of macromolecules (**39** and **40**, respectively), were achieved by simple iterative growth with **37** as the building block (Fig. 18). Cyclic voltammetry showed an expected positive shift of the reduction potentials of the anthraquinonoid moieties going from diamine (**35**) to monoamide (**36**) to diamide (**38**). UV-VIS spectroscopy helped monitor the facile, reversible chemical reduction processes of these electroactive dendrimers using $NaBH_4$. Interestingly, third generation dendrimer (**40**) is soluble even in petroleum ether unlike **38** and **39**, which could be attributed to the effect of the growing dendritic shell. Modification of *branched* building block **36** afforded electroactive dendrimers (**41, 42**, Fig. 19) with reactive functional groups on the inside [108]. Use of these electroactive macromolecules as potential cation transporting switches is envisioned.

Towards understanding biological electron transfer processes many researchers have reported the synthesis of dendrimers with electroactive cores (e.g., porphyrin). Dendrimers with organic dendrons attached tetrahedrally around an inorganic, electroactive iron-sulfur core were reported by Gorman and coworkers [109]. These are the first examples of dendrimers with a hybrid

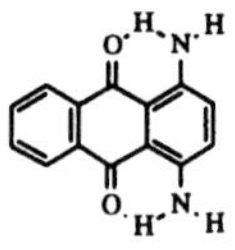

35

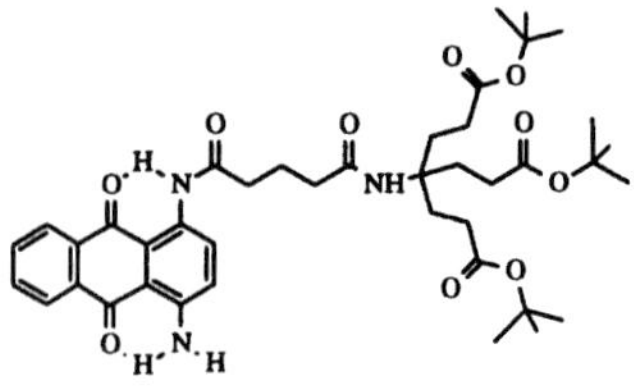

36

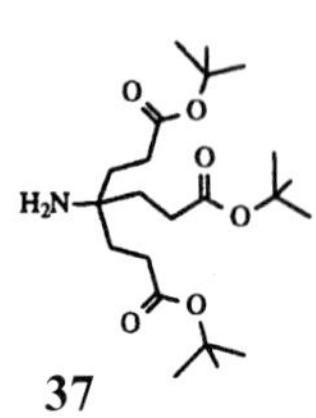

37

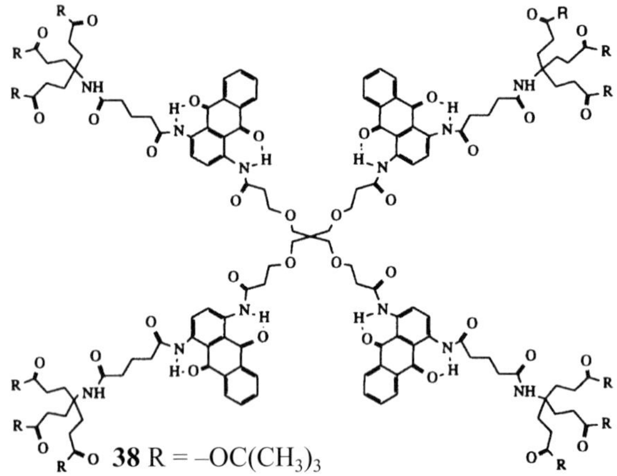

38 R = $-OC(CH_3)_3$

39 R = $-NHC(CH_2CH_2CO_2C(CH_3)_3)_3$

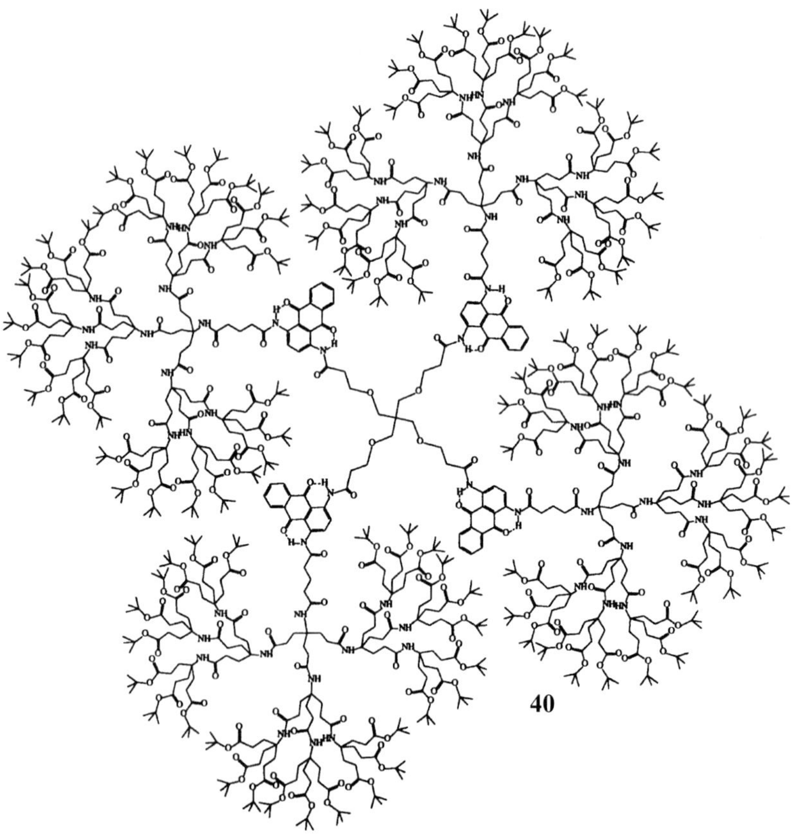

40

Fig. 18. Newkome's diaminoanthraquinone-based electroactive dendrimers

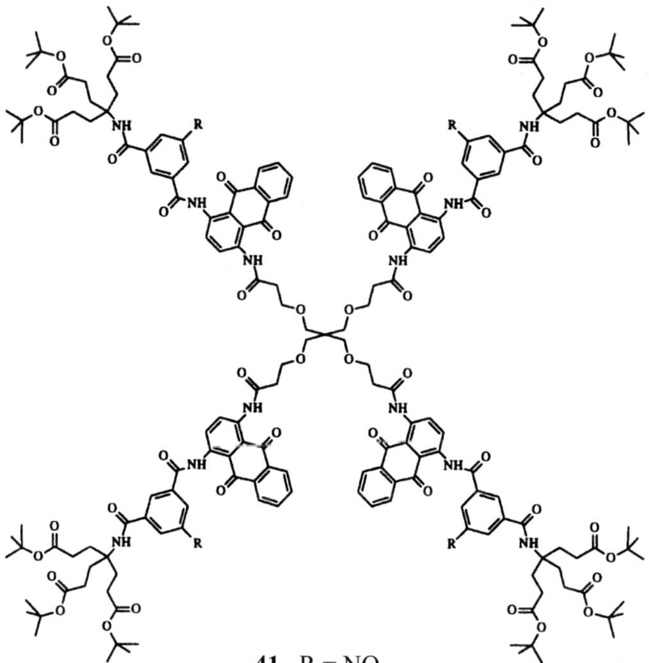

41 R = NO_2
42 R = NH_2

Fig. 19. Anthraquinonoid dendrimers with internal modifiable functional groups

organic/inorganic architecture synthesized via a ligand exchange wherein aromatic thiols were used to displace aliphatic thiols from $(n\text{-}Bu_4N)_2[Fe_4S_4(S\text{-}t\text{-}Bu)_4]$. As with other related electroactive dendrimers it was noted that, as the generation number of the dendrimers increased, electrochemical reduction became increasingly irreversible.

Organometallic dendrimers have been constructed to act as potential electro- or photo-active materials, the synthesis of which will be discussed in the following section. Apart from the examples discussed above, surface modification of dendrimers with a variety of functional groups has afforded novel redox active materials [110–116].

5 Self-Assembly Processes in Dendritic Systems

The concept of micelles consists of aggregation of amphiphilic molecules that contain polar and non-polar moieties, which associate in a manner that minimizes hydrophobic and lipophilic interactions. However, a cascade molecule consisting of an internal lipophilic framework and a external hydrophilic surface would effectively be a "unimolecular micelle" [59] capable of hosting molecular guest(s).

5.1 Self-Aggregation Assisted by Non-Directional Forces

5.1.1 *Self-Assembly Processes Related to Dendritic Amphiphiles*

In our earlier efforts to synthesize dendritic amphiphiles, we described a tridendron (**43**) which was prepared by a two-step (alkylation-amidation or triester-tris) reaction sequence applied to 1,3,5-tris(bromomethyl)benzene [117]. TEM and light scattering experiments suggested that **43** aggregated by stacking of its hydrophilic exterior into a spherical array of ca. 20 nm (diameter) reminiscent of globular micelles.

We further reported [118–121] the use of simple, $1 \rightarrow 3$ building blocks in the construction of novel amphiphiles, in general, and bolaamphiphiles, in particular. The two sequentially employed monomers – a trialkyl methanetricarboxyate and trihydroxymethylaminomethane (Tris) – resulted in polyalcohols such as **44** [122]. Spacer units ranging from a simple aliphatic chain or spirane or biphenyl units were used. The length and rigidity of the spacer unit were found to influence the ability of these compounds to form spaghetti-like aggregates. These aggregates formed thermally reversible gels which were characterized by electron microscopy as well as light scattering experiments. This novel automorphogenisis [123] (e.g., **44**, n=6) was also verified by fluorescence experiments. The hydrophobic dye chlortetracycline (CTC), which is soluble and nonfluorescent in aqueous media, fluoresces only in a hydrophobic environment. Though in very dilute solutions, fluorescence from individual molecules of **44** was undetectable, at higher concentrations it aggregated, forming a gel, and a CTC fluorescence was observed from the hydrophobic cores. This shows that the CTC dye must have intercalated into the lipophilic region of these cylindrically shaped aggregates. It was proposed that the dumbbell shaped arborols (**44**) stack up in an orthogonal array based on the lipophilic-lipophilic and hydrophilic-hydrophilic interactions. Bolaamphiphiles similar to **44** (Fig. 20) but with a central triple bond [124] in the spacer unit (**45**) were synthesized. Aqueous solution of **45** formed a gel analogous to that formed by **44**. Electron micrograph of **45** indicated formation of a helical superstructure which was attributed to the central alkyne moiety which induces a less than orthogonal chain alignment.

Jørgenson et al. reported [125] the synthesis of bolaamphiphile **46** in which a tetrathiafulvalene (TTF), a substrate currently of importance in fields such as organomagnetism and molecular electronics, was incorporated within the central lipophilic region. Gel formation of **46** (Fig. 20) in either DMF-water or ethanol-water solution was noted which bear similarities with the above findings of Newkome and coworkers.

In their efforts to construct stimuli-responsive, supramolecular amphiphiles, Fréchet et al. [126–129] reported the synthesis of a novel series of AB and ABA block copolymers via the Williamson ether synthesis (e.g., **47**, Fig. 21). Polyethylene glycols (PEGs) of different lengths were used as the linear hydrophilic B block while polyaryl ether dendrons of different sizes were used as the hydrophobic A block. These copolymers were characterized by optical microscopy,

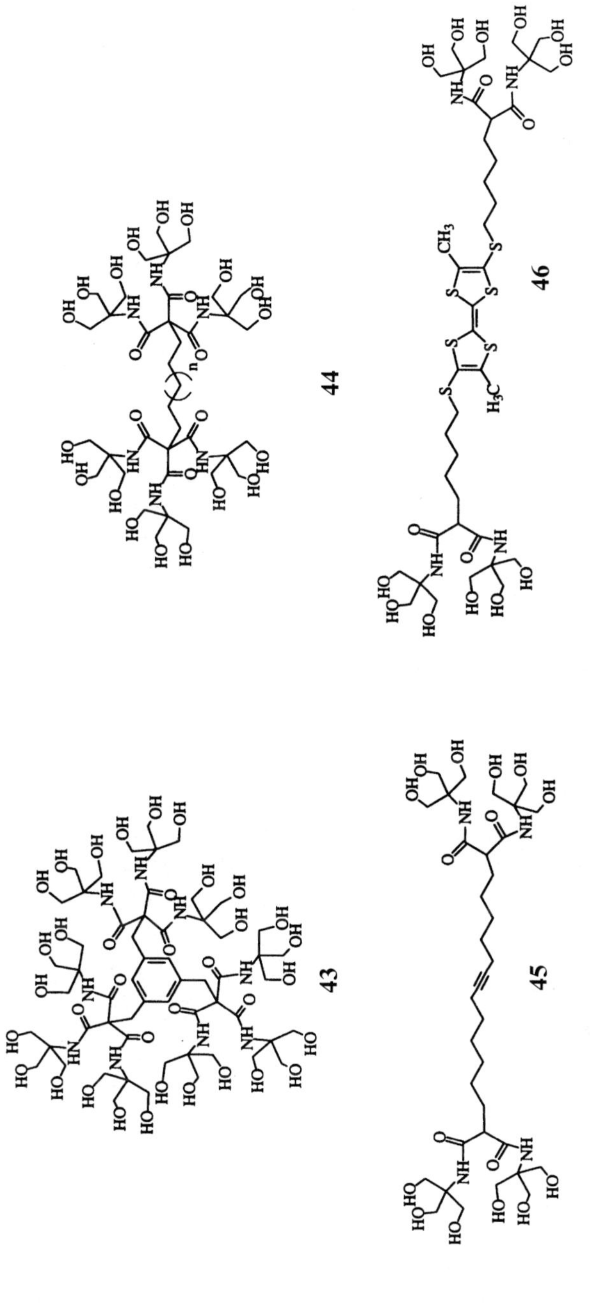

Fig. 20. Self-assembling arborols reported by Newkome and Jørgenson

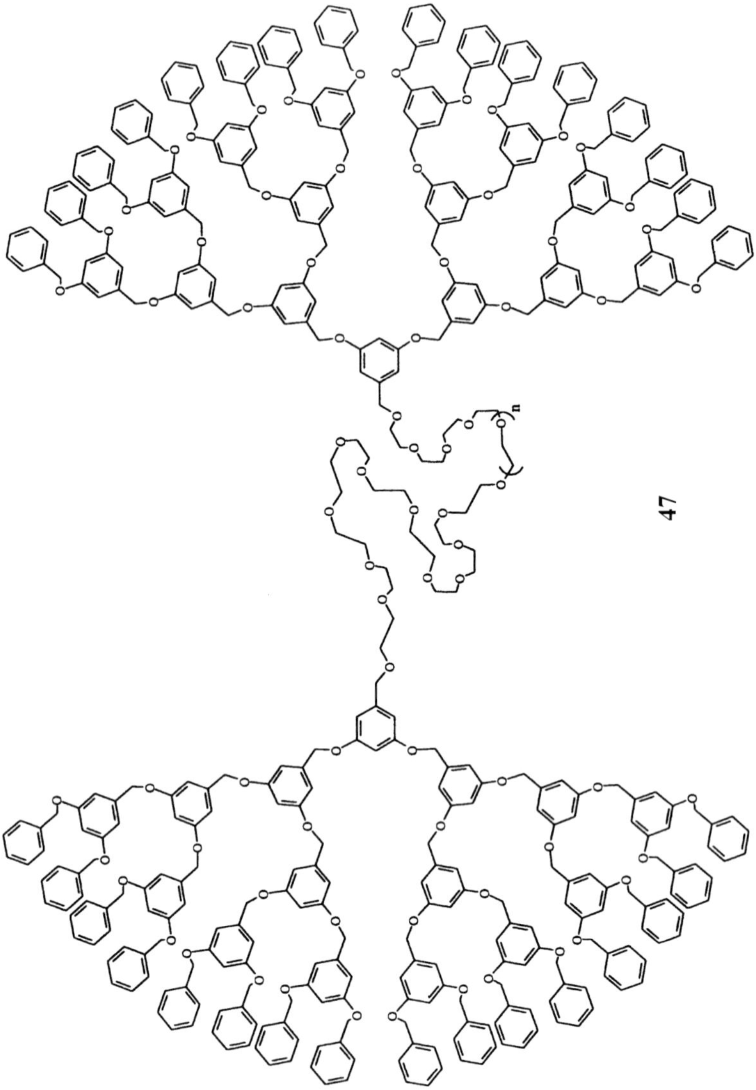

47

Fig. 21. Fréchet's dendritic copolymer

size exclusion chromatography (SEC), coupled with a viscometric detector (VISC) and ^{1}H-NMR studies. Variations in the NMR signals of the PEG unit with respect to those of the dendrimer units were useful in explaining the changes in the confirmations of the copolymers in different solvents. Additionally, aggregation behavior of these copolymers in solution was observed and studied using SEC/VISC experiments. The results also indicated that, in general, copolymers containing low generation dendrons (e.g., G1) tended to form unimolecular micelles; whereas, G2 and G3 formed polymolecular micelles which could be explained as aggregation facilitated by π-π interactions between the aryl ether units.

More recently Fréchet and Gitsov [130] used a similar approach as above and synthesized a novel series of dendritic copolymers derived from a central pentaerythritol core unit. These "hybrid star" molecules behaved as unimolecular micelles with different core-shell conformational-structures as a response to the polarity of the solvent used.

Chapman et al. [131] reported the synthesis of poly(ethylene oxide) (PEO) supported dendritic *t*-BOC-poly(α, ε-L-lysines). These dendritic polymers termed as "hydramphiphiles" formed foams possessing good temporal stability in aqueous solution. Scrimin et al. [132] synthesized a three-directional polypeptide having uses in membrane permeability modulation. Decapeptide fragments were linked to TREN [tris(2-aminoethyl)amine] core.

Meijer et al. [133] reported a novel class of amphiphilic macromolecules such as **48** synthesized by combining lipophilic linear polymer polystyrene with different generations of hydrophilic poly(propylene imine) dendrimers (Fig. 22). The nature of the micelles formed was dependent on the size of the dendrimer group as observed using techniques such as dynamic light scattering, conductivity measurements, and TEM techniques. For example, copolymers attached to first and second generation dendrimers (i.e., PS-dend-G1 and PS-dend-G2) showed inverted micellar behavior while PS-dend-G3 formed vesicular structures. However, PS-dend-G4 formed rod-like micelles while PS-dend-G5 formed spherical micelles. These observations also agreed well with the theoretical predictions of Israelachvili [134].

Meijer et al. [135] further studied aggregation phenomena of amphiphilic copolymers obtained by modifying the termini of the dendrimer units of the above series with carboxylic acid groups (**49**). TEM experiments indicated that all block copolymers formed large aggregates except third generation copolymer which formed worm-like micelles.

Host-guest systems made from dendritic materials have potential in the areas of membrane transport and drug delivery [68, 84, 85]. In a recent report [136] Tomalia and coworkers investigated structural aspects of a series of PAMAM "bolaamphiphiles" (e.g., **50**) with a hydrophobic diaminododecane core unit. Fluorescence emission of added dye (nile red) was significantly enhanced in an aqueous medium in the presence of **50** unlike the cases when **51** and **52** were added (Fig. 23). Addition of anion surfactants to this mixture generated supramolecular assemblies which enhanced their ability (ca. by 10-fold) to accommodate nile red (**53**). Further increase in emission was noted by decreasing the pH from the normal value of 11 for PAMAM dendrimers to 7. At lower pH values the

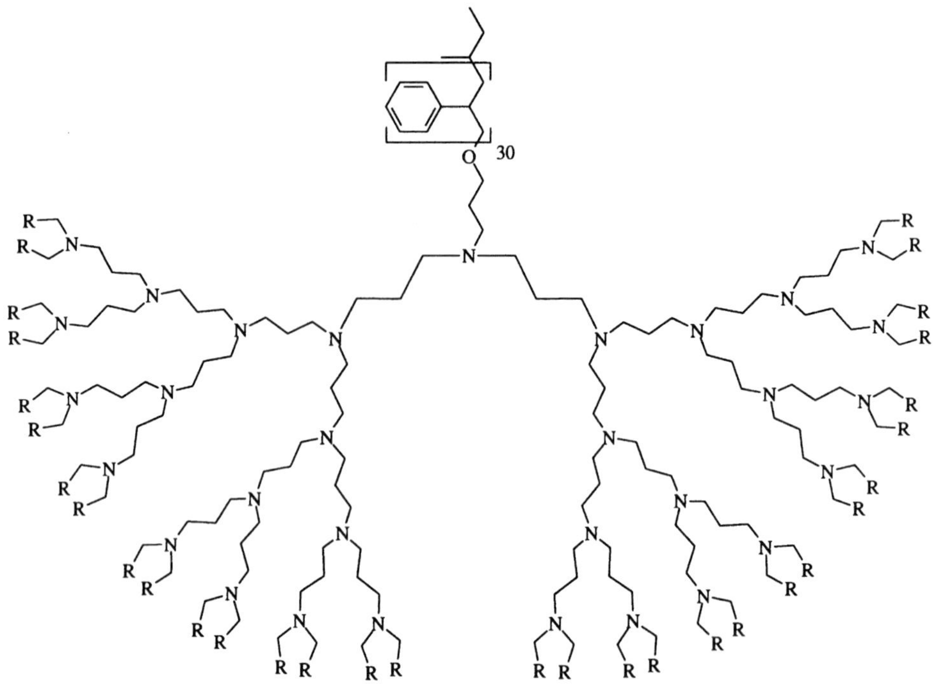

48 R = $CH_2CH_2NH_2$
49 R = COOH

Fig. 22. Amphiphilic copolymers reported by Meijer et al.

expanded form of the dendrimer is manifested since the positively charged NH_4^+ termini repel each other, thus allowing easy access of **53** to the central hydrocarbon unit.

5.1.2
Self-Assembly via π-π Interactions

The first example of a donor-acceptor molecular complex was noted in 1949 by Bensei and Hildebrand [137] in their studies involving charge transfer complexes between benzene and molecular iodine. Subsequently such complexes were studied by Mulliken [138] and now more recently have been used by Stoddart et al. [16, 139] in designing novel self-assembling systems.

In their first step towards building dendritic rotaxanes Stoddart et al. [140] reported the construction of such materials via a so-called "slipping" method (Fig. 24). Thus, treatment of bisparaphenylene-34-crown-10 (BPP34C10) with tris(bipyridinium) compound **54** at 50 °C in acetonitrile for 10 days afforded mono-, di-, and tris-rotaxanes **55**, **56**, and **57**, respectively. ES-MS was used to determine the molar masses of these macromolecules while upfield shifts in the

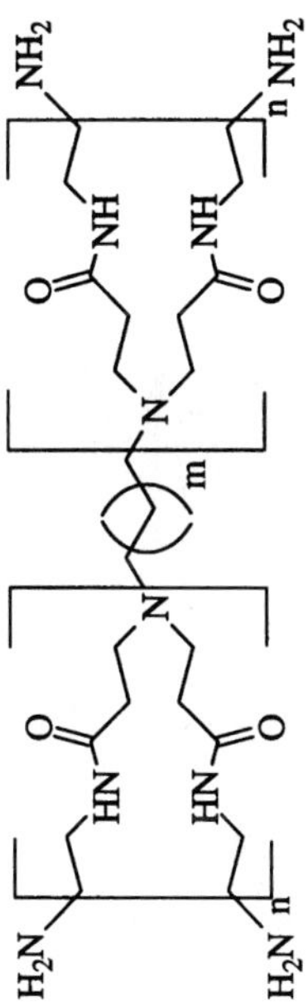

50 m = 10 ⟶ G_n(C12)

51 m = 0 ⟶ G_n(C2)

C = Chainlength = 12

G_n = Generaton Number

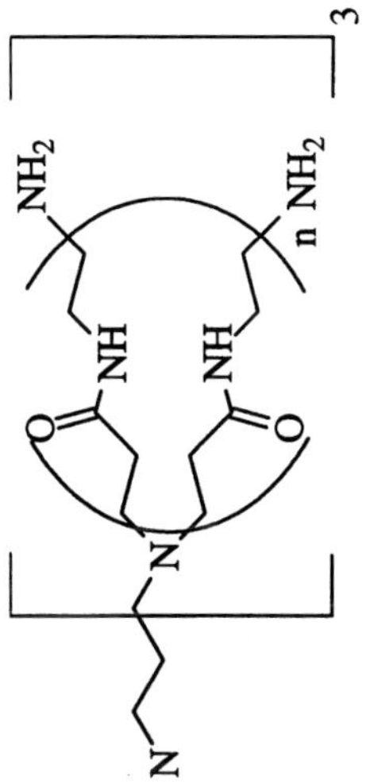

52 $G_n(NH_3)$

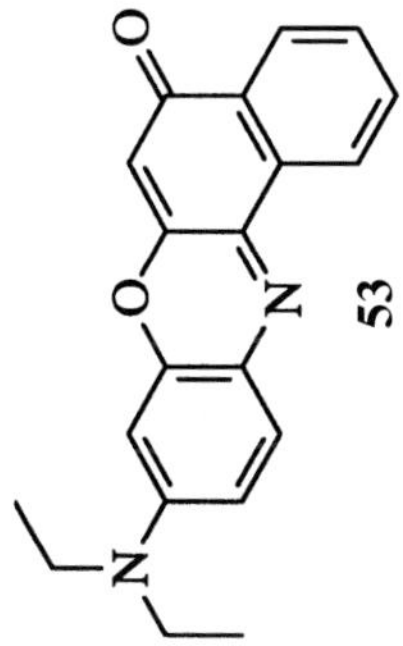

53

Fig. 23. Dendritic bolaamphiphiles reported by Tomalia et al.

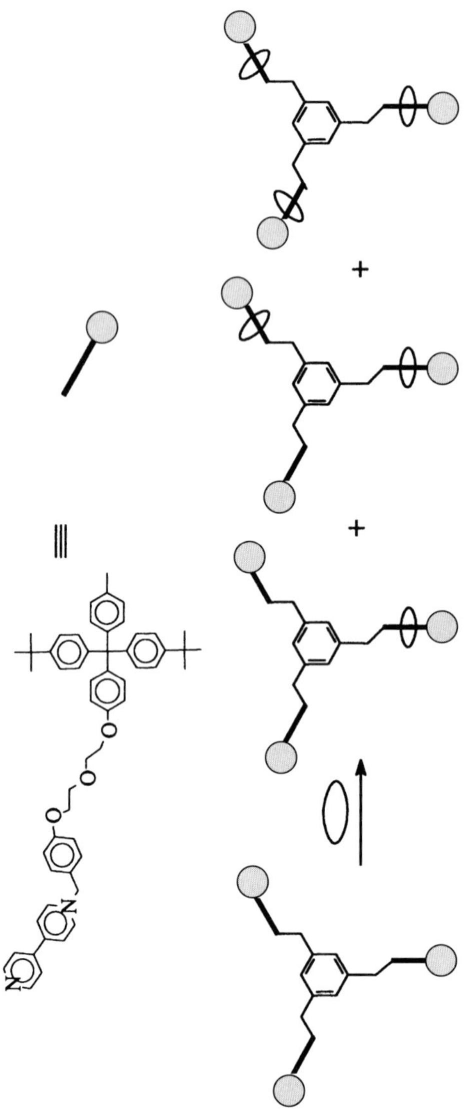

54 · PF$_6$ **55 · PF$_6$** **56 · PF$_6$** **57 · PF$_6$**

Fig. 24. Dendritic rotaxanes reported by Stoddart et al.

^{1}H NMR spectra for both hydroquinone and bipyridinium protons indicated π-π interactions.

In a recent report [141] Stoddart et al. reported a new class of rotaxanes with dendritic stoppers by using a so-called "threading" approach (Fig. 25). Alkylation of bipyridinium based units with Fréchet's third tier *branched* aryl ethereal dendron, in the presence of BPP34C10 afforded **58** as one of the products. Variable temperature ^{1}H-NMR spectroscopy in different NMR solvents helped determine the novel "shuttling" process of BPP34C10 from one bipyridinium unit to the other in **58**. The dendritic framework of **58** assists in its solubility in a wide range of solvents.

5.1.3
Liquid Crystalline Dendrimers

One of the early examples of hyperbranched polymers exhibiting liquid crystalline behavior was reported by Kim [142]. Recently, in a series of reports, Percec et al. [143–145] synthesized a novel class of thermotropic, liquid crystalline polymers possessing tertiary *C*-branching centers. "Willow-like" dendritic construction was facilitated by the synthesis of AB_2 type of novel *branched* building blocks **59–61**, which could exist in *gauche* or *anti* conformations (Fig. 26). Synthesis of polyethereal dendrimers **62** in a one-pot reaction was conducted by using traditional phenoxide alkylation methods (Bu_4NHSO_4, NaOH). The phenolic terminal groups were further capped with either alkyl or benzyl groups. *Branched* monomer **61** with a more rigid aromatic spacer, was the most effective in increasing the temperature range of the nematic meosphase. Percec et al. [146, 147] further reported the synthesis of non-spherical dendrimers using a convergent approach. The formation of a nematic phase for all dendrimers suggested that even higher generation macromolecules do not exhibit a spherical or disc shape. Figure 27 shows an idealized representation of the isotropic and nematic states **62a** and **62b** respectively.

Recently [148] Percec et al. reported the synthesis of four generations of monodendrons based on $1 \rightarrow 3$ building block, methyl 3,4,5-trishydroxybenzoate. These conical or hemispherical shaped monodendrons (e.g., **63**) self-assembled into supramolecular dendrimers displaying a spherical shape, and exhibited a novel thermotropic cubic liquid crystalline phase (Fig. 28). Liquid-like short-range interactions of the cubic phase provided the long range periodicity necessary for detailed electron mapping by X-ray diffraction, thus making definitive structural assignments possible.

Ringsdorf et al. [149] reported the preparation of a hyperbranched macromolecule **64** utilizing an AB_2-type monomer **65** possessing a mixed anhydride and two mesogenic biphenylacetate moieties and a chiral capping agent **66**, possessing a 3,4-disubstituted benzoyl chloride (Fig. 29). Polarimetry was used to prove macromolecular chirality of **64** ($[\alpha]_D = +6.8°$ in CH_2Cl_2) and it exhibited liquid crystalline behavior. Recently, introduction of mesogenic units as end groups into carbosilane based dendrimers has also been reported [150, 151].

Metallomesogens are unique compounds which combine the properties of liquid crystals with that of metal complexes to generate novel electric or

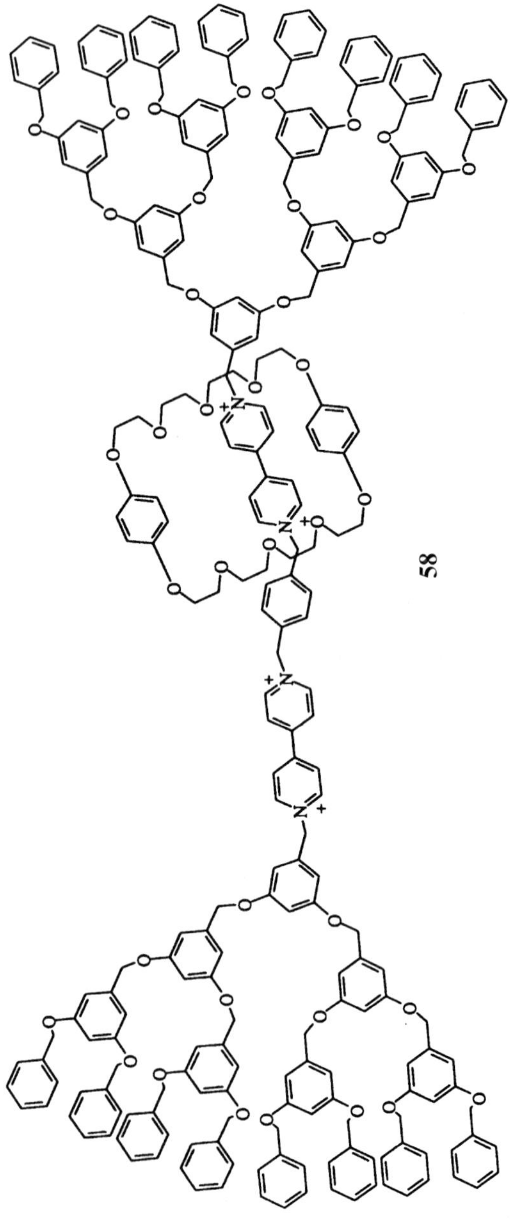

58

Fig. 25. Rotaxanes with dendritic stoppers synthesized by Stoddart et al.

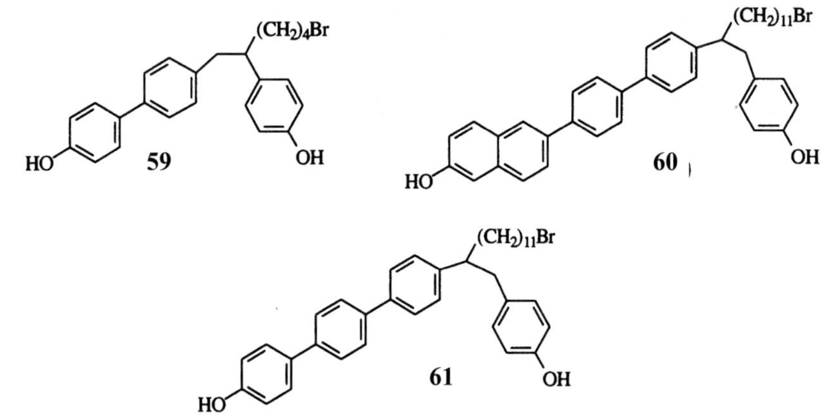

59 60 61

Fig. 26. Percec's AB_2 monomers for the synthesis of hyperbranched liquid crystalline dendrimers

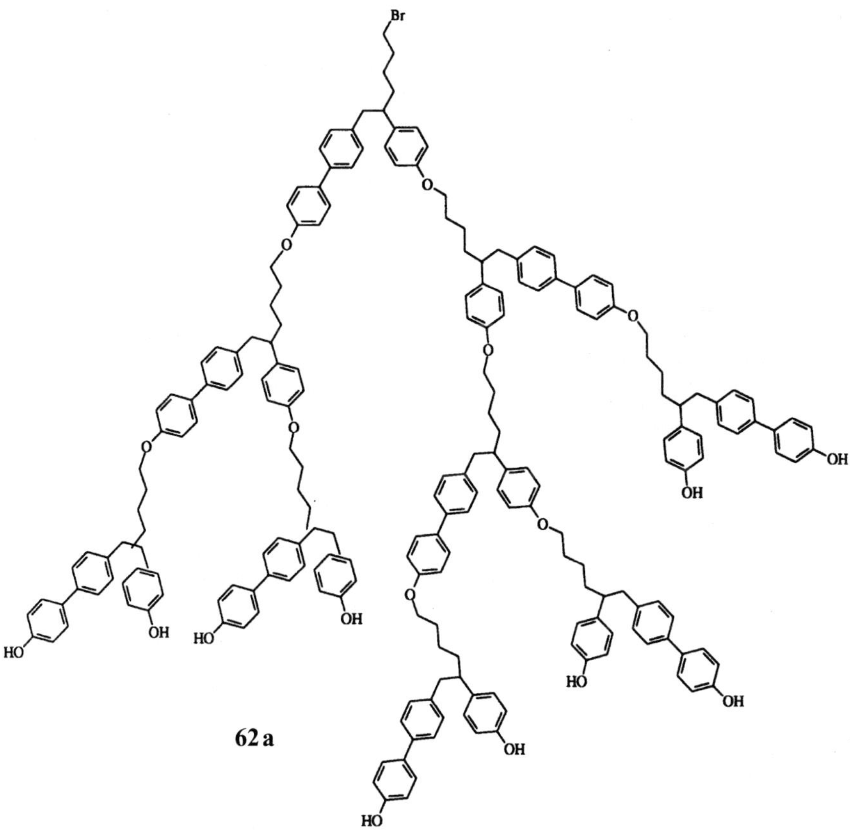

62 a

Fig. 27. Percec's hyperbranched liquid crystalline dendrimer that forms isotropic and nematic phases based on monomeric gauche and anti conformations

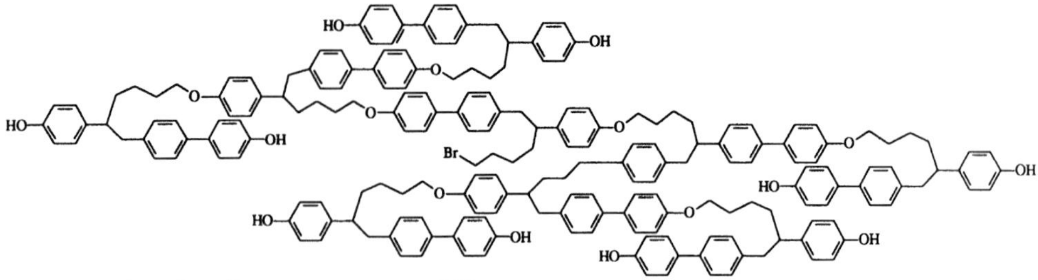

Fig. 27 (continued) **62b**

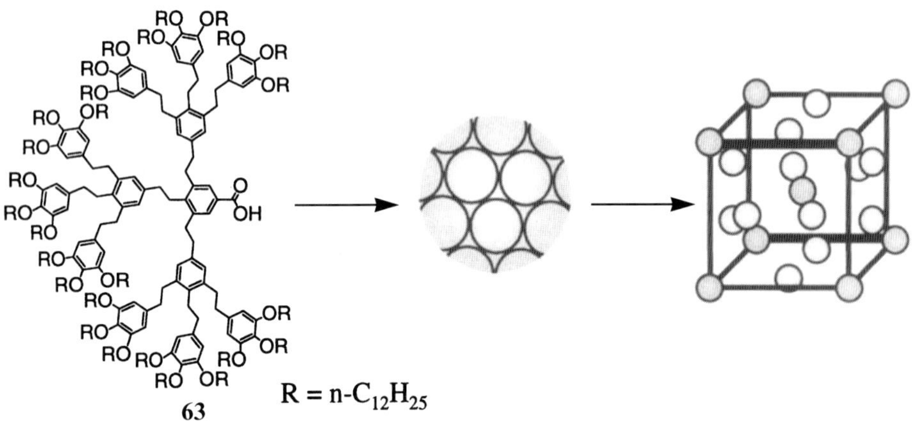

Fig. 28. Self-assembling dendrons exhibiting a thermotropic cubic liquid crystalline phase reported by Percec et al.

magnetic properties [152]. Lattermann et al. [153] reported the synthesis and properties of dendritic metallomesogens based on a three directional core units derived from tris(2-aminoethyl)amine (TREN). All complexes showed mesomorphic behavior with relatively low glass transition temperatures as determined by polarization microscopy and differential calorimetry.

Mesomorphic dendrimers containing electroactive units have potential for construction of dendrimer based molecular switches. Deschenaux et al. reported [154] the synthesis and liquid-crystalline properties of a novel dendrimer containing six mesomorphic ferrocene units. Apart from exhibiting a broad enantiotropic smectic A phase as determined by polarized optical microscopy, DSC, and XRD studies, thermogravimetry revealed the excellent thermal stability of the macromolecule.

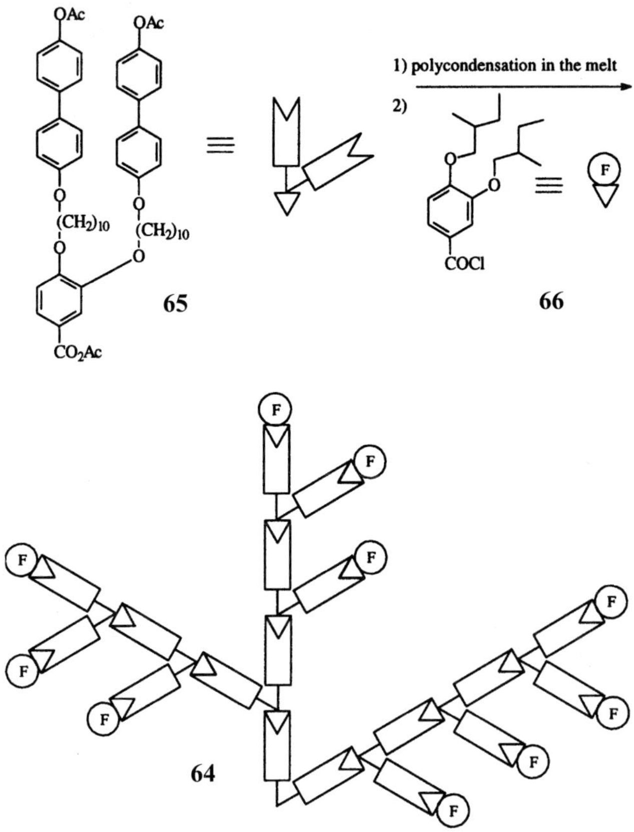

Fig. 29. Ringsdorf's liquid crystalline dendrimers

5.2 Self-Assembly via Intermolecular *H*-Bonding

In biological systems one of the primary modes of molecular recognition processes occurs via *H*-bond formation. Research concerning design and synthesis of molecular components that can self-assemble via *H*-bonding interactions has been reported [90, 155].

Probably the best example of the manifestation of self-assembly processes in dendritic systems via *H*-bonds is seen in the recent work of Zimmerman et al. [156]. Dendritic wedges possessing tetraacid moieties (**67**) self-assembling into a hexameric, disc-like framework (**68**) was confirmed by SEC and ^{1}H-NMR studies. The tetraacid unit (**69**) is known to form cyclic as well as linear aggregates in solution via carboxylic acid dimerization (Fig. 30). However, with incorporation of larger dendritic wedges on **69**, the hexamer form is preferred.

Recently, Kraft and Osterod [157] reported the synthesis of poly(aramide) dendrimers possessing either electron-deficient 1,3,4-oxadiazole (**70**) or aromatic systems (**71**) linked by amide units to a central triphenylmethane unit (Fig. 31).

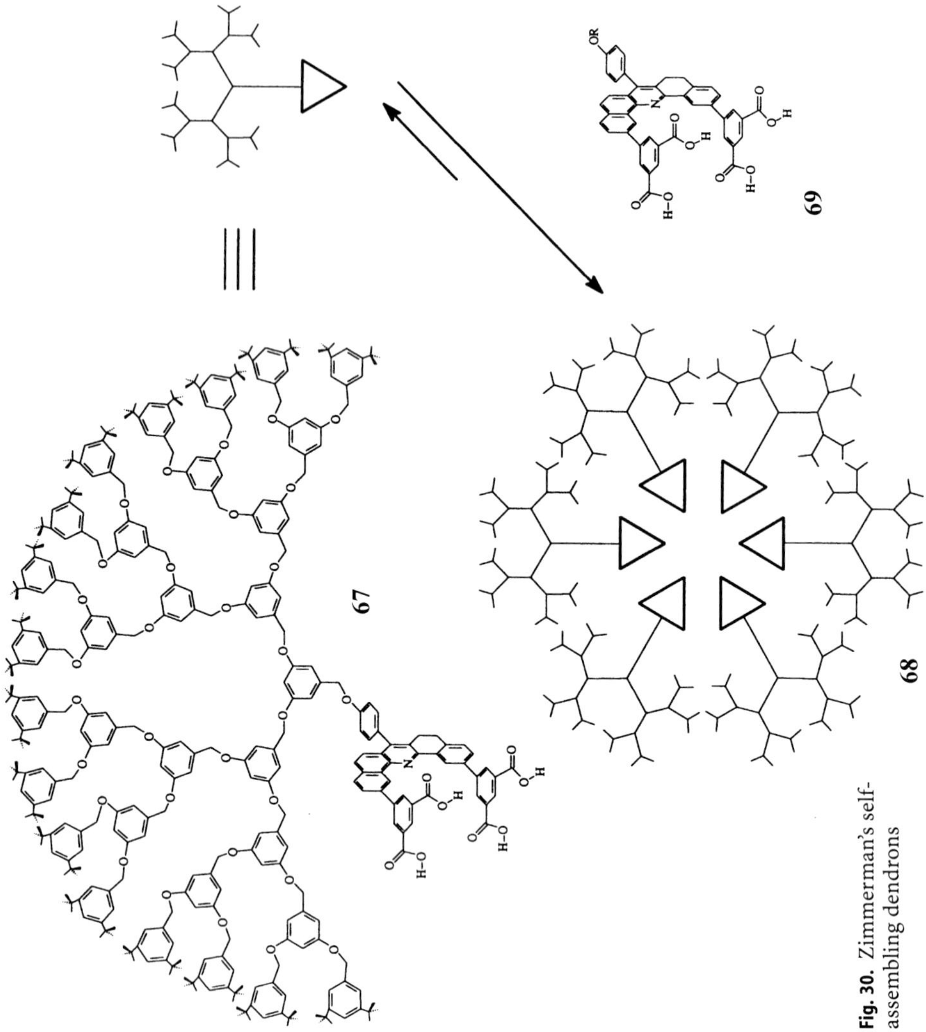

67

68

69

Fig. 30. Zimmerman's self-assembling dendrons

Viscosity and molar mass measurements for **70** and **71** supplemented with broad ^{1}H-NMR signals which depended on concentration, temperature, and solvent but independent of NMR frequency, strongly suggested self-association of these macromolecules (in a $CHCl_3$ solution).

5.3
Mono- and Multi-Layer Dendrimer Self-Assembly

Due to their well defined geometrical sizes, globular shapes, and uniform multiple surface functional groups, dendrimers are promising candidates for the construction of mono- or multi-layer films.

Regen and Watanabe [158] fabricated multi-layers close to a thickness of 800 Å by using fourth or sixth generation PAMAM dendrimers with 16 or 12 cycles, respectively. The construction techniques involved activation with K_2PtCl_4 on a silicon wafer containing amino groups on the surface, which was followed by deposition of the dendrimer layer. However, elimination of Pt^{2+} layer resulted in absence of layer growth, as examined by X-ray photoelectron spectroscopy.

In their efforts towards developing novel chemical sensors Crooks and Wells [159] reported the first example of a covalently bound dendrimer layer. PAMAM dendrimers were chemically linked to a mercaptoundecanoic acid (MUA) self-assembled monolayer (SAM) via amide bond formation, as indicated by FTIR external reflection spectroscopy (ERS). Five different generations of PAMAM dendrimers were used (G0, G2, G4, and G8) and the magnitude of the ellipsometric film thickness and dendrimer diameter could be well correlated to the total amide peak area in the IR spectrum. Michael addition to methyl acrylate by the primary amine groups of the dendrimer layer was measured by monitoring the ester peak in the IR spectrum, thus proving the presence of a reactive surface. Lower generation dendrimers (G0, G2) do not show surface reactivity due

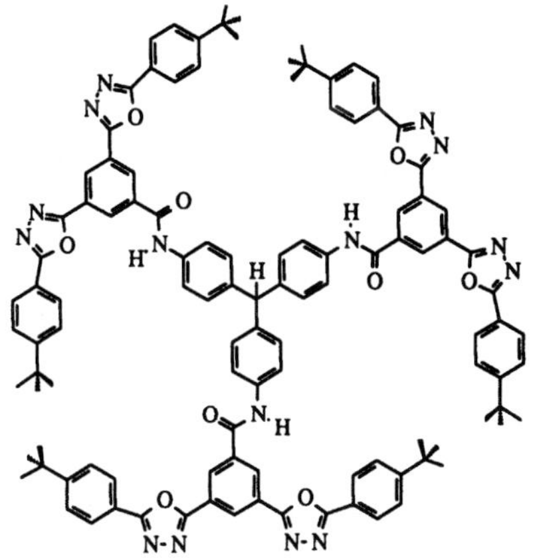

70

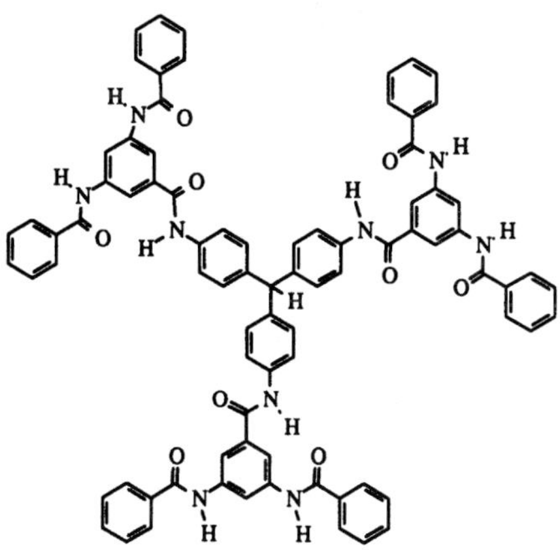

71

Fig. 31. Self-assembling aramide dendrimers reported by Kraft et al.

to complete reaction with SAM. Rapid, reversible responses and excellent signal to noise ratios were observed for these dendrimer-modified SAMs when treated with several volatile organic compounds in a surface acoustic wave (SAW) mass balance.

In a concurrently published report [160] Crooks and coworkers reported similar MUA-SAMs modified by covalent linking of hyperbranched macromolecules. These films containing a high density of surface carboxylic acid groups could selectively bind metal ions or undergo chemical modification.

Numerous modern applications require highly ordered, ultra-thin coatings with fine tunable surface properties. In their efforts towards building such systems Tsukruk et al. [161] reported construction of mono and multi-layer films by self-assembly of PAMAM dendrimers. Fabrication by self-assembly of dendrimers of two adjacent generations was conceived by electrostatic layer-by-layer deposition. Dendritic multilayers of the type (AB)x were assembled on a SiO_2 surface (where $A=G_n$, and $b=G_{n-1/2}$, G_n and $G_{n-1/2}$ being the amine terminated and carboxylic acid terminated dendrimers of adjacent generations respectively; and x (=20) being the number of molecular layers). Strong interactions between the terminal groups of adjacent dendritic layers resulting in compact, compressed layers were studied using scanning probe microscopy (SPM) and X-ray reflectivity techniques.

5.4 Self-Assembly of Metallo-Dendrimers

Metal ions have desirable properties as constituents of supramolecular systems and linking elements for self-assembly processes. Metal-ligand interactions display (a) inherent coordination geometries which are well-defined, (b) a range of weak to strong binding strengths, and (c) novel photo- and electro-chemical properties. This section deals with the different types of dendritic structures possessing metal complexes. Examples dealing with dendrimers based on metallo-porphyrin cores and metal-complexing, crown-ether units have been discussed in earlier sections.

5.4.1 *Dendrimers with Metal Centers as Linkers and Branching Points*

Balzani et al. reported [162, 163] one of the early examples of self-assembling metallo-dendrimers (Fig. 32). Thus, dendrimers with metal-binding sites acting as key links and branching points were synthesized by a divergent protection/deprotection scheme. Ruthenium(II)-polypyridine complexes were constructed using 2,3- or 2,5-bis(2-pyridyl)pyrazine (dpp, **72**) as the bridging ligand and 2,2′-bipyridine (**73**), as terminating ligand. Balzani et al. prefer to call this methodology as the "complexes-as-ligands/complexes-as-metals" strategy. $[Ru(2,3\text{-dpp})_3]^{2+}$ (**74**) which possesses three, free metal complexing sites was used as the core, while $[Ru(2,3\text{-Medpp})_2Cl_2]^{2+}$ (**75**) was used as the building block. The reaction between **74** and **75** generated the first generation tetranuclear metal complex (**76**). After removal of the methyl groups from the peripheral

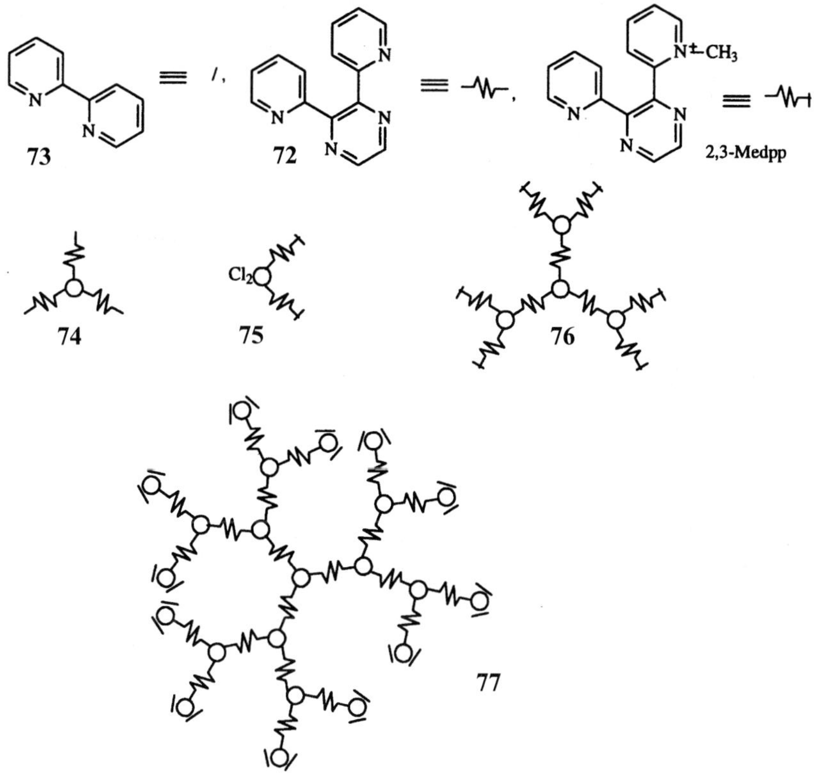

Fig. 32. Dendritic ruthenium complexes reported by Balzani et al.

nitrogens (Dabco in refluxing DMF), **76** is reacted with **75** to generate second generation dendrimer. These researchers eventually constructed the third generation dendrimer (**77**) containing 22 metal ions (MW: 10,890 amu). It was further proposed that novel dendritic-architectures, involving layers of ligands, complexed with different transition metal ions (e.g., Os^{2+}), could be self-assembled in a controlled manner, thus creating species with potentially useful photochemical properties. In their efforts towards such goals, Balzani and coworkers [164] recently investigated a series of luminescent dendrimers with an $[Os(2,3\text{-}dpp)_3]^{2+}$ core and **75** as the building block in a divergent approach to generate novel mixed metal macromolecular complexes.

5.4.2
Connectivity of Dendritic Fragments to Metallic-Cores

Recently, Newkome et al. [165] described the synthesis of dendritic lock and key complexes, such as **78**, which is based on the well-researched *bis*terpyridine-ruthenium (—⟨Ru⟩—) coordination chemistry (Fig. 33). Dendritic terpyridines of the first and second generation were reacted with $RuCl_3$ and this formed the

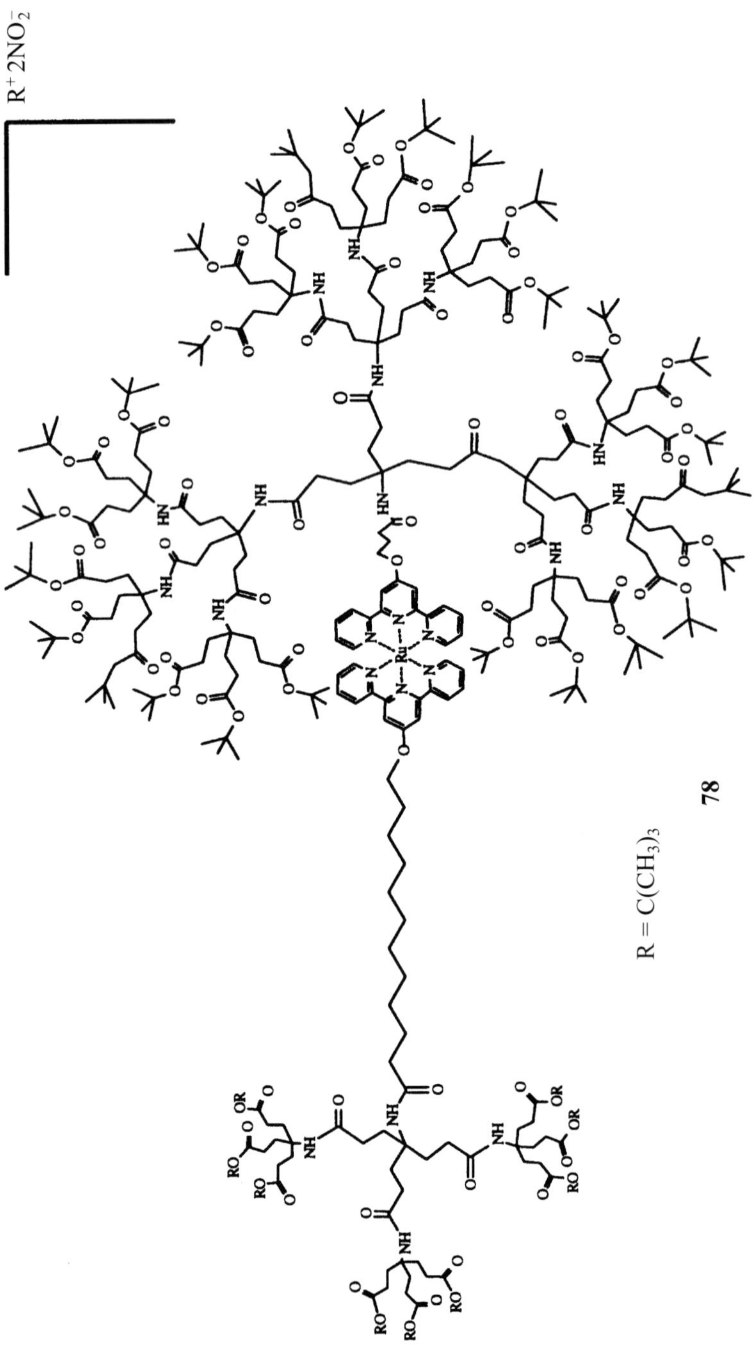

$R = C(CH_3)_3$

78

Fig. 33. Newkome's "lock and key" dendritic Ru-complexes metal terminated "keys." Dendritic terpyridines from the first through fourth generation were synthesized and used as the "locks." The lock and key complexation generated different dendritic metal complexes (e.g., **78**). Cyclic voltammetry, interestingly, exhibited irreversible redox processes as the generation number increased. Ineffective electron transfer due to the isolation of the redox center from the electrode surface by the dendritic shell or destabilization of redox products due to steric hindrance are the possible reasons for such irreversible electrochemistry.

Chow et al. [166] reported similar electrochemical results with dendrimers possessing *bis*(terpyridine) iron(II) complexes.

An alternative approach to bypass multi-step procedures involved in both convergent- and divergent-growth approach to dendrimers was reported by Tor and Tzalis [167]. Dendrimers with transition metal ions, as the core, were built with a novel 1,10-phenanthroline ligand, symmetrically substituted at the 3,8-positions with branched functional groups. Three units of the phenanthroline ligand complexed in an octahedral geometry when treated to a methanolic solution of aqueous $FeSO_4 \cdot 7H_2O$ followed by addition of excess KPF_6. ES-MS analysis was used to characterize the octahedral dendrimer complex $[Fe(Ligand)_3]^{2+}(PF_6^-)_2$. The ligand was also treated with $(CH_3CN)_4Cu^+PF_6^-$ in acetonitrile at 25 °C to afford the tetrahedral complex $[Cu(Ligand)_2]^+PF_6^-$ in high yields. Besides characterization of this complex by ES and FAB-mass spectroscopic methods, MLCT bands in the UV spectrum typical to this type of complex were also noted at 464 nm.

Due to their easily accessible redox potentials and low-energy excited states, dendrimers containing metals have important applications. For example, a long lifetime, luminescent excited state is important for immunoassay applications. This is because the signal of the label can be read after the decay of the background fluorescence of the sample, whose lifetime is usually in the nanosecond timescale. Vögtle and coworkers [168] recently reported the synthesis and characterization of a novel series of dendritic molecules based on a $[Ru(bpy)_3]^{2+}$ core. An approach similar to Tor and Tzalis [167] was followed in constructing dendrimers with a metal ion core. Fluorescence data revealed that second generation dendritic complex **79** had a more intense emission and a longer excited-state lifetime (>1 μs). It was also shown that the shielding effect of the dendrimer branches on the $[Ru(bpy)_3]^{2+}$ core of **79** strongly decreased the quenching effect of molecular oxygen by almost a factor of 12 compared to that of $[Ru(bpy)_3]^{2+}$ (Fig. 34). Thus desirable properties are bestowed on the Ru metal complex due to the "dendritic effect;" although the third generation bipyridine dendron (G3) was unable to form the trimer complex with Ru^{2+}. Conversion of the transoid structure of the free ligand (G3) to the cisoid conformation needed for metal complexation could be hindered by the dendritic branches or by the donor moieties in the dendritic part of the ligand competing with the bipyridine nitrogen atoms.

5.4.3
Dendrimers Assembled via Metal Connectivity at Multiple Sites

Amongst the multitude of metals that could be employed as connecting units in dendritic macromolecular assembly, ruthenium is by far the most commonly

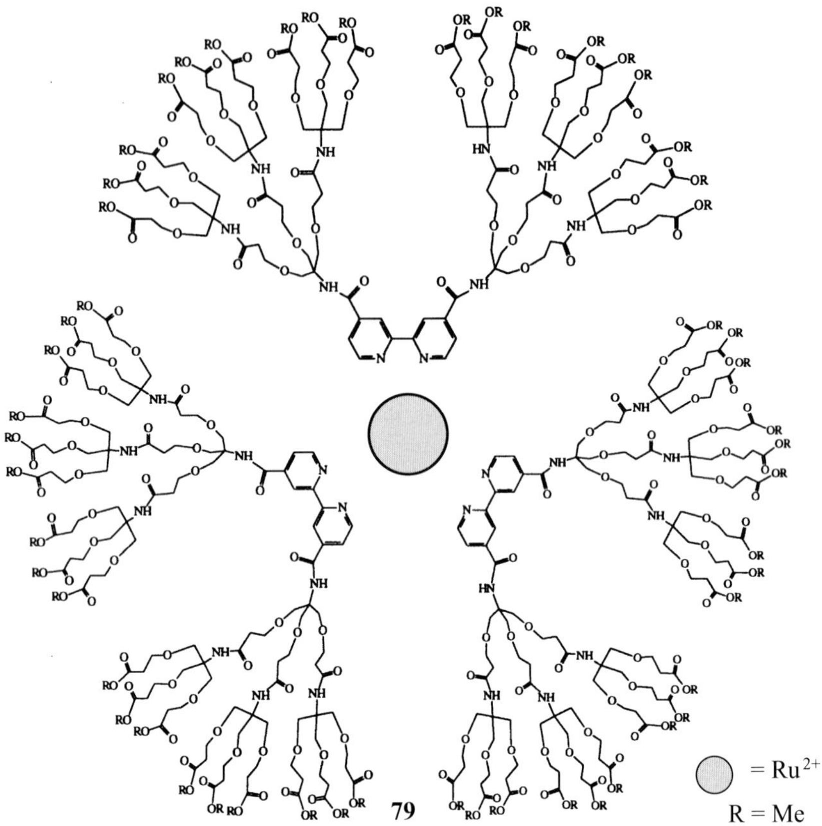

Fig. 34. A fluorescent dendrimer with a ruthenium core reported by Vögtle and Balzani

preferred metal, attributable to its unique chemistry [169] and highly desirable properties.

For instance, some time ago Newkome et al. reported the synthesis of ruthenium based dendrimers [170]. A dendrimer (**80**) with twelve peripheral terpyridine ligands was built around a central quaternary carbon-based core. In the final step complexation between the terminal ligand of the dendrimer and a terpyridinyl ruthenium chloride building block afforded the dodecaruthenium cascade molecule **80** (Fig. 35). Thus, preconstructed cores and dendritic fragments were linked by Ru^{2+} as the connecting unit and this mode of connectivity could be denoted by [—⟨Ru⟩—].

In a recent report [171] Newkome and He extended this concept and described the use of two ruthenium centers per appendage [—⟨Ru⟩—⟨x⟩—⟨Ru⟩—] towards construction of a four-directional dendrimer (e.g., **81**, Fig. 36). A combination of convergent and divergent approaches, hence, allowed the stepwise construction of metallodendrimers via controlled metal complexation.

Constable has described [172] the construction of a variety of metallodendrimers using both convergent and divergent approaches wherein the metal

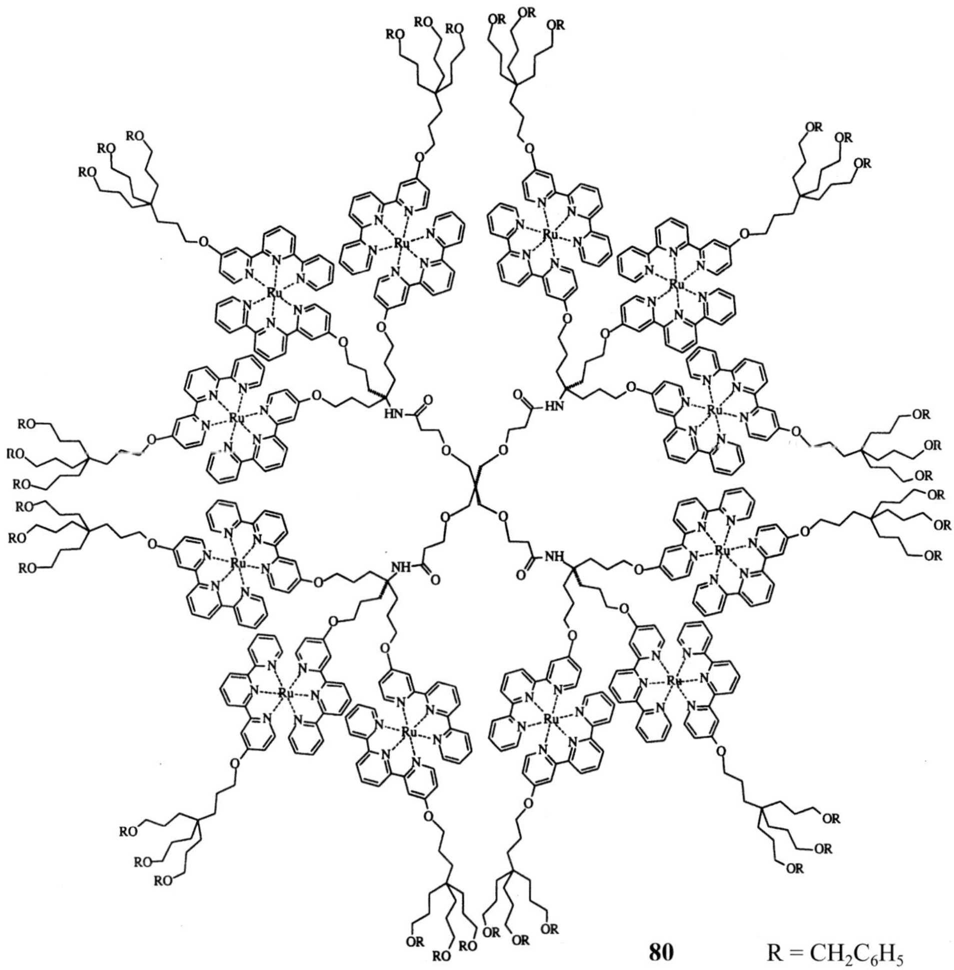
80 R = $CH_2C_6H_5$

Fig. 35. A dodeca-ruthenium dendrimer reported by Newkome et al.

ions used as key linking units have been described as a "supramolecular glue". In a related report [173], he and others described the synthesis of a pentaerythritol based dendrimer (**82**, Fig. 37) containing sixteen Ru ions embedded in a symmetric framework facilitated by the use of a metal-binding 2,2′:6′,2″-terpyridine ligand. Potential uses of these and related high nuclearity metallo-dendrimers as magnetic, electronic or photo-optical materials are envisioned.

Reinhoudt et al. [174] reported the self-assembly of an AB_2 type of monomer **83** with two Pd centers. Labile coordinating ligand, acetonitrile, when removed by heating from a solution of nitromethane in vacuo initiated an intramolecular coordination of benzylnitrile group with the tridentate Pd-centers leading to a hyperbranched structure. Coordination of the benzylnitrile group was moni-

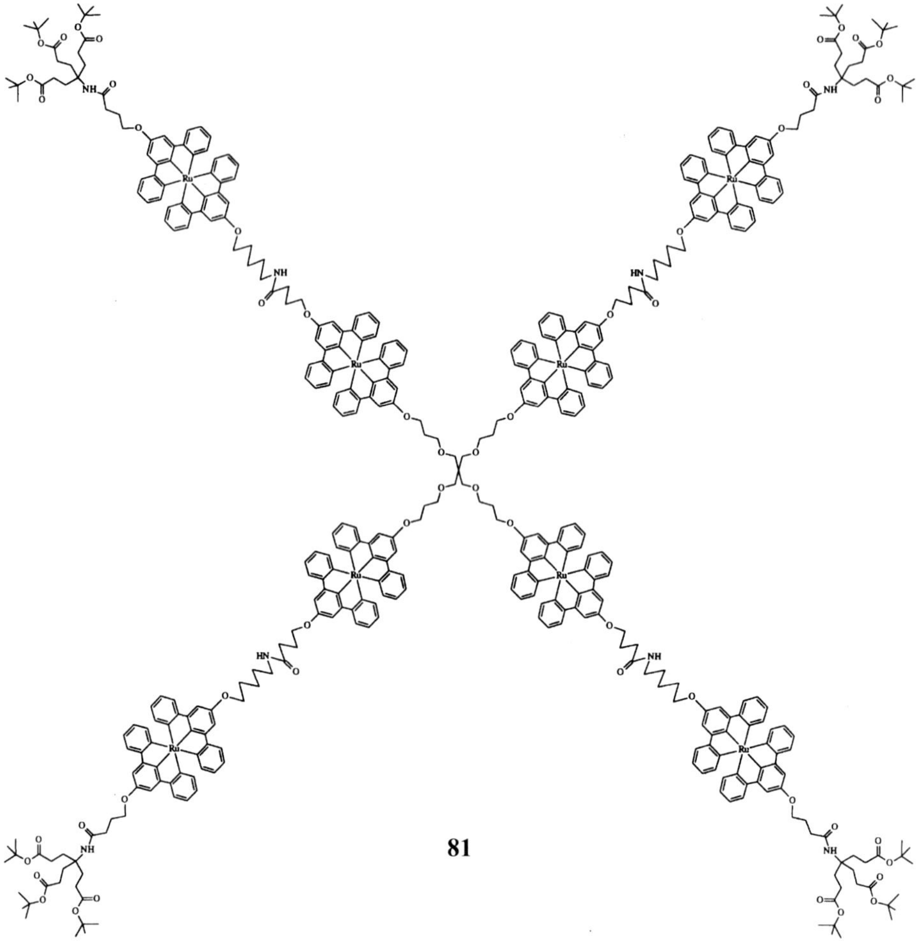

81

Fig. 36. Newkome's dendritic complex with —⟨Ru⟩—⟨x⟩—⟨Ru⟩— connectivity

tored by FTIR spectroscopy. The self-assembled polymeric structure was shown to disassemble by addition of acetonitrile. Using ^{1}H-NMR studies particles of polymeric aggregates measuring 200 nm were characterized using quasi-elastic-light-scattering (QELS), atomic-force-microscopy (AFM), and transmission-electron-microscopy (TEM). These researchers also noted that an analogous AB type monomer did not self-assemble into similar globular structures.

Reinhoudt et al. [175] in a later report described how they synthesized (Pd-based) metallo-dendrimers using a controlled divergent methodology. Building block **83** contains a labile coordinating cyano group and two kinetically inert, tridentate, (S-C-S)-palladium complexes wherein the Pd centers are temporarily protected by coordinating chloride ions. Core **84** on treatment with $AgBF_4$ (Fig. 38) afforded activated Pd centers which on further treatment with three equivalents of **83** gave the first generation dendrimer **85** formed by

Fig. 37. A pentaerythritol based polyruthenium dendrimer reported by Constable et al.

nitrile-palladium complexation. Repetitive addition of $AgBF_4$ and **83** afforded dendrimers up to the third generation. All metal dendritic palladium complexes were characterized by usual spectroscopic methods and ES-MS was used to determine the molecular masses.

In a very recent report [176] Reinhoudt et al. further demonstrated the usefulness of the above metal-based dendritic building blocks in the synthesis of bigger nanostructures which for the first time combined both coordination chemistry and *H*-bonding interactions. For example, a dendron containing a barbituric acid residue capable of *H*-bonding when treated with *N*-octadecanyl-*N′*-(2-*N*-tBoc-amino)phenylmelamine (NPM) formed a hexameric rosette. ^{1}H-NMR spectra at temperatures between −60 and −30 °C indicated the formation of a supramolecular complex via *H*-bonds between the barbituric acid groups of the dendron and NPM; however, the ^{1}H-NMR signals between 11 and 15 ppm for the *H*-bonded aggregates disappeared above −20 °C. Peak assignments for the different *H*-bonded protons were assigned on the basis of 2-D

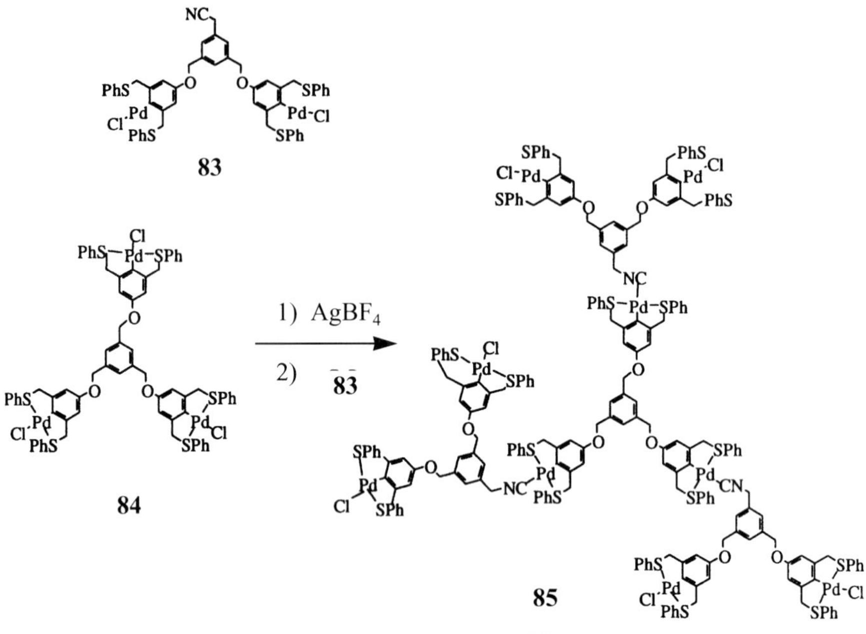

Fig. 38. Reinhoudt's Pd-based dendrimers

NOESY and TOCSY experiments. It was noted that while the monodendrons are soluble only in CH_3NO_2 the assemblies readily dissolve in CH_2Cl_2 on addition of NPM.

Puddhephatt et al. [177, 178] reported a convergent approach in the synthesis of a novel series of alkyl-platinum dendrimers (Fig. 39). The synthetic strategy was based on two known reactions, displacement of SMe_2 ligands from $[Pt_2Me_4(\mu\text{-}SMe_2)_2]$ (**87**) by diimine (2,2′-bipyridine) and oxidative addition of the resultant product $[PtMe_2$(2,2′-bipyridine)] (86) to 5,5′-*bis*(bromomethyl)-2,2′bipyridine (**88**) to yield **89**. Apart from notable color changes, formation of products was monitored by growth and decay of MLCT bands in the UV-VIS spectra. Additionally ^1H NMR and GPC provided useful data on structural features and molecular weights, respectively. Dendrimers up to the fourth generation possessing 28 Pt moieties were constructed. Reaction between **87** and **88**, however, afforded an insoluble, yellow, hyperbranched polymer which could not be characterized.

5.4.4
Dendrimers with Metal-Binding Sites on the Surface

There is a plethora of examples representing dendrimers with metal-binding sites on the periphery. Polycarboxylates were the simplest, early examples [179], although dendrimers possessing more complicated ligand complexing centers have invariably been built via a convergent approach.

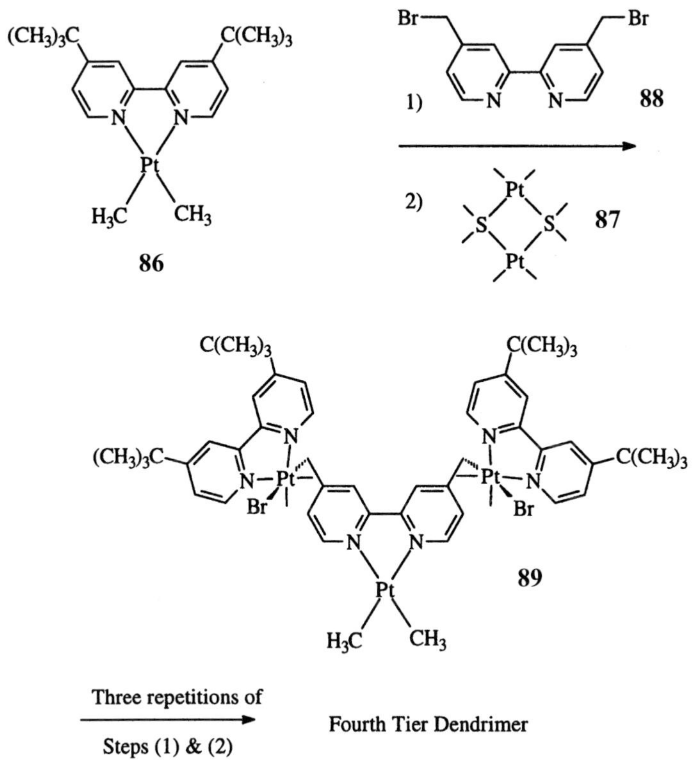

Fig. 39. Platinum based dendrimers constructed by Puddephatt et al.

van Koten et al. [180] reported the synthesis of polysilane dendrimers with either four or twelve (**90**) diamino aryl-nickel(II) complexes on their periphery (Fig. 40). These dendritic nickel complexes were used as catalysts in the Karasch addition reaction of polyhalogenoalkanes to carbon-carbon double bonds. Despite the slightly lower reactivity, these dendrimers exhibited excellent regiospecificity when compared to the corresponding mono-nickel catalyst. The additional advantage of these dendrimers is their easy separation from the reaction mixture by filtration procedures.

Seyferth and coworkers [181] introduced ethynyl groups onto the periphery of carbosilane dendrimers by displacement of chloride from the terminal silicon groups. They further treated these ethynyl terminated silicon dendrimers with $Co_2(CO)_8$ to afford the corresponding acetylenedicobalt hexacarbonyl dendritic complexes.

Moss and Liao [182, 183] reported a metallo-dendrimer built by a convergent strategy. An approach similar to that of Fréchet was used to synthesize a fourth generation dendrimer containing 48 peripheral ruthenium atoms.

Dubois et al. [184] reported the synthesis of phosphorus based dendrimers terminated with multivalent palladium complexes. Iterative free radical addition

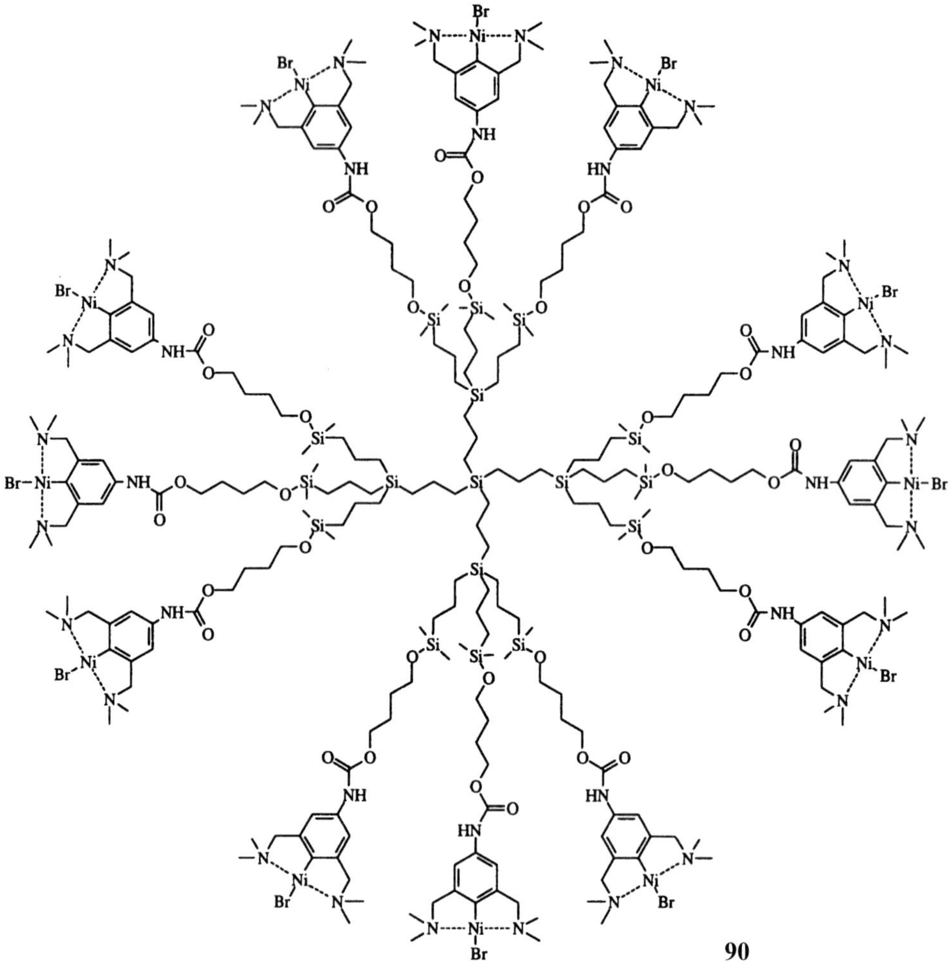

90

Fig. 40. van Koten's dendritic Ni-complex

of diethylvinylphosphonate to a primary phosphine followed by reduction of the resulting phosphonate with $LiAlH_4$ afforded dendrimers of different generations. These dendrimers were treated with $[Pd(CH_3CN)_4](BF_4)_2$ to produce metallated dendrimers which were shown to catalyze the electrochemical reduction of CO_2 to CO.

Complexation of gold ions, [Au(I)], with peripheral phosphine groups of a *P*-based dendrimer was reported by Majoral et al. [185]. Transmission electron microscopy (TEM) was used to analyze the large aggregates formed by the dendritic gold complexes and a direct correlation was observed between the size of the particles and the dendrimer generation number. In a recent report [186], Majoral et al. further demonstrated that up to 48 diphosphino groups could be anchored to the surface of dendrimers and various dendritic metal-complexes

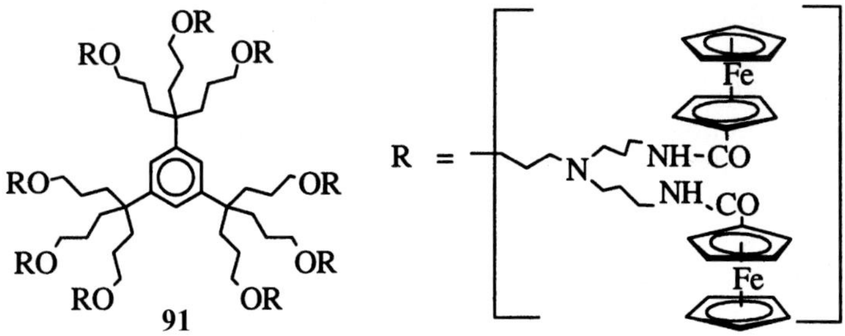

Fig. 41. Polyamido ferrocene dendrimers synthesized by Astruc et al.

were obtained (Pd, Pt, Rh), and subsequently used as potentially useful organometallic catalysts.

Astruc et al. [187] reported a nonairon sandwich complex by treating a nonaol with $\{[C_5H_5]Fe(\eta^6\text{-}p\text{-}MeC_6H_4F)(PF_6)\}$. In a subsequent report [188], Astruc and Marvaud reported the synthesis of "aromatic star" molecules with or without a central $Fe(\eta^5\text{-}C_5H_5)^+$ group. These bipyridine and terpyridine terminated dendrimers were further capped with $[Ru(bipy)_2Cl_2]$ and $[Ru(terpy)Cl_3]$, respectively, to afford the corresponding hexa or hepta nuclear complexes.

Recently, Astruc et al. [189] reported novel amido-ferrocene dendrimers (e.g., **91**) which were shown to act as supramolecular redox sensors for the recognition of small inorganic ions (Fig. 41). It was further observed that as the dendrimer generation number increased the sensitivity to the guest molecules also increased as observed by cyclic voltammetry experiments.

Cuadrado et al. [190] modified the terminal Si-Cl groups of silicon based dendrimers by reacting with $[Fe(C_5H_5)(C_5H_5Li)]$ or $[Fe(C_5H_5)(C_5H_5CH_2CH_2NH_2)]$ to obtain first and second generation ferrocenyl dendrimers. Platinum and indium-tin-oxide electrode surfaces were modified with electroactive films of these novel dendrimers, thus, affording potentially useful multi-electron redox catalysts.

6 Conclusions

The growing interest in dendrimer research has sky-rocketed and shall predictably continue unabated as chemists use this simple iterative technology to approach the meso- and macromolecular frontiers. Dendrimers, in a unique way, combine characteristics such as monodisperse definite composition typically to small organic molecules, with attributes unique to traditional polymers, such as high molecular weights. This has generated a plethora of new properties for these macromolecules which are currently being continually tapped by chemists to achieve utilitarian goals.

At present, improved methods of dendrimer construction have made possible the commercial availability of these macromolecules with a variety of molecular weight ranges and surface functional groups. Instead of testing the limits in building dendrimers of higher molecular weights, the focus has now largely shifted to exploiting the novel properties of these macromolecules towards finding new useful commercial applications.

The overwhelming number of dendrimer-related reports flooding the chemical arena, particularly, in the last five years, has made it a difficult task to summarize all important developments in one treatise. The restricted scope of this chapter – supramolecular chemistry within dendritic materials – denotes the utilitarian character to the unique infrastructure of these materials. Surface coatings and attachments to molecular spheres should possess a common theme respective of their frameworks, and thus there should be less differentiation between the mode of construction but rather what is the surface functionality.

Dendrimers, for example, are being tested for their potential use in biological and biomedical applications. Recent reports describe the ability of amine terminated PAMAM dendrimers [191–194] possessing positively charged surfaces under physiological pH conditions to bind with polyanionic DNA and function as in vitro gene-transfer agents. Attachment of multiple copies of peptide sequences to small dendritic scaffolds has afforded novel peptide dendrimers [195–197] with increased immunogenic effects, and one case even acts as a promising synthetic AIDS vaccine [198]. A sialic acid dendrimer which inhibits the infuenza A virus haemaglutinin is a fine example of the rapidly growing field of glycodendrimers and glycopolymers [199]. From sugar-binding receptors (application in diabetes treatment) [200, 201] to polygadolinium chelates (magnetic resonance imaging contrast agents) [202], dendrimer surfaces have been modified with a variety of application-based units. Surface modification of dendrimers with varied functional units has generated utilitarian dendrimers having potential applications as novel catalysts [203–206], radiotherapeutic agents [207], and in chromatographic separations [208–210]. Recently asymmetric catalysis was achieved using a chiral hyperbranched macromolecule affording high yields and good enantioselectivity for the resultant product [211].

Recent reports of dendrimers with a versatile C_{60} core [212] allowing an unusual core branching multiplicity of 12, and reaching dense-packing limits already in the first generation, continues to prove that there are many more unique dendrimers with varied architectures and associated interesting properties that need to be discovered.

As new synthetic methods are pursued in future to synthesize dendrimers more efficiently and cheaply on an industrial scale, one of the most promising areas for these macromolecules is seen in the construction of higher order assemblies or "dendritic networks" [213].

Combinatorial approach to unsymmetrically tiered macromolecules [214] is a brand new area of research which would allow "dendrimerization" of materials (e.g., glass, classical polymers, fibers) and thus enable fine tuning of macromolecular properties. For example, treatment of an amine terminated dendrimer with a "*mixture*" of complementary, isocyanate-based monomers [215, 216] affords a heterogeneous surfaced dendrimer. Selective transformation of the surface nitrile moieties via metal-catalyzed reduction to obtain a new polyamine dendrimer allows further *combinatorial-based* elaboration as illustrated in Fig. 42.

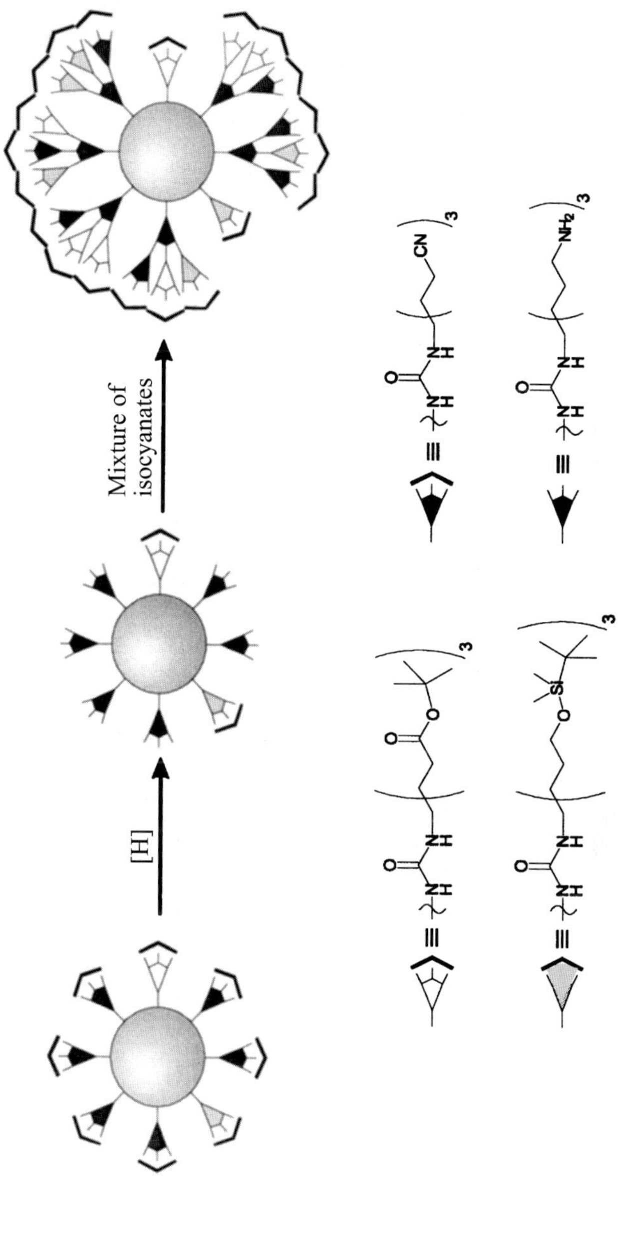

Fig. 42. Combinatorial tier construction with isocyanate based dendritic monomers

Dendrimers produced in this way will necessarily possess unique cavities, clefts, and void regions thereby facilitating the investigation of novel, dissymmetric architectures [polycelles = poly(micelles)] [214], and thus add the next chapter to this ever expanding field of supramolecular chemistry.

Thus, the concept and manifestation of molecular recognition mediated supramolecular self-assembly of small chemical units, be it an AB_n type building block or a dendrimer molecule, have proved to be vital factors in bridging the gap between small molecules and novel "new-age" materials.

Acknowledgement. We wish to acknowledge the National Science Foundation and NATO for partial support of this work.

7 References

1. Vögtle F (1991) Supramolecular chemistry. Wiley, Chichester
2. Lehn J-M (1995) Supramolecular chemistry: concepts and perspectives. VCH Publishing, Weinheim, Germany
3. Lawrence DS, Jiang T, Levett M (1995) Chem Rev 95:2229
4. Smith DR (1994) Supramolecular chemistry. Chemistry & Industry, p 14
5. Newkome GR, Moorefield CN (1996) In: Reinhoudt DN, Lehn J-M (eds) Comprehensive supramolecular chemistry, vol 10, chap 25. Springer, Berlin Heidelberg New York, p 777
6. Newkome GR, Moorefield CN (1996) Design, syntheses, and supramolecular chemistry of smart cascade polymers. In: Müller A, Dress A, Vögtle F (eds) From simplicity to complexity in chemistry and beyond, pt 1. Vieweg & Sohn, Braunschweig/Wiesbaden, p 127
7. Zeng F, Zimmerman SC (1997) Chem Rev 97(5):1681
8. Lehninger AL (1976) Biochemistry chap 36. Worth Publishers, New York
9. Pederson CJ (1988) Angew Chem Int Ed Engl 27:7027
10. Cram DJ (1988) Angew Chem Int Ed Engl 27:1009
11. Lehn JM (1988) Angew Chem Int Ed Engl 27:89
12. Stupp SI, LeBonheur V, Walker K, Li LS, Huggins KE, Kesser M, Amsturz A (1997) Science 276:384
13. Gardner GB, Venkataraman D, Moore JS, Lee S (1995) Nature 374:792
14. (a) Seto CT, Whitesides GM (1991) J Am Chem Soc 113:712; (b) Seto CT, Whitesides GM (1992) J Am Chem Soc 115:905; (c) Seto CT, Whitesides GM (1993) J Am Chem Soc 115:1330; (d) Zerkowski JA, Seto CT, Whitesides GM (1992) J Am Chem Soc 114:5473; (e) Seto CT, Mathias J-P, Whitesides GM (1993) J Am Chem Soc 115:1321
15. (a) Lehn J-M, Mascal M, DeCian A, Fischer J (1990) Chem Commun 479; (b) Lehn J-M (1988) Angew Chem Int Ed Engl 27:89; (c) Lehn J-M (1990) Angew Chem Int Ed Engl 29:1304; (d). Kotera M, Lehn J-M, Vigneron J-P (1994) Chem Commun 197
16. Philp D, Stoddart JF (1996) Angew Chem Int Ed Engl 35:1154
17. Percec V, Heck J, Johansson G, Tomazos D, Kawasumi M, Chu P (1994) J Macromol Sci Pure Appl Chem A31(11):1719
18. Branda N, Grotzfeld RM, Valdes C, Rebek J (1995) J Am Chem Soc 117:85
19. Ghadiri MR, Granja JR, Milligan RA, McRee DE, Khazanovich N (1993) Nature 366:324
20. Newkome GR, Moorefield CN, Vögtle F (1996) Dendritic molecules: concepts, synthesis, perspectives. VCH Publishing, Weinheim, Germany
21. Adroin N, Astruc D (1996) Bull Soc Chim Fr 132:875
22. Voit BI (1995) Acta Polym 46:87
23. Issberner J, Moors R, Vögtle F (1994) Angew Chem Int Ed Engl 106:2507
24. Fréchet JMJ (1994) Science 263:1710
25. Tomalia DA, Durst HD (1993) Top Curr Chem 165:193
26. Newkome GR, Yao Z-Q, Baker GR, Gupta VK (1985) J Org Chem 50:2003

27. Tomalia DA, Naylor AM, Goddard WA III (1990) Angew Chem Int Ed Engl 29:138
28. Buhleier E, Wehner W, Vögtle F (1978) Synthesis 155
29. Tomalia DA, Baker H, Dewald JR, Hall M, Kallos G, Martin S, Roeck J, Ryder J, Smith P (1985) Polymer J 17:117
30. Newkome GR (1998) Royal Society of Chemistry 1998 National Congress, Durrham, England
31. See [32] and [33] for details regarding the term"site-specific"
32. Newkome GR, Moorefield CN (1993) Polym Preprints 34 : 75
33. Newkome GR, Moorefield CN (1994) Macromol Symp 77:63
34. Flory PJ (1953) Principles of polymer chemistry. Cornell University Press
35. Worner C, Mülhaupt R (1993) Angew Chem Int Ed Engl 32:1306
36. de Brabander-van den Berg EMM, Meijer EW (1993) Angew Chem Int Ed Engl 32:1308
37. Tomalia DA (1993) Aldrichimica Acta 26:91
38. Tomalia DA, Swanson DR, Klimash JW, Brothers HM III (1993) Polym Preprints 34(1):52
39. Kim C, Park E, Jung I (1996) J Korean Chem Soc 40:347
40. Slany M, Bardaji M, Caminade AM, Chaudret B, Majoral JP (1997) Inorg Chem 36:1939
41. Moucheron C, Mesmaekar AK-D, Dupont-Gervais A, Leize E, Dorsselaer AV (1996) J Am Chem Soc 118:12,834
42. Denkewalter RG, Kolc J, Lukasavage WJ (1981) US Pat 4,289,872
43. Denkewalter RG, Kolc J, Lukasavage WJ (1982) US Pat 4,360,646
44. Denkewalter RG, Kolc JF, Lukasavage WJ (1983) US Pat 4,410,688
45. Young JK, Baker GR, Newkome GR, Morris KF, Johnson Jr CS (1994) Macromolecules 27:3464
46. Hawker CJ, Fréchet JMJ (1990) Chem Commun 1010
47. Fréchet JMJ, Hawker CJ, Phillippides AE (1991) US Pat 5,041,516
48. Miller TM, Neenan TX (1990) Chem Mater 2:346
49. Xu ZF, Moore JS (1993) Angew Chem Int Ed Engl 32:246
50. Xu ZF, Moore JS (1993) Angew Chem Int Ed Engl 32:1354
51. L'abbé G, Forier B, Dehaen W (1996) Chem Commun 2143
52. Zeng FW, Zimmerman SC (1996) J Am Chem Soc 118:5326
53. Kim YH (1992) Adv Mater 4:764
54. Maciejewski M (1982) J Macromol Sci-Chem A17:689
55. Naylor AM, Goddard WAI, Kiefer GE, Tomalia DA (1989) J Am Chem Soc 111:2339
56. de Gennes PG, Hervet H (1983) J Phys Lett 44:351
57. Menger FM, Takeshita M, Chow JF (1981) J Am Chem Soc 103:5938
58. Shinkai S, Mori S, Koreishi H, Tsubaki T, Manabe O (1986) J Am Chem Soc 108:2409
59. Newkome GR, Moorefield CN, Baker GR, Saunders MJ, Grossman SH (1991) Angew Chem Int Ed Engl 30:1178
60. Newkome GR, Moorefield CN, Baker GR, Johnson AL, Behera RK (1991) Angew Chem Int Ed Engl 30:1176
61. Newkome GR, Moorefield CN (1992) US Pat 5,154,853
62. For a review see: Ringsdorf H, Schlarb B, Venzmer J (1988) Angew Chem Int Ed Engl 27:113
63. Menger FM, Portnoy CE (1967) J Am Chem Soc 89:4698
64. Menger FM, Littau CA, (1991) J Am Chem Soc 113:1451
65. Hawker CJ, Wooley KL, Fréchet JMJ (1993) J Chem Soc Perkin Trans 1 21:1287
66. Stevelmans S, van Hest JCM, Jansen JFGA, van Boxtel DAFJ, de Brabander-van den Berg EMM, Meijer EW (1996) J Am Chem Soc 118:7398
67. Jansen JFGA, de Brabander-van den Berg EMM, Meijer EW (1995) J Am Chem Soc 117:4417
68. Jansen JFGA, de Brabander-van den Berg EMM, Meijer EW (1994) Science 266:1226
69. Jansen JFGA, de Brabander-van den Berg EMM, Meijer EW (1996) Macromol Symp 102:27
70. Jansen JFGA, Janssen RAJ, de Brabander-van den Berg EMM, Meijer EW (1995) Adv Mater 7:561
71. Newkome GR, Moorefield CN, Keith JM, Baker GR, Escamilla GH (1994) Angew Chem Int Ed Engl 33:666

72. Nicholas KM, Pettit R (1971) Tetrahedron Lett 3475
73. Exon C, Magnus P (1983) J Am Chem Soc 105:2477
74. Knudsen MJ, Schore NE (1984) J Org Chem 49:5025
75. Magnus P, Lewis RT, Huffman JC (1988) J Am Chem Soc 110:6921
76. Nicholas K (1987) Acct Chem Res 29:207
77. Newkome GR, Moorefield CN (1994) US Pat 5,376,690
78. Newkome GR, Moorefield CN (1995) US Pat 5,422,379
79. Newkome GR, Moorefield CN (1996) US Pat 5,516,810
80. Newkome GR, Moorefield CN (1996) US Pat 5,585,457
81. Nagasaki T, Kimura O, Ukon M, Arimori S, Hamachi I, Shinkai S (1994) J Chem Soc Perkin Trans 1:75
82. For a comprehensive review see Smid J (1981) Makromol Chem Suppl 5:203
83. Newkome GR, Lin X (1991) Macromolecules 24:1443
84. Mattei S, Seiler P, Diederich F, Gramlich V (1995) Hel Chim Acta 78:1904
85. Wallimann P, Seiler P, Diederich F (1996) Hel Chim Acta 79:779
86. Tomoyose Y, Jiang D-L, Jin R-H, Aida T, Yamashita T (1996) Macromolecules 29:5236
87. Jiang D-L, Aida T (1996) Chem Commun 1523
88. Collman JP, Fu L, Zingg A, Diederich F (1997) Chem Commun 193
89. MacDonald JC, Whitesides GM (1994) Chem Rev 94:2383
90. Newkome GR, Wooseley BD, He E, Moorefield CN, Güther R, Baker GR, Escamilla GH, Merill J, Luftmann H (1996) Chem Commun 2737
91. Newkome GR, Moorefield CN, Baker GR (1997) US Pat 5, 620, 101
92. Dandliker PJ, Diederich F, Gross M, Knobler CB, Louati A, Sanford EM (1994) Angew Chem Int Ed Engl 33:1739
93. Dandliker PJ, Diederich F, Gisselbrecht J-P, Louati A, Gross M (1995) Angew Chem Int Ed Engl 34:2725
94. Mekelburger H-B, Rissanen K, Vögtle F (1993) Chem Ber 126:1161
95. Junge DM, McGrath DV (1997) Chem Commun 857
96. Jiang D-L, Aida T (1997) Nature 388:454
97. Xu Z, Moore JS (1994) Acta Polymer 45:83
98. Young JK, Devadoss C, Zhu Z, Wang P-W, Moore JS (1995) Polym Mat Sci Eng 224
99. Kamlet MJ, Taft RW (1976) J Am Chem Soc 98:377
100. Johnson BP, Khaledi MG, Dorsey JG (1986) Anal Chem 58:2354
101. Hawker CJ, Wooley KL, Fréchet JMJ (1993) J Am Chem Soc 115:4375
102. Hanson JE, Murphy WR, Riley JM, Tyler TL, Kelley SO, Makarewicz A (1995) Polym Mat Sci Eng 358
103. Stewart GM, Fox MA (1996) J Am Chem Soc 118:4354
104. Newkome GR, Narayanan VV, Patri AK, Groß J, Moorefield CN, Baker GR (1995) Polym Mater Sci Eng 73:222
105. Narayanan VV, Newkome GR, Echegoyen L, Pérez-Cordero E (1996) Polym Preprints 37(2):419
106. Newkome GR, Narayanan VV, Echegoyen L, Pérez-Cordero E, Luftmann H (1997) Macromolecules 30(17):5187
107. Newkome GR, Behera RK, Moorefield CN, Baker GR (1991) J Org Chem 56:7162
108. Narayanan VV, Newkome GR (unpublished results)
109. Gorman CB, Parkhurst BL, Su WY, Chen K-Y (1997) J Am Chem Soc 119:1141
110. Bryce MR, Devonport W, Moore AJ (1994) Angew Chem Int Ed Engl 33:1761
111. Duan RG, Miller LL, Tomalia DA (1995) J Am Chem Soc 117:10,783
112. Miller LL, Hashimoto T, Tabakovic I, Swanson DR, Tomalia DA (1995) 7:9
113. Bryce MR, Devonport W (1996) Synthetic Metals 76:305
114. Tabakovic I, Miller LL, Duan RG, Tully DC, Tomalia DA (1997) Chem Mater 9:736
115. Castro R, Cuadrado I, Alonso B, Casado CM, Morán M, Kaifer AE (1997) J Am Chem Soc 119:5760
116. Cuadrado I, Morán M, Casado CM, Alonso B, Lobete F, García B, Ibisate M, Losada J (1996) Organometallics 15:5278

117. Newkome GR, Yao Z-Q, Baker G, Gupta VK, Russo PS, Saunders MJ (1986) J Am Chem Soc 108:849
118. Escamilla GH, Newkome GR (1994) Angew Chem Int Ed Engl 33:1937
119. Newkome GR, Baker GR, Saunders MJ, Russo PS, Gupta VK, Yao Z-Q, Miller JE, Bouillion KJ (1986) Chem Commun 752
120. Newkome GR, Lin X, Chen Y, Escamilla GH (1993) J Org Chem 58:3123
121. Escamilla GH (1995) In: Newkome GR (ed) Advances in dendritic macromolecules, vol 2, chap 6. JAI Press, Greenwich, Connecticut, p 157
122. Newkome GR, Baker GR, Arai S, Saunders MJ, Russo PS, Theriot KJ, Moorefield CN, Rogers LE, Miller JE, Lieux TR, Murray ME, Phillips B, Pascal L (1990) J Am Chem Soc 112:8458
123. Lehn JM (1990) Angew Chem Int Ed Engl 29:1304
124. Newkome GR, Moorefield CN, Baker GR, Behera RK, Escamilla GH, Saunders MJ (1992) Angew Chem Int Ed Engl 31:917
125. Jørgensen M, Bechgaard K, Bjørnholm T, Somme-Larsen P, Hansen LG, Schaumburg K (1994) J Org Chem 59:5877
126. Gitsov I, Wooley KL, Fréchet JMJ (1992) Angew. Chem. Int Ed Engl 31:1200
127. Gitsov I, Wooley Kl, Hawker CJ, Ivanova PT, Fréchet JMJ (1993) Macromolecules 26:5621
128. Gitsov I, Fréchet JMJ (1993) Macromolecules 26:6536
129. (a) Gitsov I, Wooley KL, Hawker CJ, Fréchet JMJ (1991) Polym Preprints 32(3):631; (b) Fréchet JMJ, Gitsov I (1995) Macromol Symp 98:441
130. Gitsov I, Fréchet JMJ (1996) J Am Chem Soc 118:3785
131. Chapman TM, Hillyer GL, Mahan EJ, Shaffer KA (1994) J Am Chem Soc 116:11,195
132. Scrimin P, Veronese A, Tecilla P, Tonellato U, Monaco V, Formaggio F, Crima M, Toniolo C (1996) J Am Chem Soc 118:2505
133. Vanhest JCM, Delnoye DAP, Baars MWPL, Vangenderen MHP, Meijer EW (1995) Science 268:1592
134. Israelachvili JN (1992) In: Intermolecular and surface forces. Academic Press, New York, p 380
135. Vanhest JCM, Baars MWPL, Elissenroman C, Vangenderen MHP, Meijer EW (1995) Macromolecules 28:6689
136. Watkins DM, Sayed-Sweet Y, Klimash JW, Turro NJ, Tomalia DA (1997) Langmuir 13:3136
137. Benesi HA, Hildebrand JH (1949) J Am Chem Soc 71:2703
138. Mulliken RS (1952) J Am Chem Soc 74:811
139. Ashton PR, Everitt SRL, Gómez-López M, Jayaraman N, Stoddart JF (1997) Tetrahedron Lett 38:5691
140. Ambalino DB, Ashton PR, Belohradsky M, Raymo FM, Stoddart JF (1995) Chem Commun 751
141. Ambalino DB, Ashton PR, Balzani V, Brown CL, Credi A, Fréchet JMJ, Leon JW, Raymo FM, Spencer N, Stoddart JF, Venturi M (1996) J Am Chem Soc 118:12,012
142. Kim YH (1992) J Am Chem Soc 114:4947
143. Percec V, Kawasumi M (1992) Polym Preprints 33(1):221
144. Percec V, Chu PW, Kawasumi M (1994) Macromolecules 27:4441
145. Percec V (1995) Pure Appl Chem 67:2031
146. Percec V, Chu PW, Ungar G, Zhou JP (1995) J Am Chem Soc 117:11,441
147. Li JF, Crandall KA, Chu PW, Percec V, Petschek RG, Rosenblatt C (1996) Macromolecules 29:7813
148. Balagurusamy VSK, Ungar G, Percec V, Johansson G (1997) J Am Chem Soc 119:1539
149. Bauer S, Fischer H, Ringsdorf H (1993) Angew Chem Int Ed Engl 32:1589
150. Ponomarenko SA, Rebrov EA, Bobrovsky AY, Boiko NI, Muzafarov AM, Shibaev VP (1996) Liq Cryst 21:1
151. Lorenz K, Holter D, Stuhn B, Mülhaupt R, Frey H (1996) Adv Mater 8:414
152. Serrano JL, (1995) Metallomesogens. VCH, Weinheim
153. Stebani U, Lattermann G, Wittenberg M, Wendorff JH (1996) Angew Chem Int Ed Engl 35:1858

154. Deschenaux R, Serrano E, Levelut A-M (1997) Chem Commun 1577
155. Lawrence DS Levett TJ (1995) Chem Rev 95:2229
156. Zimmerman SC, Zeng FW, Reichert DEC, Kolotuchin SV (1996) Science 271:1095
157. Osterod F, Kraft A (1997) Chem Commun 1435
158. Watanabe S, Regen SL (1994) J Am Chem Soc 116:8855
159. Wells M, Crooks RM (1996) J Am Chem Soc 118:3988
160. Zhou Y, Bruening ML, Bergbreiter DE, Crooks RM, Wells M (1996) J Am Chem Soc 118:3773
161. Tsukruk VV, Rinderspacher F, Bliznyuk VN (1997) Langmuir 13:2171
162. Serroni S, Denti G, Campagna S, Juris A, Ciano M, Balzani V (1992) Angew Chem Int Ed Engl 31:1493
163. Juris A, Venturi M, Pontoni L, Resino IR, Balzani V, Serroni S, Campagna S, Denti G (1995) Can J Chem 1875
164. Serroni S, Juris A, Venturi M, Campagna S, Resino IR, Denti G, Credi A, Balzani V (1997) J Mater Chem 7:1227
165. Newkome GR, Güther R, Moorefield CN, Cardullo F, Echegoyen L, Pérez-Cordero E, Luftmann H (1995) Angew Chem Int Ed Engl 34:2023
166. Chow HF, Chan IYK, Chan DTW, Kwok RWM (1996) Chem Eur J 2:1085
167. Tzalis D, Tor Y (1996) Tetrahedron Lett 37:8293
168. Issberner J, Vögtle F, De Cola L, Balzani V (1997) Chem Eur J 3:706
169. Seddon KR (1996) Platinum Metals Rev 40:128
170. Newkome GR, Cardullo F, Constable EC, Moorefield CN, Thompson AMWC (1993) Chem Commun 925
171. Newkome GR, He E (1997) J Mater Chem 7:1237
172. Constable EC (1997) Chem Commun 1073
173. Cattalini M, Constable EC, Housecroft CE, Eich O, Lazzarini C, Phillips D, Pohl-Ferry C (1997) Chimia 51:602
174. Huck WTS, van Veggel FCJM, Kropman BL, Blank DHA, Keim EG, Smithers MMA, Reinhoudt DN (1995) J Am Chem Soc 117:8293
175. Huck WTS, van Veggel FCJM, Reinhoudt DN (1996) Angew Chem Int Ed Engl 35:1213
176. Huck WTS, Hulst R, Timmerman P, van Veggel FCJM, Reinhoudt DN (1997) Angew Chem Int Ed Engl 36:1006
177. Achar S, Puddephatt RJ (1994) Angew Chem Int Ed Engl 33:847
178. Achar S, Vittal JJ, Puddephatt RJ (1996) Organometallics 15:43
179. Newkome GR, Lin X, Young JK (1992) Synlett 53
180. Knapen JWJ, Vandermade AW, Dewilde JC, Vanleeuwen PWNM, Wijkens P, Grove DM, van Koten G (1994) Nature 372:659
181. Seyferth D, Kugita T, Rheingold AL, Yap GPA (1995) Organometallics 14:5362
182. Liao YH, Moss JR (1993) Chem Commun 1774
183. Liao YH, Moss JR (1995) Organometallics 14:2130
184. Miedaner A, Curtis CJ, Barkley RM, Dubois DL (1994) Inorg Chem 33:5482
185. Slany M, Bardaji M, Casanove MJ, Caminade A-M, Majoral J-P, Chaudret B (1995) J Am Chem Soc 117:9764
186. Bardaji M, Kustos M, Caminade A-M, Majoral J-P, Chaudret B (1997) Organometallics 16:403
187. Moulines F, Djakovitch L, Boese R, Gloaguen B, Thiel W, Fillaut J-L, Delville MH, Astruc D (1993) Angew Chem Int Ed Engl 32:1075
188. Marvaud V, Astruc D (1997) Chem Commun 773
189. Valério C, Fillaut J-L, Ruiz J, Guittard J, Blais J-C, Asruc D (1997) J Am Chem Soc 119:2588
190. Alonso B, Cuadrado I, Moran M, Losada J (1994) Chem Commun 2575
191. Haensler J, Szoka FC Jr (1993) Bioconjugate Chem 7:372
192. Tang M, Redemann CT, Szoka FC Jr (1996) Bioconjugate Chem 7:703
193. Kukowska-Latallo JF, Bielinska AU, Johnson J, Spindler R, Tomalia DA, Baker JRJ (1996) Proc Natl Acad Sci USA 93:4897
194. Bielinska A, Kukowska-Latallo JF, Johnson J, Tomalia DA, Baker JRJ (1996) Nucleic Acids Res 24:2176

195. Tam JP (1988) Proc Natl Acad Sci USA 85:5409
196. Posnett DN, McGrath H, Tam JP (1988) J Biol Chem 263:1719
197. Tam JP, Lu Y-A (1989) Proc Natl Acad Sci USA 86:9084
198. Defroot J-P, Nardelli B, Huang W, Ho DD, Tam JP (1992) Proc Natl Acad Sci USA 89:3879
199. Roy R, Zanini D, Meunier SJ, Romanowska A (1993) Chem Commun 1869
200. James TD, Sandanayake KRAS, Iguchi R, Shinkai S (1995) Nature 374:345
201. James TD, Shinmori H, Takeuchi M, Shinkai S (1996) Chem Commun 705
202. Wiener EC, Brechbiel MW, Brothers H, Magin RL, Gansow OA, Tomalia DA, Lauterbur PC (1994) Magn Reson Med 31:1
203. Issberner J, Bohme M, Grimme S, Nieger M, Paulus W, Vögtle F (1996) Tetrahedron Asymmetry 7:2223
204. Suh J, Hah SS, Lee SH (1997) Bioorganic Chem 25:63
205. Chow H-F, Mak CC (1997) J Org Chem 62:5116
206. Bardají M, Caminade A-M, Majoral J-P, Chaudret B (1997) Organometallics 16:3489
207. Martin VV, Ralston WH, Hynes MR, Keana JFW (1995) Bioconjugate Chem 6:616
208. Kudzal SA, Monnig CA, Newkome GR, Moorefield CN (1994) J Am Chem Soc 119:2255
209. Castagnola M, Cassiano L, Lupi A, Messana I, Patamia M, Rabino R, Rossetti DV, Giardina B (1995) J Chromatogr 694:463
210. Muijselaar PGHM, Claessens HA, Cramers CA, Jansen JFGA, Meijer EW, de Brabander-van den Berg EMM, Vanderwal S (1995) HRC J High Res Chromat 18:121
211. Bolm C, Derrien N, Seger A (1996) Synlett 387
212. Camps X, Schönberger H, Hirsch A (1997) Chem Eur J 3:561
213. Newkome GR, Moorefield CN, Vögtle F (1996) Dendritic molecules: concepts, synthesis, perspectives. VCH Publishing, Weinheim, Germany, p 223–239
214. Newkome GR, Weis CD, Moorefield CN, Baker GR, Childs BJ, Epperson J (1998) Angew Chem Int Ed Engl 37:307
215. Newkome GR, Weis CD, Childs BJ (1998) Designed Monomers and Polymers, in press
216. Newkome GR, Weis CD (1997) US Pat 5, 703, 271

Divergent Approaches to Phosphorus-Containing Dendrimers and their Functionalization

Jean-Pierre Majoral · Anne-Marie Caminade

Laboratoire de Chimie de Coordination, CNRS, 205 route de Narbonne, F-31077 Toulouse cedex 4, France. *E-mail: majoral@lcctoul.lcc-toulouse.fr*

Divergent approaches to phosphorus-containing dendrimers are described. One of them allows one to prepare the dendrimer of the highest generation known up to now (generation 12, theoretical molecular weight >3000000). The reactivity on the surface of these macromolecules is detailed. Emphasis is made on a concept of multiplurifunctionalization allowing one to graft at the periphery a large number of sets of two, three or four functional groups. Reactivity into the internal voids of phosphorus-containing dendrimers is also reported showing that it is possible to develop a macromolecular chemistry into the cavities of a given dendrimer. Lastly the essential contribution of phosphorus in dendrimer chemistry is pointed out.

Keywords: Dendrimer, phosphorus, multiplurifunctionalization, complexation, dipole moments.

1 Introduction

Although the first publication on dendrimers appeared in 1978 [1] it is only in 1990 that the first report on phosphorus-containing dendrimers was published [2] i.e. after numerous papers devoted to the synthesis of organic dendrimers and practically at the same time than the description of a silicon containing dendrimer.

In part this is probably due to experimental difficulties encountered in phosphorus chemistry, i.e. the complexity and low yields of reactions and the sensitivity towards oxidation and hydrolysis of phosphorus(III) derivatives.

In view of the widespread use of phosphorus and specifically of tertiary phosphines and polyphosphines in coordination chemistry and catalysis the search of methods of synthesis of phosphorus-containing dendrimers appears to be one of the most exciting challenge in main group element chemistry during the last decade. Moreover phosphorus polymers such as polyphosphazenes of general formulae $[R_2P=N-]_n$ are known to present interesting properties as special rubbers, flame resistant materials, polymer conductors, liquid crystal polymers, adhesives, photocutting polymers, binders in paint formulation, flame retardant polymer additives, etc. Use of these polyphosphazenes for biomedical applications is also reported. Therefore one can expect that well defined phosphorus-containing dendrimers will present some of these properties as well as new ones because of the intrinsic properties of dendritic macromolecules.

The aim of this review is to present the state-of-the-art concerning the synthesis and the chemical properties of phosphorus-containing dendrimers with emphasis made on the specificity introduced by phosphorus.

2 Synthesis

2.1 Phosphorus on the Surface

α-(Diphenylphosphino) acetic acid and *p*-(diphenylphosphino) benzoic acid have been employed as spacers to functionalize dendritic polyamines of second and third generation such as diaminobutane-poly(trimethylene amine) dendrimers DAB-PPI$(NH_2)_{16}$ and DAB-PPI$(NH_2)_{32}$ with up to 32 peripheral diphenylphosphino groups [3] (Scheme 1). These polynuclear gold complexes are expected to be valuable in biochemical diagnostic and imaging as well as in medicine as antiinflamatory and antitumour drugs.

2.2 Phosphorus on the Core, Within the Cascade Structure and on the Surface

Dubois et al. [4] describe the synthesis of organophosphine dendrimers via the sequential addition of diethylvinyl phosphonate to primary phosphines followed by reduction with lithium aluminum hydride (Scheme 2). Metallation of

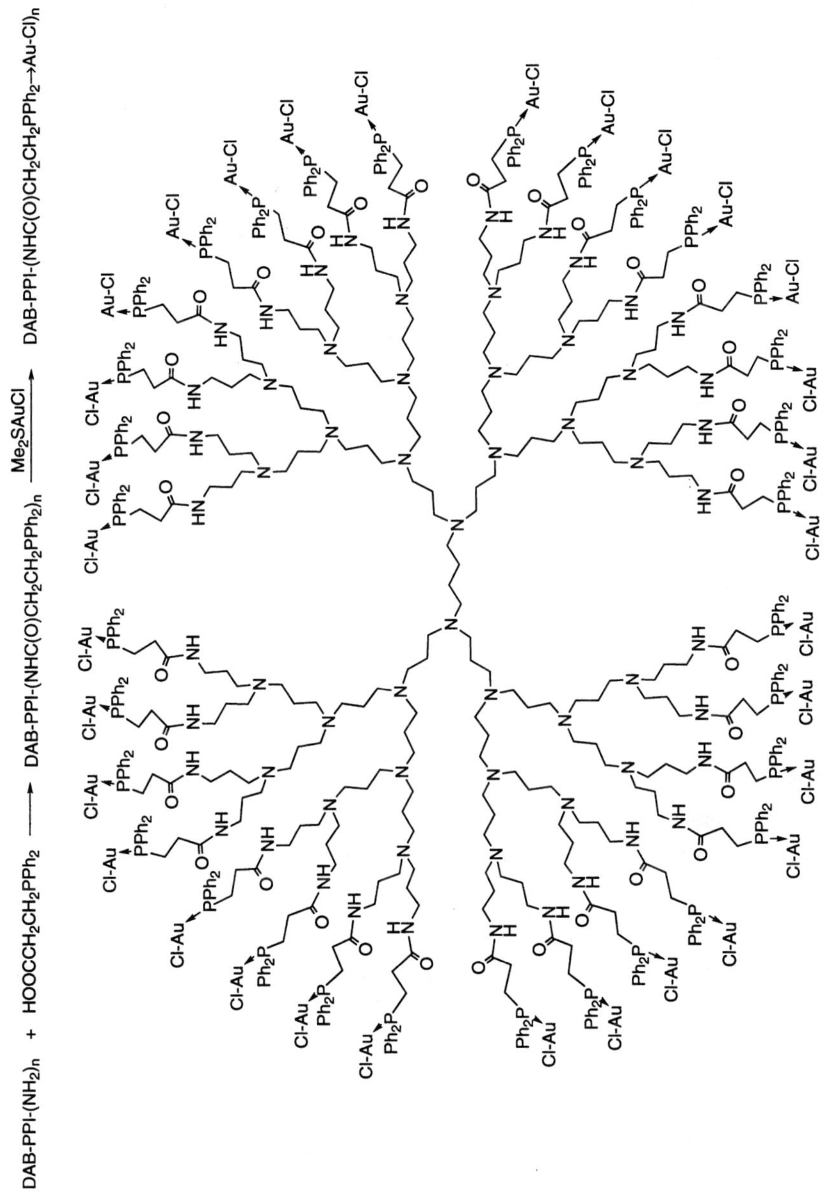
DAB-PPI-(NH2)n + HOOCCH2CH2PPh2 → DAB-PPI-(NHC(O)CH2CH2PPh2)n
Me2SAuCl
DAB-PPI-(NHC(O)CH2CH2PPh2→Au-Cl)n

Scheme 1

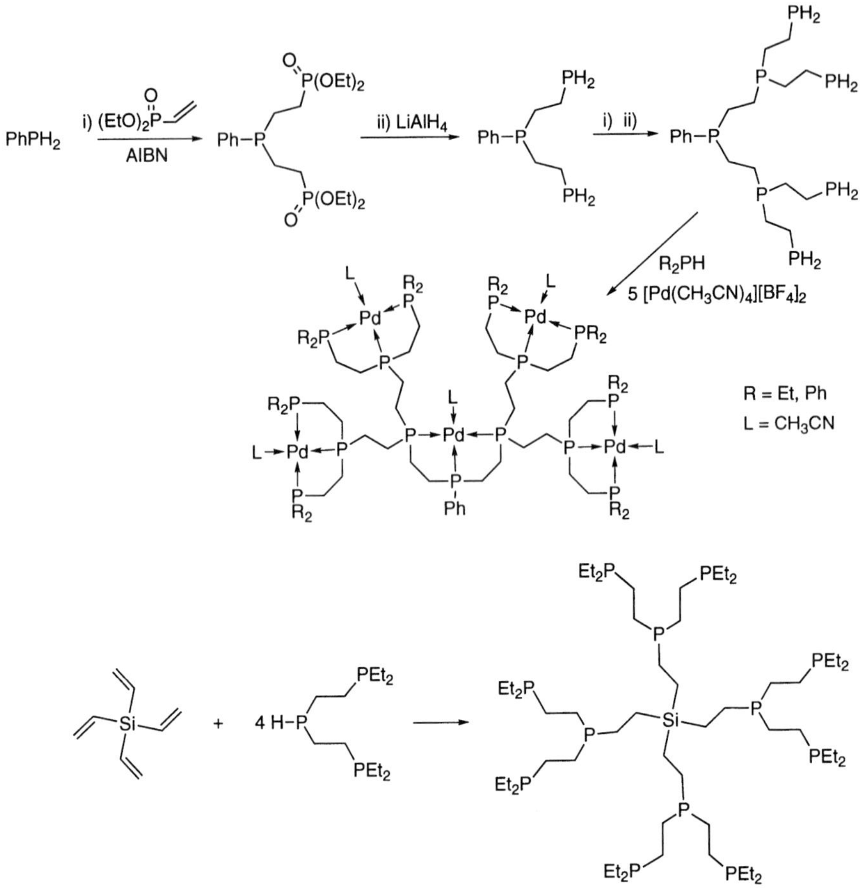

Scheme 2

some of these dendrimers containing 12 and 15 phosphorus atoms with $[Pd(CH_3CN)_4]$ $(BF_4)_2$ gives rise to complexes exhibiting catalytic activity for electrochemical CO_2 reduction. A similar approach for synthesizing a phosphorus-containing dendrimer with a silane core is applied and consists of the reaction of tetravinylsilane with a triphosphine ligand (Scheme 2).

Engel and Rengan report the synthesis of polycationic dendrimers bearing phosphorus at the core and at branching points [2, 5]. Tri-, tetra- and pentadirectional dendrimers have been generated from a phosphino, a phosphonium or a phosphorane core. The elaboration of all these macromolecules begins with tris (*p*-methoxymethyl)phenylphosphine $P(C_6H_4CH_2OMe)_3$ **1** which can be used as it is or as the corresponding oxide or sulfide. **1** can be quaternized either by reaction with an alkylhalide to generate the primary core $[RP(C_6H_4CH_2OMe)_3]^+$ **2** for a tridirectional development of the dendrimer or with *p*-methoxymethylbromobenzene in the presence of nickel bromide to generate the primary core $[P(C_6H_4CH_2OMe)_4]^+$ **3** for a tetradirectional growing up of the dendritic structure. A neutral quinquedirectional phosphorus core $P(C_6H_4CH_2OMe)_5$ **4** is also prepared from the phosphonium salt **3** which is treated with 4-(methoxymethyl)phenyllithium (Scheme 3).

The selective and facile cleavage of the benzylic ether linkages of **1**, **2**, **3** or **4** is accomplished by treatment with iodotrimethylsilane to form the corresponding benzylic iodide. Further addition to these iodide derivatives of **1** affords dendrimers of generation 1 with phosphonium ion sites at the periphery. Such a strategy is conducted up to generation 3 with a phosphine or a phosphonium core (Scheme 3).

The same reactions can be conducted from $(O)P(C_6H_4CH_2OMe)_3$. The phosphine oxide core of the polycationic dendrimer of generation 2 is reduced using trichlorosilane and coordinated to gold (Scheme 4). Both the reduction and the gold complexation reactions occur in high yield. The electronic absorption spectrum of the gold complex exhibits a slight bathochromic shift from that of the corresponding complex involving triphenylphosphine.

A D_{3h} cyclophosphazene core **6** elaborated via aminolysis of $N_3P_3Cl_6$ **5** by long chain diamines $H_2N(CH_2)_nNH_2 (n \geq 6)$ constitutes a useful starting reagent for the design of spherical cyclophosphazene dendrimers built up to the fifth generation [6]. This process involves the repetition of the two steps procedure outlined in Scheme 5: grafting of $P_3N_3Cl_5$ moieties on the amino end groups of **6**

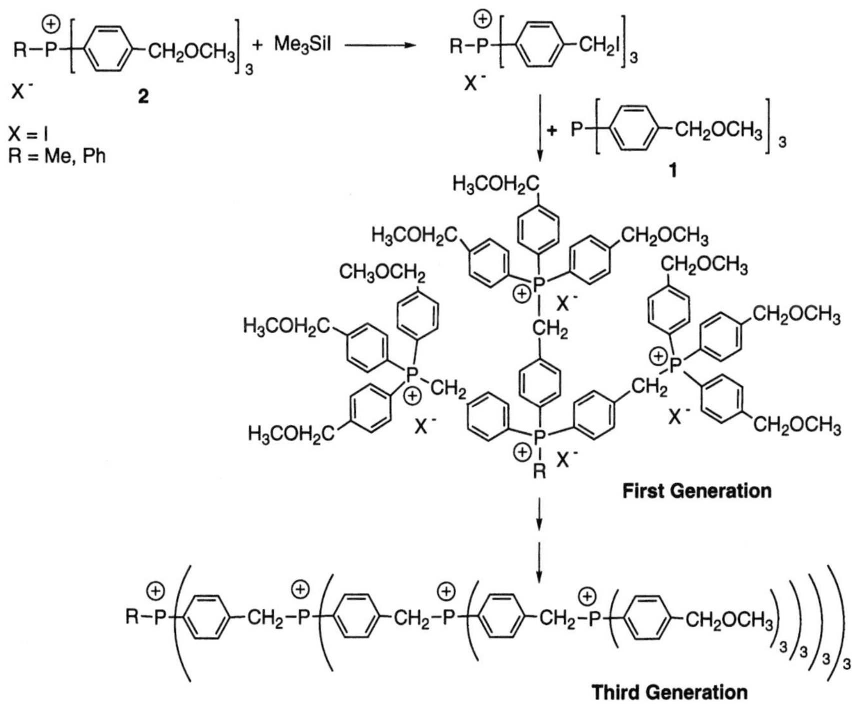

Scheme 3

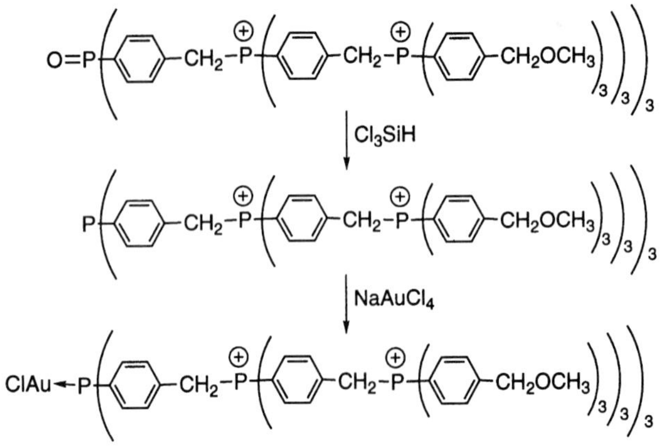

Scheme 4

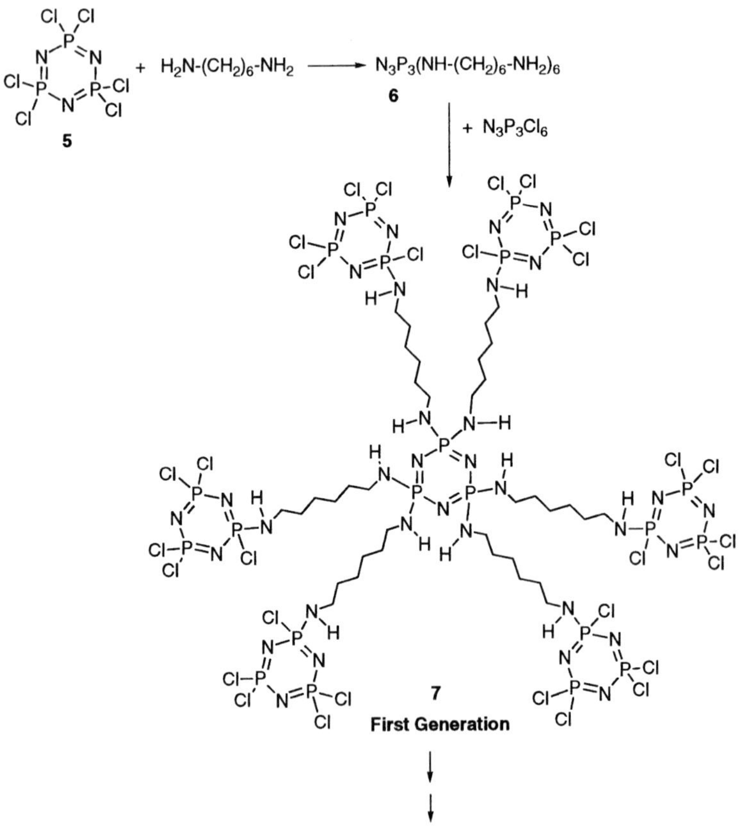

$N_3P_3\left[NH\text{-}(CH_2)_6\text{-}NH\text{-}N_3P_3\left(NH\text{-}(CH_2)_6\text{-}NH\text{-}N_3P_3[NH\text{-}(CH_2)_6\text{-}NH\text{-}N_3P_3(\,NH\text{-}(CH_2)_6\text{-}NH\text{-}N_3P_3(\,NH\text{-}(CH_2)_6\text{-}NH\text{-}N_3P_3Cl_5)_5\}_5]_5\right)_5\right]_5$

Fifth Generation

Scheme 5

to give **7** and substitution of all the remaining chlorine atoms of **7** with $H_2N(CH_2)_nNH_2$.

A general convergent growth procedure for the synthesis of nucleic acid dendrimers has been proposed by Damha and Hudson [7]. This strategy involves thymidines anchored to a long chain alkylamine-controlled pore glass. Chain extension is obtained with an automated DNA synthesizer. A tetrazole-activated adenosine 2′, 3′-bis (phosphoramidite) derivative **8** allows the coupling of two adjacent polymer-bound nucleotide chains (Scheme 6). Repetitive chain elongation and branching steps led to the formation of various dendrimers based on thymidine and adenosine building blocks, including an 87-unit-long dendrimer having a molecular weight of around 25000 with six branch points and twelve terminal ends.

Most of the methods of synthesis reported above present some limitations and constraints: lack of solubility for polycationic phosphorus-containing dendrimers from generation 3 [2], dendrimers of weak molecular weight [4], difficulties to overcome cross linking reactions [6], possibility only to incorporate phosphorus groups on the surface of dendrimers [3], etc.

All these drawbacks can be avoided by using a simple and general strategy for the synthesis of neutral phosphorus dendrimers elaborated up to generation 12, in which the core and subsequent branching points are pentavalent phosphorus atoms.

Numbering of compounds hereafter will be done as follows: **x-[G_n]** or **x-[G'_n]** in which x is the number attributed to a compound, n is the number of generation, G corresponds to a dendrimer with terminal PCl_2 bonds, G′ to a dendrimer with terminal aldehyde groups.

The first step consists of the reaction of trichlorophosphine sulfide (1 equiv.) with the sodium salt of 4-hydroxybenzaldehyde **9** to give **10-[G'_0]**. Subsequent addition of the hydrazine derivative $H_2NN(CH_3)P(S)Cl_2$ **11** to **10-[G'_0]** leads quantitatively to **10-[G_1]** a first generation dendrimer possessing six P-Cl bonds (Scheme 7). Further elaboration to second, third and up to generation 12 is accomplished by repetition of these two steps [8, 9]. In each reaction cycle terminal aldehyde groups and then terminal dichlorothiophosphoryl groups are introduced quantitatively and remain available for continued reactions. No protection/deprotection procedure is necessary. Noteworthy is the fact that the only by-products are sodium chloride and water. All the dendrimers including those of the twelfth generation (theoretical molecular weight >3000000, 12288 terminal P-Cl bonds) are stable and remain perfectly soluble in a variety of organic solvents (THF, CH_2Cl_2). However the dendrimer of generation 12, **10-[G_{12}]** appears poorly soluble in these solvents, preventing further construction to higher generations. All the dendrimers were characterized by NMR, IR spectroscopy and elemental analysis. Mass spectrometry (FAB) is useful for dendrimer up to the third generation.

^{31}P NMR was found to be an extraordinary tool to follow rigourously the construction of these derivatives. All phosphorus sites are distinguishable at each generation by their chemical shift and by the intensities of the different signals (see Fig. 1 for illustration).

A dendrimer of generation 1 is also characterized by X-ray diffraction studies (Fig. 2). Three main features can be emphasized. First each $OC_6H_4CH{=}N\text{-}N(Me)P$

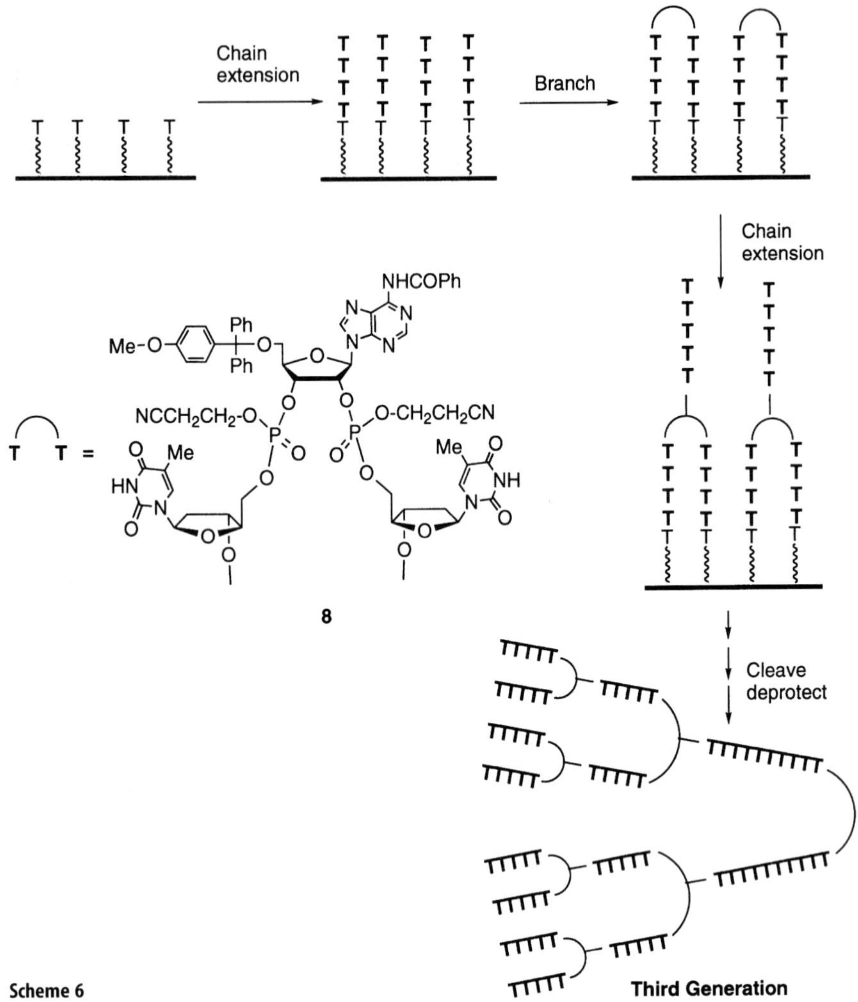

Scheme 6

unit is planar, and the molecule as a whole looks like a three blade propeller when examined in the direction of terminal P=S groups. Second, the distance between terminal P=S groups is very large (17 Å). Last, the length of each arm is 9 Å. All these data suggest that steric congestion does not disturb the construction of dendrimers even for the highest generations and that terminal groups are readily available for further reactions.

The method of synthesis of dendrimers outlined in Scheme 7 is not limited to the preparation of macromolecules bearing exclusively P=S units. Indeed construction of dendrimers possessing both P=S and P=O groups has been performed leading for instance to species shown in Scheme 8: (i) a dendrimer of generation 4 bearing alternatively P=S and P=O groups at each generation [10]; (ii) a dendrimer of generation 4 with P=S units on the core and within the

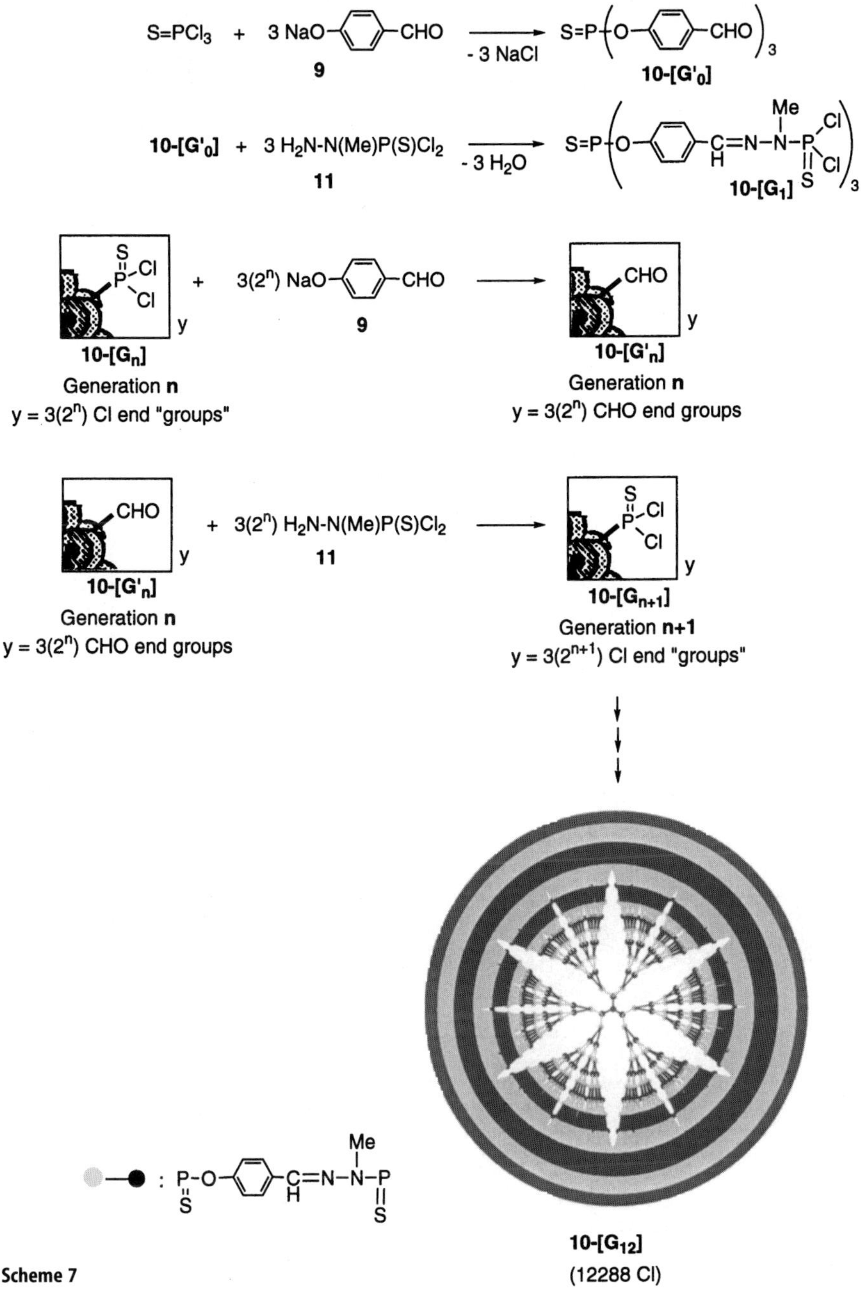
S=PCl3 + 3 NaO–C6H4–CHO
9
- 3 NaCl
10-[G'0]
10-[G'0] + 3 H2N-N(Me)P(S)Cl2
11
- 3 H2O
10-[G1]
3(2n) NaO–C6H4–CHO
10-[Gn]
Generation n
y = 3(2n) Cl end "groups"
10-[G'n]
Generation n
y = 3(2n) CHO end groups
3(2n) H2N-N(Me)P(S)Cl2
10-[Gn+1]
Generation n+1
y = 3(2n+1) Cl end "groups"
10-[G12]
(12288 Cl)

Scheme 7

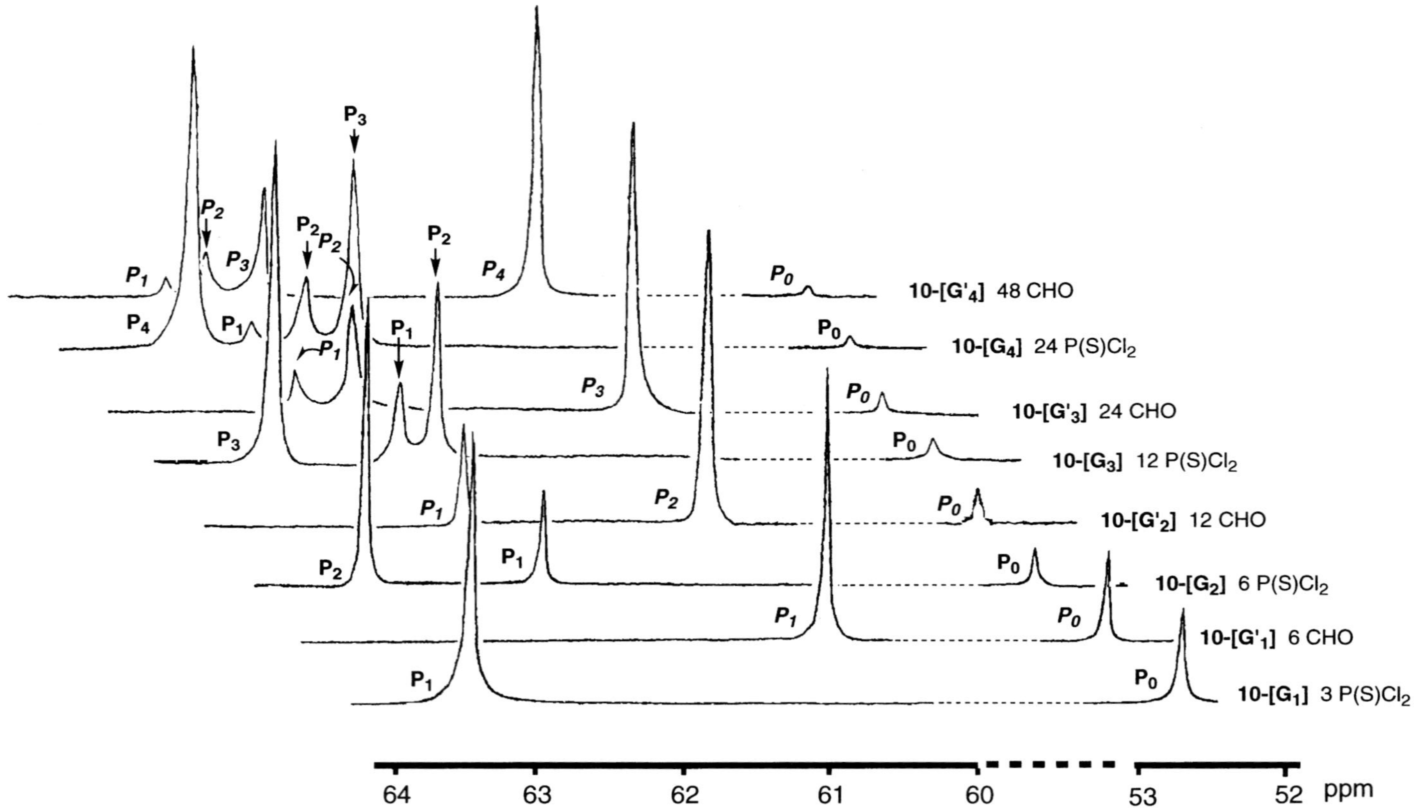

Fig. 1. $^{31}P\ \{^1H\}$-NMR spectra (THF) of dendrimers of generation 1, 2, 3 and 4 **10-[G$_1$]**, **10-[G$_1'$]**, **10-[G$_2$]**, **10-[G$_2'$]**, **10-[G$_3$]**, **10-[G$_3'$]**, **10-[G$_4$]**, **10-[G$_4'$]**. P_0 phosphorus of the core, P_1 phosphorus of generation 1, P_2 phosphorus of generation 2, P_3 phosphorus of generation 3, P_4 phosphorus of generation 4

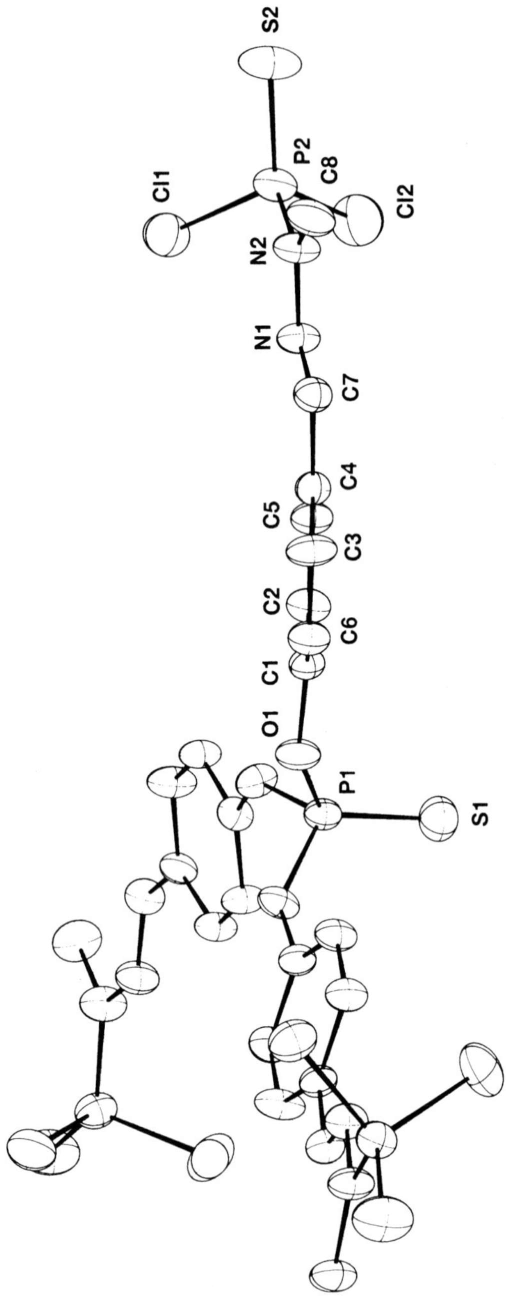

Fig. 2. ORTEP view of 10-[G_1]

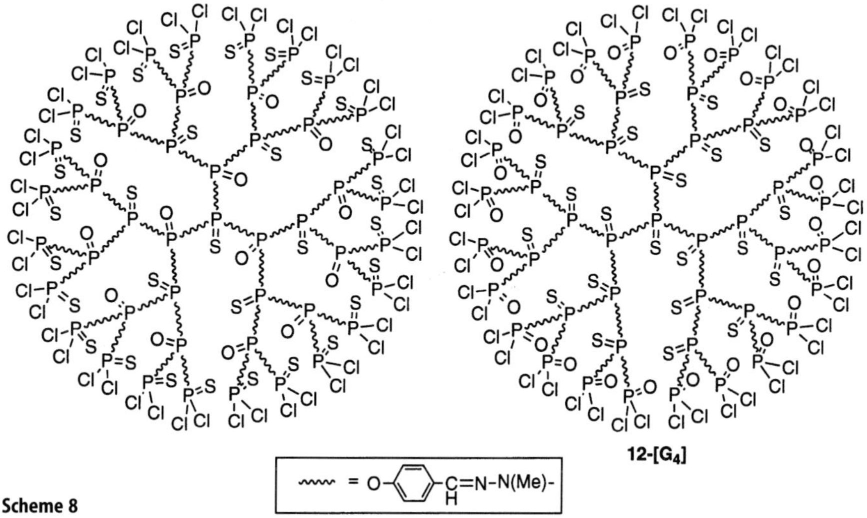

Scheme 8

cascade structure, and P=O units on the surface 12-[G_4]. Indeed all combinations are possible, proving the usefulness of this methodology. Moreover a variety of cores can be used. As an example a dendrimer of generation 8 **10′-[G'_8]** possessing 1536 aldehyde chain ends was built from hexachlorocyclotriphosphazene $N_3P_3Cl_6$ [11] (Scheme 9].

Other ways of preparation of neutral phosphorus-containing dendrimers using hydrazono derivatives have been proposed in the recent past but none presents the same advantage as the one described above. One can mention the formation of dendrimers up to generation 3 using the reiiteration of a sequence of three reactions [12]. The core is the hexapodant $N_3P_3(OC_6H_4CHO)_6$, and the first step of the elaboration of the dendrimer consists of the treatment of this derivative with methylhydrazine giving rise to the hexahydrazone **13**. Addition of a chlorophosphine such as chlorodiphenylphosphine to **13** in the presence of base affords the hexahydrazonophosphine **14**. The last step involves a Staudinger type reaction between **14** and the azide $N_3P(S)(OC_6H_4CHO)_2$ leading to the dendrimer of the first generation **15-[G_1]** (Scheme 10). This iterative reaction sequence allows one to prepare dendrimers of generation 2 and 3 possessing 12 and 24 terminal phosphino groups respectively. Analogous experiments can be done with chlorodiazaphospholane instead of chlorodiphenylphosphine [12]. Similarly the same dendrimer construction using $(S)P(OC_6H_4CHO)_3$ was investigated and led also to the formation of dendrimers of generation 1, 2 and 3 (Scheme 11).

Such a way offers significant advantages including a high initiator core multiplicity allowing one to reach quickly a large external shape, and the possibility to introduce inorganic P=N-P linkages which are useful probes for controlling branch cells assembling. Moreover this procedure leading to electrophilic or nucleophilic surfaces permits the introduction of three different reactive groups

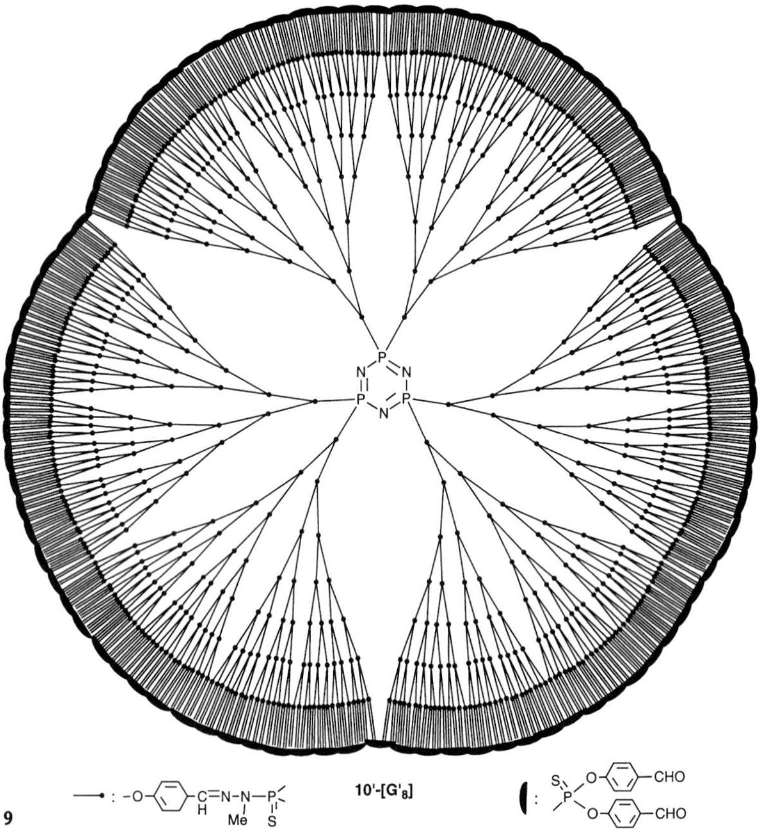

Scheme 9

at the periphery such as aldehydes, hydrazones and hydrazonophosphines. However the lack of solubility of dendrimers of generation 3 prevents from elaboration of higher generations.

3 Reactivity on the Surface

3.1 Chiroptical Properties of Dendrimers with Chiral End Groups

A few papers reported the chiroptical properties of dendrimers with chiral substituents on the surface. Newkome et al. have shown that the molecular ellipticity increases proportionately to the number of chiral groups up to generation 2 [13], whereas Meijer et al. showed that some of the amino acid terminated dendrimers they studied undergo a decrease to almost zero of their specific optical rotation on going from the first to the fifth generation [14]. The use of modified amino acids with less N-H bonds gives dendrimers whose specific optical rotation remains constant for all generations. Vögtle also reported a nearly constant

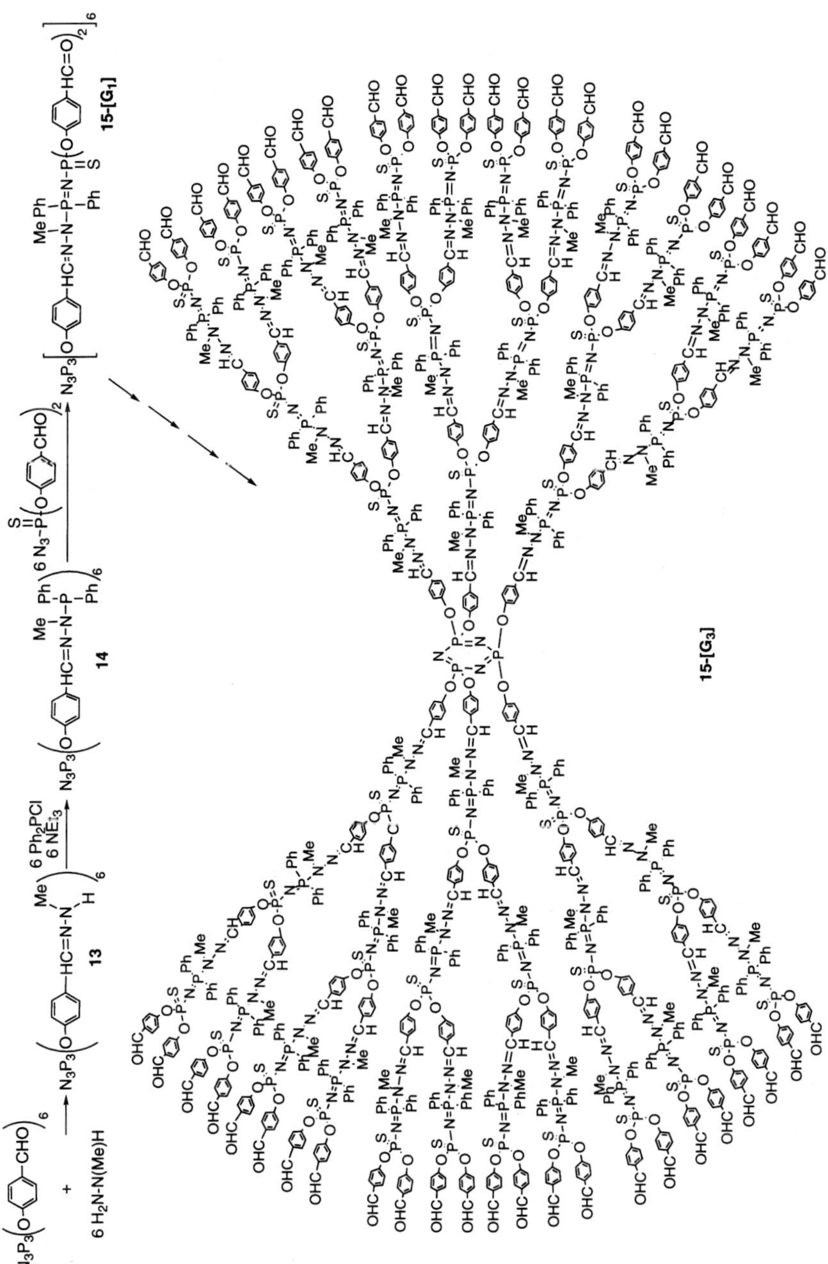

Scheme 10

ture. A neutral quinquedirectional phosphorus core $P(C_6H_4CH_2OMe)_5$ **4** is also prepared from the phosphonium salt **3** which is treated with 4-(methoxymethyl)phenyllithium (Scheme 3).

The selective and facile cleavage of the benzylic ether linkages of **1**, **2**, **3** or **4** is accomplished by treatment with iodotrimethylsilane to form the corresponding benzylic iodide. Further addition to these iodide derivatives of **1** affords dendrimers of generation 1 with phosphonium ion sites at the periphery. Such a strategy is conducted up to generation 3 with a phosphine or a phosphonium core (Scheme 3).

The same reactions can be conducted from $(O)P(C_6H_4CH_2OMe)_3$. The phosphine oxide core of the polycationic dendrimer of generation 2 is reduced using trichlorosilane and coordinated to gold (Scheme 4). Both the reduction and the gold complexation reactions occur in high yield. The electronic absorption spectrum of the gold complex exhibits a slight bathochromic shift from that of the corresponding complex involving triphenylphosphine.

A D_{3h} cyclophosphazene core **6** elaborated via aminolysis of $N_3P_3Cl_6$ **5** by long chain diamines $H_2N(CH_2)_nNH_2 (n \geq 6)$ constitutes a useful starting reagent for the design of spherical cyclophosphazene dendrimers built up to the fifth generation [6]. This process involves the repetition of the two steps procedure outlined in Scheme 5: grafting of $P_3N_3Cl_5$ moieties on the amino end groups of **6**

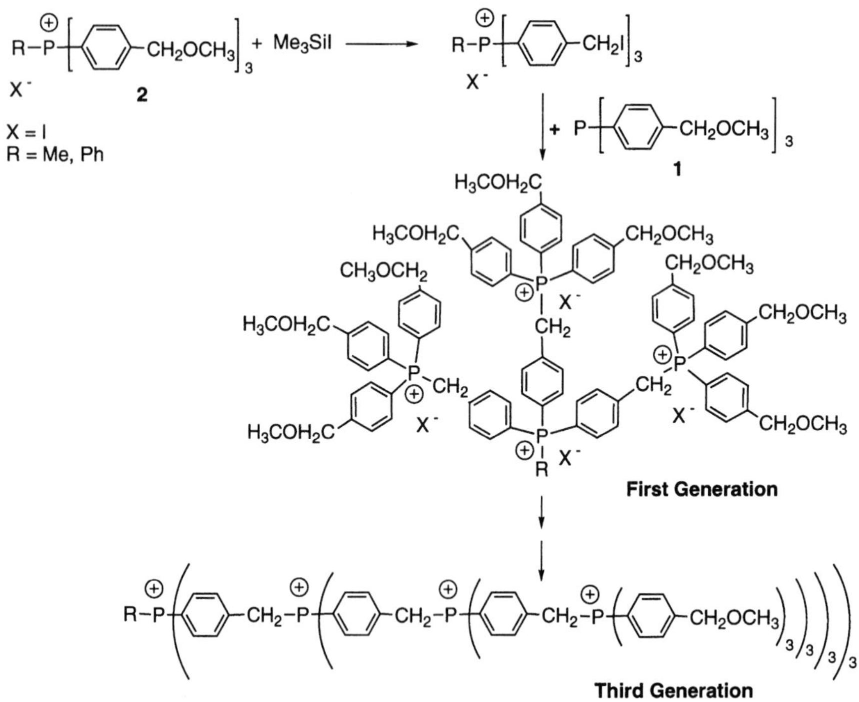

Scheme 3

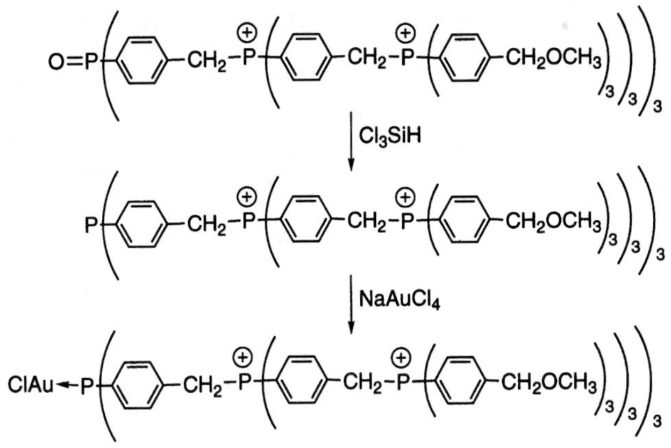

Scheme 4

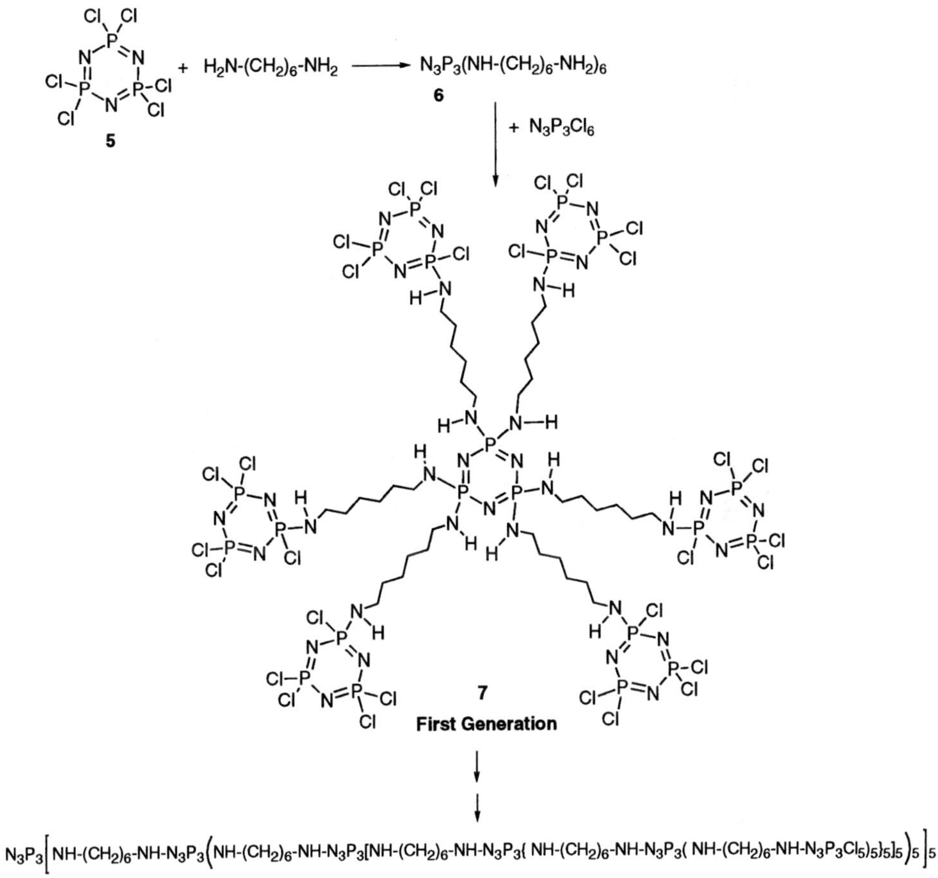

Scheme 5

to give **7** and substitution of all the remaining chlorine atoms of **7** with $H_2N(CH_2)_nNH_2$.

A general convergent growth procedure for the synthesis of nucleic acid dendrimers has been proposed by Damha and Hudson [7]. This strategy involves thymidines anchored to a long chain alkylamine-controlled pore glass. Chain extension is obtained with an automated DNA synthesizer. A tetrazole-activated adenosine 2′, 3′-bis (phosphoramidite) derivative **8** allows the coupling of two adjacent polymer-bound nucleotide chains (Scheme 6). Repetitive chain elongation and branching steps led to the formation of various dendrimers based on thymidine and adenosine building blocks, including an 87-unit-long dendrimer having a molecular weight of around 25000 with six branch points and twelve terminal ends.

Most of the methods of synthesis reported above present some limitations and constraints: lack of solubility for polycationic phosphorus-containing dendrimers from generation 3 [2], dendrimers of weak molecular weight [4], difficulties to overcome cross linking reactions [6], possibility only to incorporate phosphorus groups on the surface of dendrimers [3], etc.

All these drawbacks can be avoided by using a simple and general strategy for the synthesis of neutral phosphorus dendrimers elaborated up to generation 12, in which the core and subsequent branching points are pentavalent phosphorus atoms.

Numbering of compounds hereafter will be done as follows: **x-[G_n]** or **x-[G'_n]** in which x is the number attributed to a compound, n is the number of generation, G corresponds to a dendrimer with terminal PCl_2 bonds, G′ to a dendrimer with terminal aldehyde groups.

The first step consists of the reaction of trichlorophosphine sulfide (1 equiv.) with the sodium salt of 4-hydroxybenzaldehyde **9** to give **10-[G'_0]**. Subsequent addition of the hydrazine derivative $H_2NN(CH_3)P(S)Cl_2$ **11** to **10-[G'_0]** leads quantitatively to **10-[G_1]** a first generation dendrimer possessing six P-Cl bonds (Scheme 7). Further elaboration to second, third and up to generation 12 is accomplished by repetition of these two steps [8, 9]. In each reaction cycle terminal aldehyde groups and then terminal dichlorothiophosphoryl groups are introduced quantitatively and remain available for continued reactions. No protection/deprotection procedure is necessary. Noteworthy is the fact that the only by-products are sodium chloride and water. All the dendrimers including those of the twelfth generation (theoretical molecular weight >3000000, 12288 terminal P-Cl bonds) are stable and remain perfectly soluble in a variety of organic solvents (THF, CH_2Cl_2). However the dendrimer of generation 12, **10-[G_{12}]** appears poorly soluble in these solvents, preventing further construction to higher generations. All the dendrimers were characterized by NMR, IR spectroscopy and elemental analysis. Mass spectrometry (FAB) is useful for dendrimer up to the third generation.

^{31}P NMR was found to be an extraordinary tool to follow rigourously the construction of these derivatives. All phosphorus sites are distinguishable at each generation by their chemical shift and by the intensities of the different signals (see Fig. 1 for illustration).

A dendrimer of generation 1 is also characterized by X-ray diffraction studies (Fig. 2). Three main features can be emphasized. First each $OC_6H_4CH{=}N\text{-}N(Me)P$

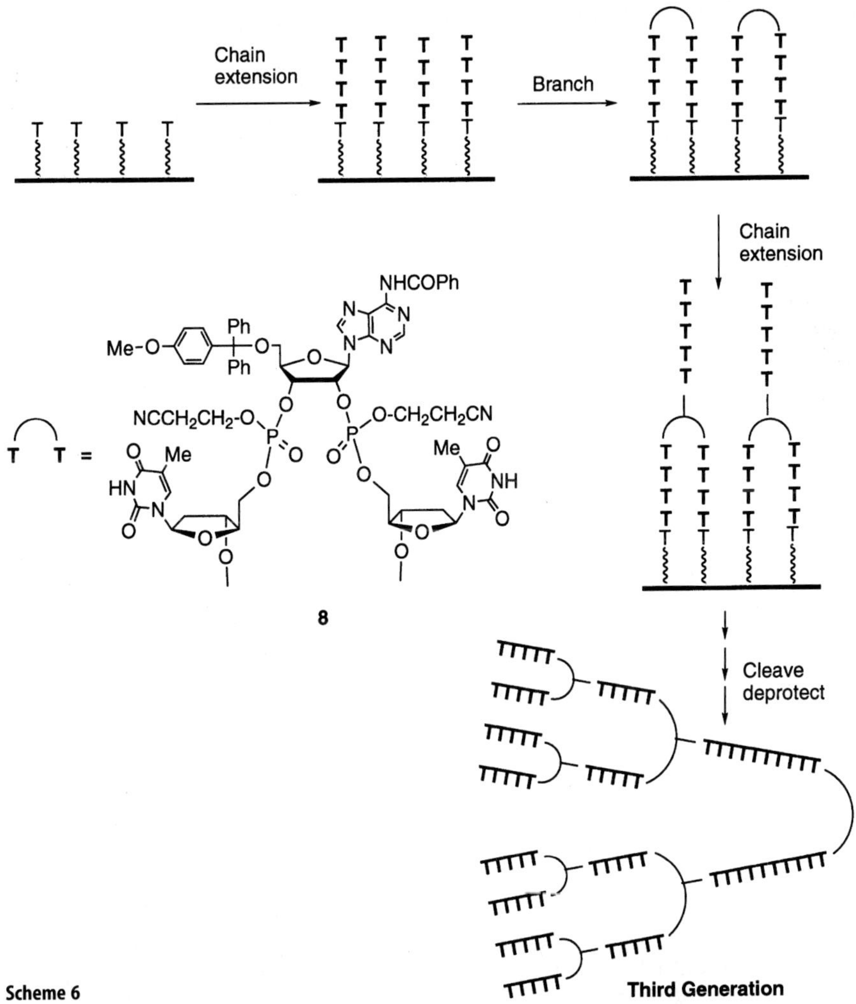

Scheme 6

unit is planar, and the molecule as a whole looks like a three blade propeller when examined in the direction of terminal P=S groups. Second, the distance between terminal P=S groups is very large (17 Å). Last, the length of each arm is 9 Å. All these data suggest that steric congestion does not disturb the construction of dendrimers even for the highest generations and that terminal groups are readily available for further reactions.

The method of synthesis of dendrimers outlined in Scheme 7 is not limited to the preparation of macromolecules bearing exclusively P=S units. Indeed construction of dendrimers possessing both P=S and P=O groups has been performed leading for instance to species shown in Scheme 8: (i) a dendrimer of generation 4 bearing alternatively P=S and P=O groups at each generation [10]; (ii) a dendrimer of generation 4 with P=S units on the core and within the

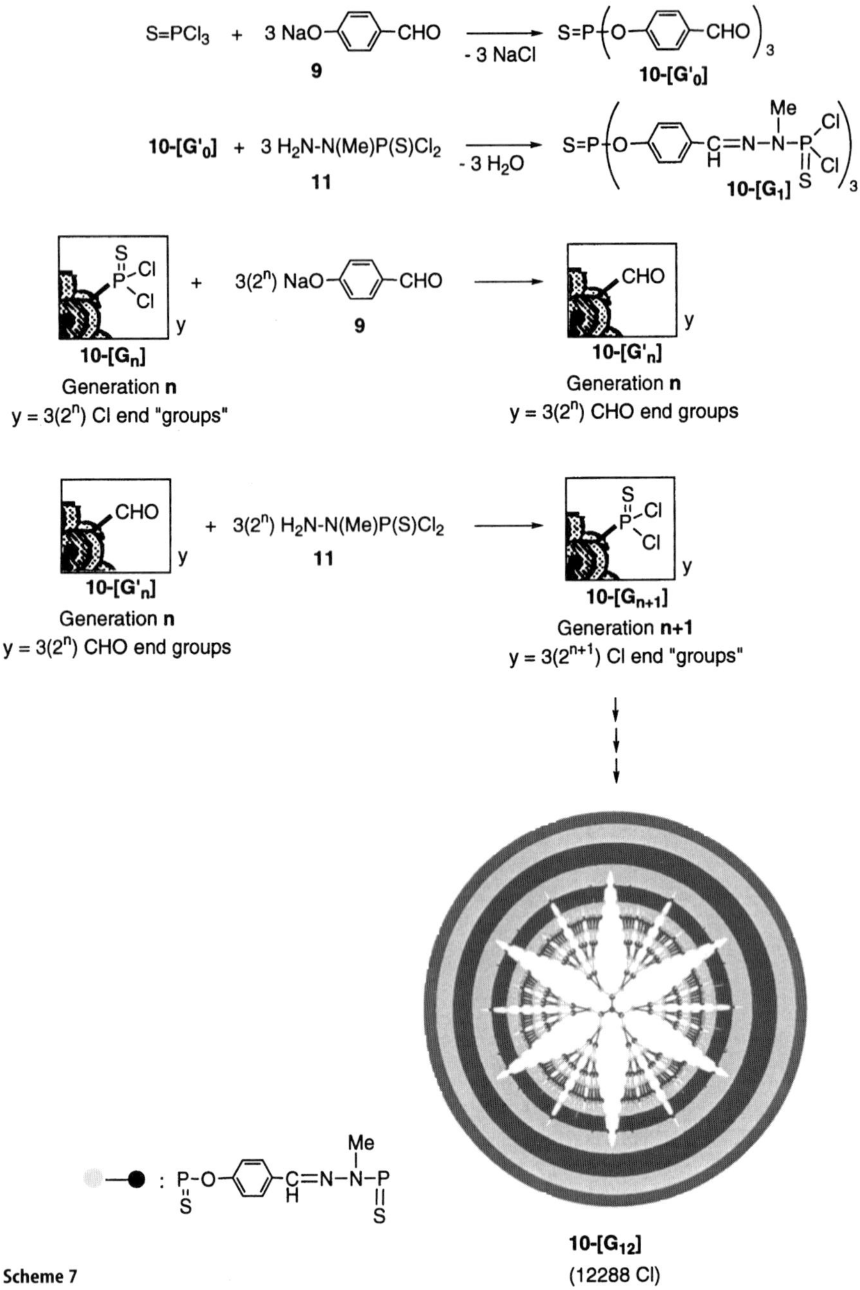
S=PCl3 + 3 NaO–C6H4–CHO → S=P(O–C6H4–CHO)3
- 3 NaCl
9
10-[G'0]
10-[G'0] + 3 H2N-N(Me)P(S)Cl2 → 10-[G1]
- 3 H2O
11
10-[G1]
+ 3(2n) NaO–C6H4–CHO
9
10-[Gn]
Generation n
y = 3(2n) Cl end "groups"
10-[G'n]
Generation n
y = 3(2n) CHO end groups
+ 3(2n) H2N-N(Me)P(S)Cl2
11
10-[G'n]
Generation n
y = 3(2n) CHO end groups
10-[Gn+1]
Generation n+1
y = 3(2n+1) Cl end "groups"
10-[G12]
(12288 Cl)

Scheme 7

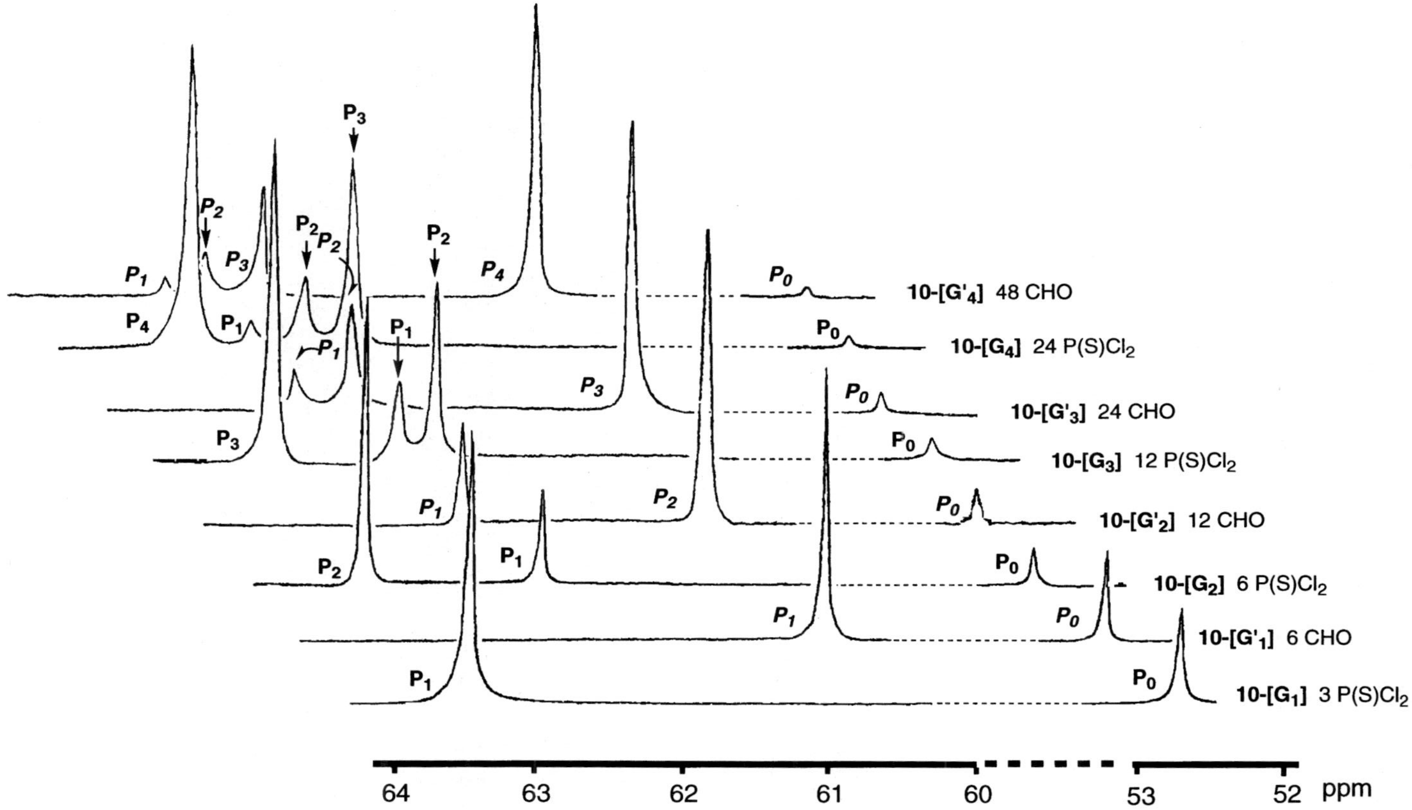

Fig. 1. $^{31}P\{^1H\}$-NMR spectra (THF) of dendrimers of generation 1, 2, 3 and 4 **10-[G$_1$]**, **10-[G$_1'$]**, **10-[G$_2$]**, **10-[G$_2'$]**, **10-[G$_3$]**, **10-[G$_3'$]**, **10-[G$_4$]**, **10-[G$_4'$]**. P_0 phosphorus of the core, P_1 phosphorus of generation 1, P_2 phosphorus of generation 2, P_3 phosphorus of generation 3, P_4 phosphorus of generation 4

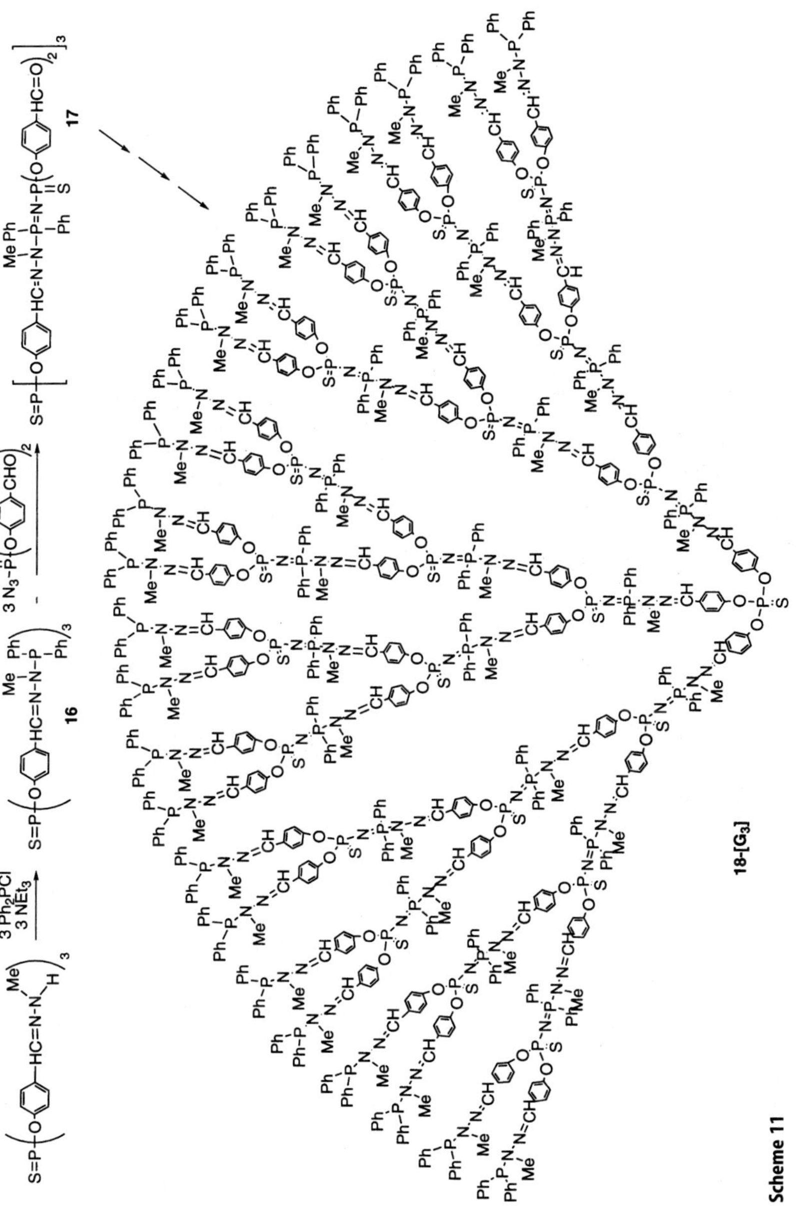

Scheme 11

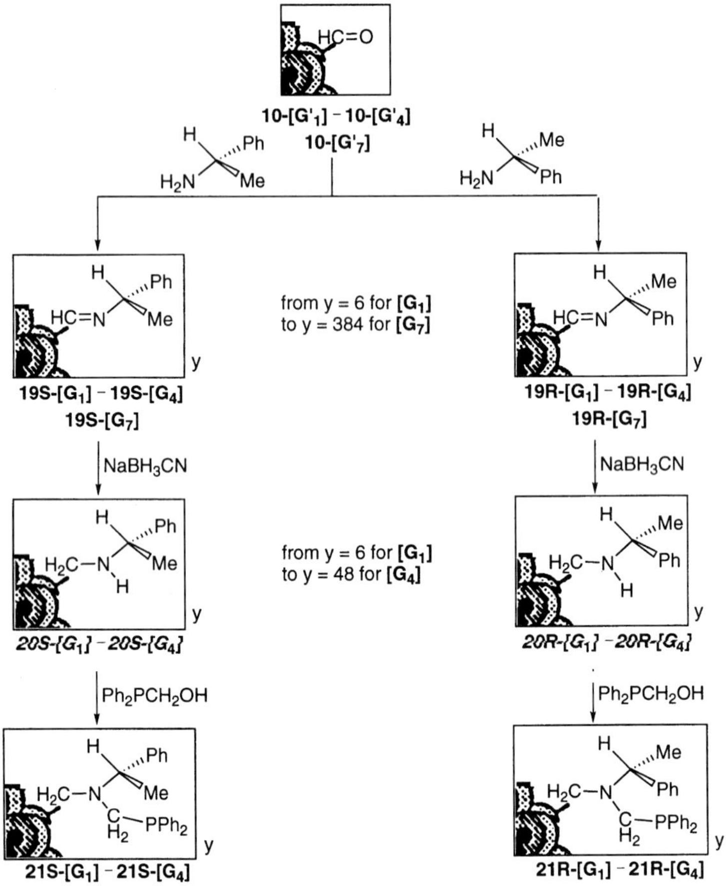

Scheme 12

optical activity with increasing generations for dendrimers bearing planar chiral paracyclophane [15].

Phosphorus-containing dendrimers with chiral end groups **19 S-[G$_n$]** and **19 R-[G$_n$]** (n=1–4, 7) are obtained from the condensation of (*S*)-(–) α-methylbenzylamine or (*R*)-(+)-α-methylbenzylamine with aldehyde terminated dendrimers **10-[G$'_n$]**. Up to 384 chiral groups (generation 7) can be grafted onto the surface [16]. The chemospecific reduction of the imine bonds thus created affords two new series of chiral dendrimers **20 S-[G$_n$]** and **20 R-[G$_n$]** (n=1–4). Then the reaction with Ph_2PCH_2OH gives a series of chiral phosphino terminated dendrimers **21 S-[G$_n$]** and **21 R-[G$_n$]** (n=1–4) (Scheme 12) (Fig. 3). Studies of the chiroptical properties of all these non-hindered chiral dendrimers indicate that each terminal group behaves independently and that the optical rotation depends only on the number of chiral groups grafted on the dendrimer whatever the generation is (Fig. 4).

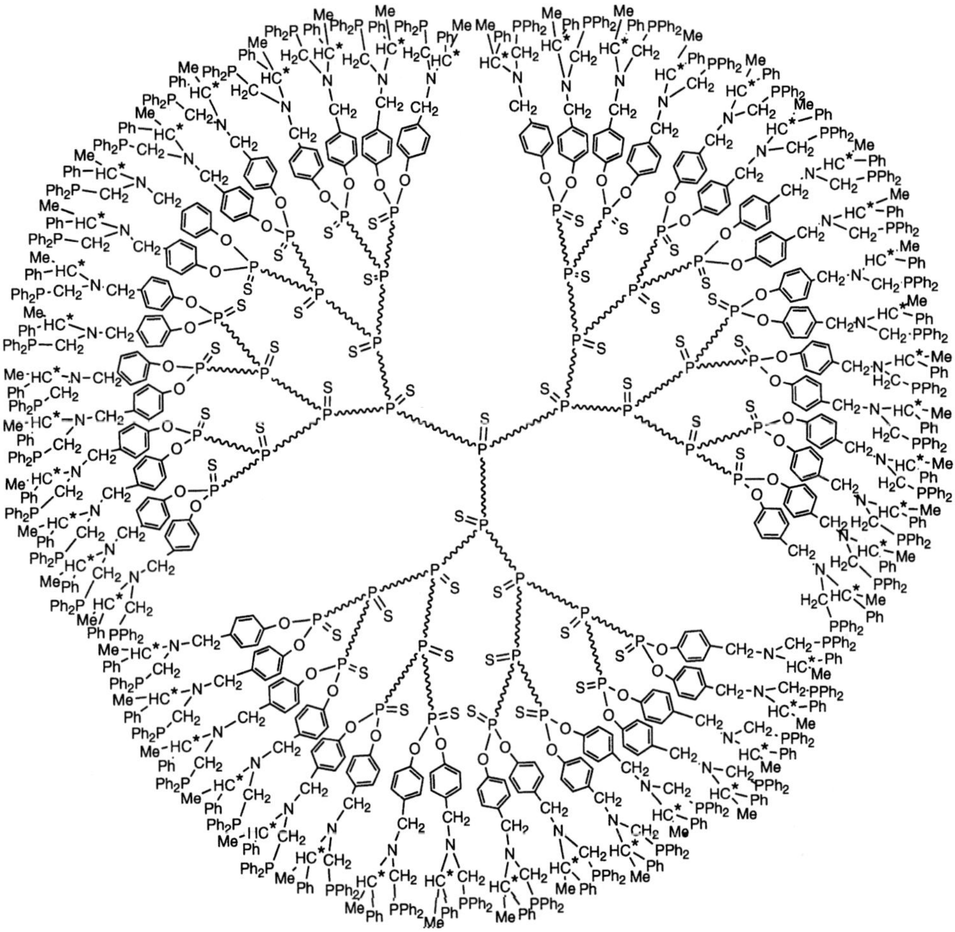

〰 = O-C₆H₄-C(H)=N-NMe

21-[G₄]

Fig. 3. Representation of the dendrimer **21**-[G_4] (generation 4, 48 terminal (*S*) or (*R*) aminophosphines)

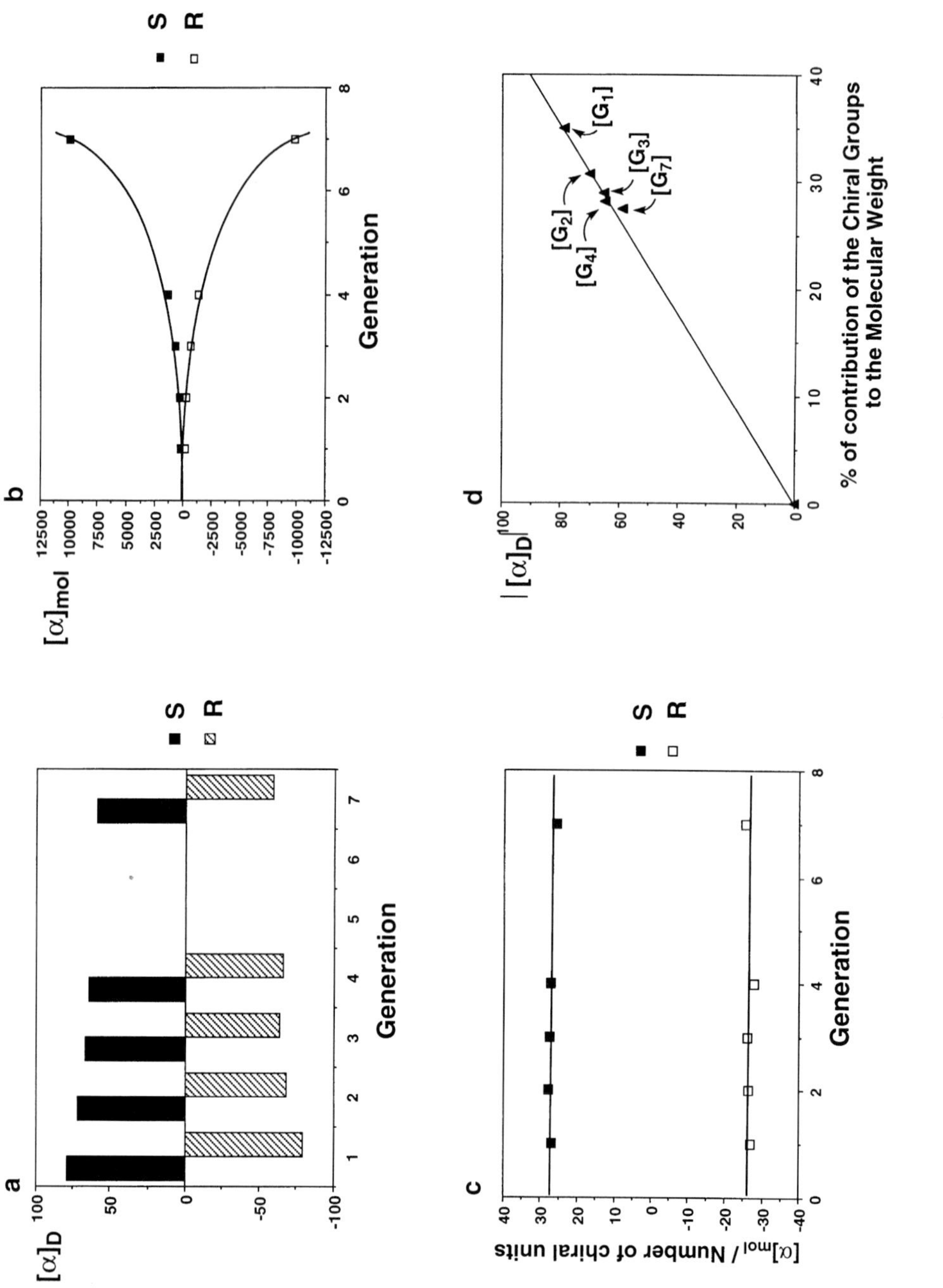

Fig. 4. **a** Specific rotation $[\alpha]_D$ vs generation for dendrimers **19** S-[G_n] and **19** R-[G_n]. **b** Molar rotation $[\alpha]_{mol}$ vs generation for dendrimers **19** S-[G_n] and **19** R-[G_n]. **c** Molar rotation $[\alpha]_{mol}$ divided by the number of chiral groups vs generation for dendrimers **19** S-[G_n] and **19** R-[G_n]. **d** Absolute value of the specific rotation $|[\alpha]_D|$ vs the percentage of contribution of the chiral groups to the molecular weight for the dendrimers **19** S-[G_n] and **19** R-[G_n]

3.2 Anchorage of Macrocycles

Various macrocycles can be grafted at the periphery of phosphorus-containing dendrimers possessing either aldehyde chain ends or terminal P-Cl bonds. A Schiff reaction between for example dendrimers of generations 1, 2 and 4 **10-[G'_1]**, **10-[G'_2]**, **10-[G'_4]** and 4-aminobenzo-15 crown-5 takes place to give macromolecules **22-[G_1]**, **22-[G_2]**, **22-[G_4]** bearing up to 48 crown ether units (Scheme 13) (Fig. 5). Formation of the dendrimer of generation 1 **22-[G_1]** (6 terminal crown ether units) necessitated one week of stirring in refluxing THF while that of the dendrimer of generation 4 **22-[G_4]** necessitated three weeks in the same experimental conditions [9a]. Therefore steric hindrance plays an important role but does not prevent the reactions from going to completion.

Grafting of nitrogen macrocycles, namely 1,4,8,11-tetraazacyclotetra or pentadecane and 1,4,8,11-tetraazacyclotetradecane-5,7 dione on the $P(S)Cl_2$ chain ends of dendrimers of generation 1 and 3 is easily achieved in the presence of K_2CO_3 [17a]. Up to 12 of these nitrogen macrocycles are thus anchored on the surface (Scheme 14). Remarkably no cross linking reactions are detected and the resulting polymacrocyclic dendritic structures **23-[G_1]–23-[G_3]**, **24-[G_1]–24-[G_3]**, **25-[G_1]–25-[G_3]** are obtained in good yields.

3.3 Phosphate, Phosphite, Ylide and Phosphonate Terminated Dendrimers

It is possible to anchor to the surface of dendritic molecules a large variety of phosphorus groups (from 6 to 96 units), each of them, potentially, of great interest in different fields. Indeed biologically important phosphates are known (nucleotides, phospholipides, nucleosides, polyphosphates, phosphate sugars); phosphite monomers are used in some catalytic processes (acrylonitrile dimerization) and in the well-known Arbusov rearrangement while phosphorus ylides play a key role in Wittig reactions. Moreover phosphonate monomers have found wide applications in general organic synthesis (Horner-Wadsworth-Emmons condensation, Diels-Alder reactions, Michael additions, etc.) and they can be used as versatile intermediates for the preparation of a number of heterocycles.

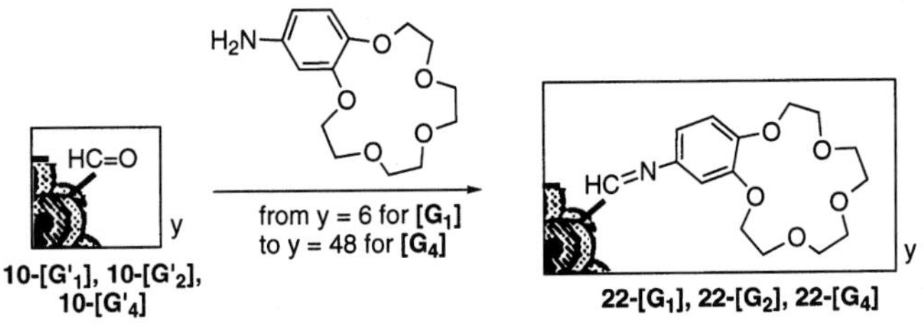

Scheme 13

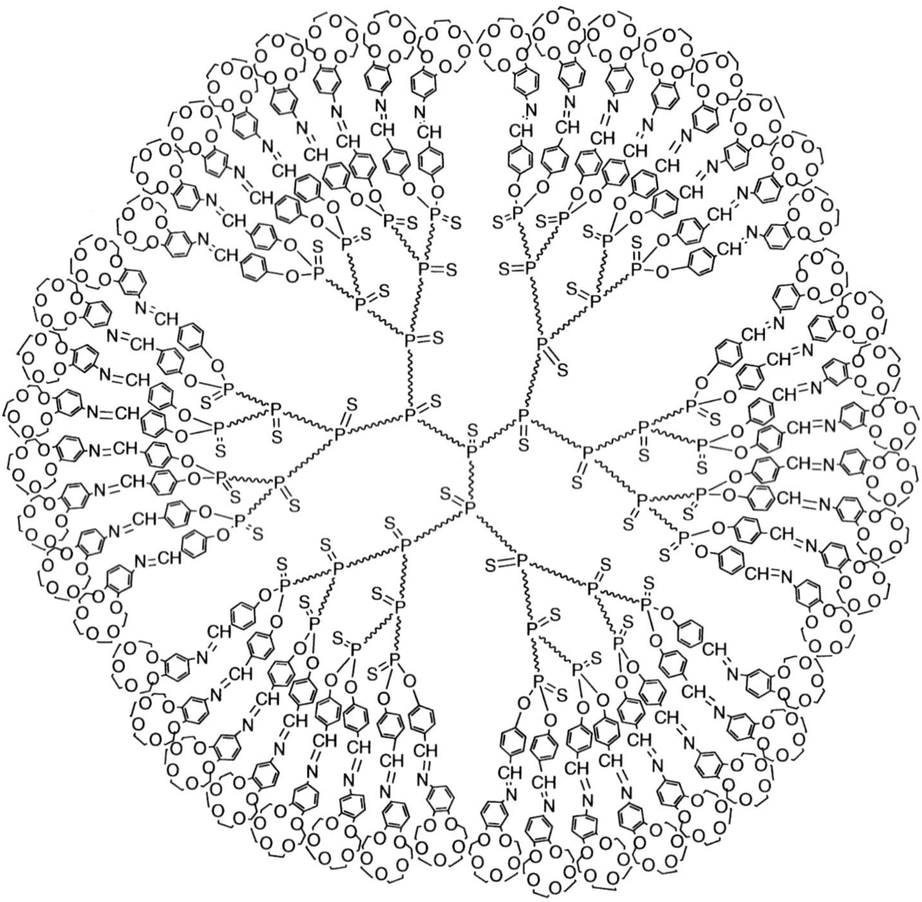

22-[G_4]

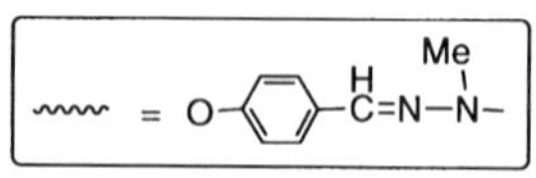

Fig. 5. Dendrimer 22-[G_4]: generation 4, 48 terminal crown ether units

All the experiments are conducted on aldehyde terminal functions. Depending on the solubility of the resulting phosphorylated dendrimers, anchorage of phosphorus moieties has been carried out on generation 1 (6 phosphate or phosphinite groups) and up to generation 5 [(96 aminophosphate (Fig. 6, Scheme 15), amino phosphite or functionalized phosphonate groups)] [17b].

Dendrimer 27-[G_1] can be used as the starting material for the grafting of ylides. Addition of triphenylphosphoranylidene ethenone $Ph_3P=C=C=O$

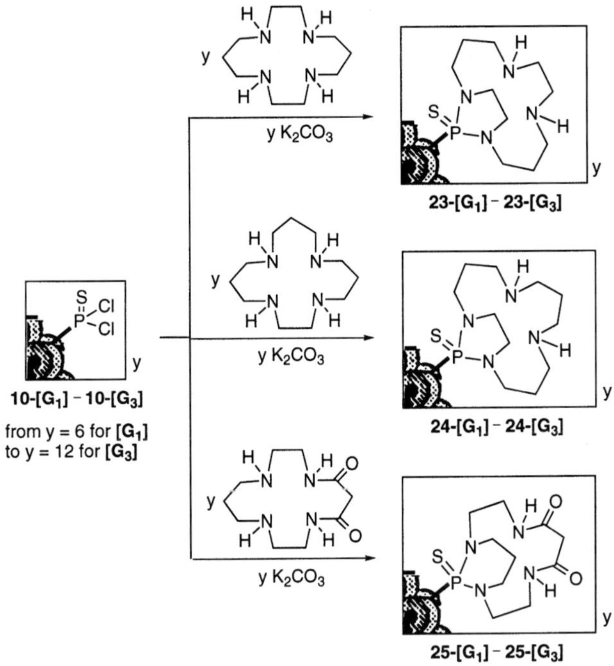

Scheme 14

affords compound **28-[G$_1$]** possessing 6 ylide chain ends. A Wittig reaction involving **28-[G$_1$]** and benzaldehyde takes place at room temperature leading to **29-[G$_1$]** obtained as two isomers (*trans/cis* 3/1). An analogous reaction is observed with crotonaldehyde and gives **30-[G$_1$]** (Scheme 16) [17b].

Higher generations of dendrimers with ylide end groups are found to be insoluble in water and a variety of organic solvents.

3.4 Concept of Multiplurifunctionalization

Most of the procedures reported above and experiments conducted on purely organic dendrimers describe the introduction of no more than two different types of functionalities at the periphery of dendrimers. Because surface modifications have a large effect on the properties of these macromolecules there is a need to develop versatile new strategies for anchoring a large variety of reactive groups at the surface. The particular behavior of terminal $P(X)Cl_2$ (X=O,S) allows the straightforward control of reactivity at the periphery of the dendrimers and the introduction of sets of two but also three or four different functional groups.

This concept will be illustrated hereafter with selected examples.

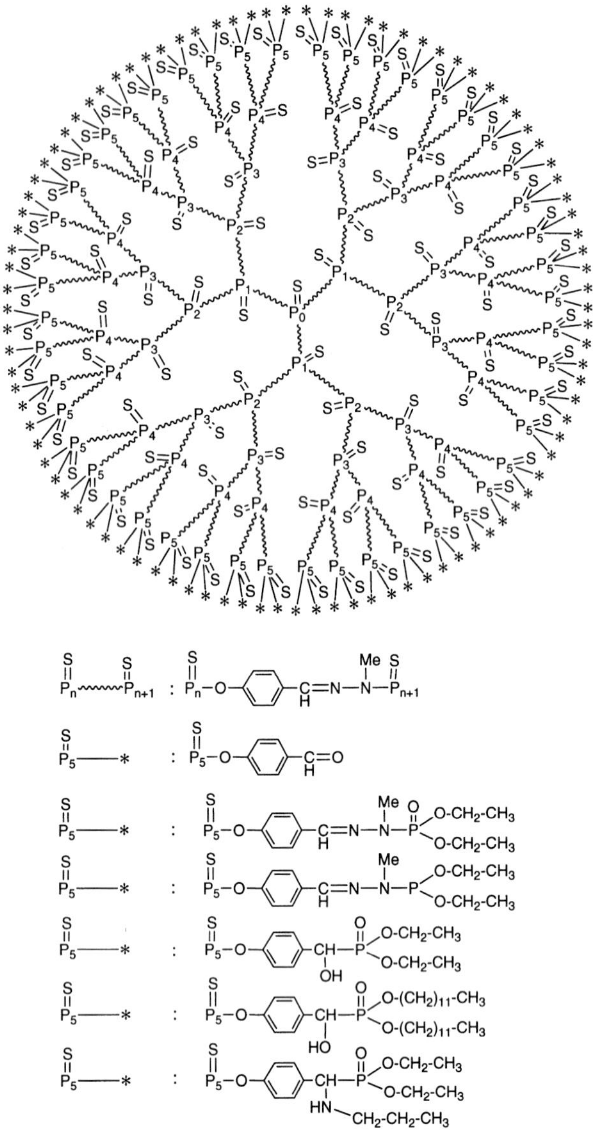

Fig. 6. Dendrimers of generation 5 with 96 various terminal phosphorus groups

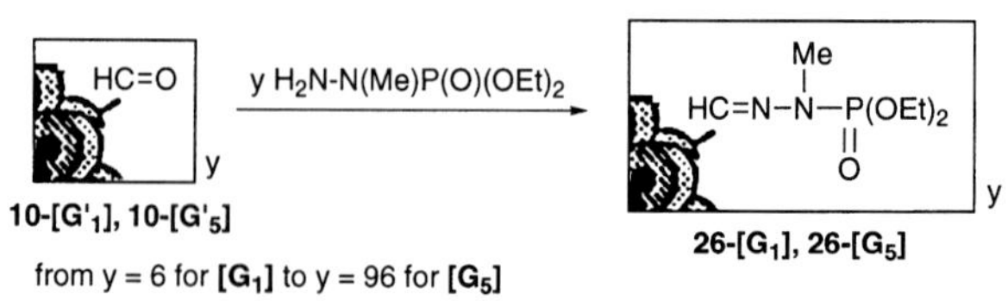

Scheme 15

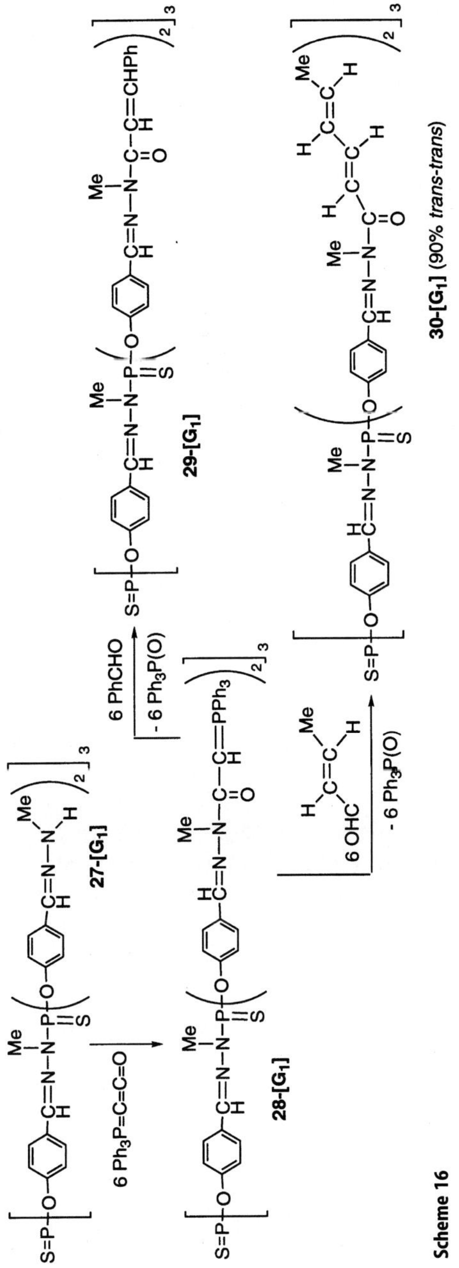
6 Ph_3P=C=C=O
27-[G_1]
28-[G_1]
6 PhCHO
- 6 $Ph_3P(O)$
29-[G_1]
6 OHC
- 6 $Ph_3P(O)$
30-[G_1] (90% *trans-trans*)

Scheme 16

3.4.1
Multidifunctionalization

Application of the Horner-Wadsworth-Emmons reaction to the functionalization of dendrimers allows one to prepare amino acid terminated macromolecules. Such a reaction conducted with dendrimers **10-[G′$_1$]**, **10-[G′$_3$]**, **10-[G′$_4$]** and phosphonates unsubstituted at the carbon α to the phosphoryl group affords in moderate yield dendrimers bearing various α,β unsaturated functional groups on the surface [18]. (Schemes 17 and 18).

Addition of bisallylamine to dendrimers **10-[G$_1$]–10-[G$_4$]**, possessing 3, 6, 12 and 24 terminal $P(S)Cl_2$ groups respectively, leads selectively to monosubstitution of the $P(S)Cl_2$ moieties, regardless of the number of bisallylamines used, with the formation of compounds **33-[G$_1$]–33-[G$_4$]** (Scheme 19) [19].

Addition of 2 equiv. of allylamine or propargylamine per $P(S)Cl_2$ chain ends takes place on dendrimers of generations 1–7 to give the multidifunctionalized dendrimers bearing up to 96 NH and either 96 allyl or propargyl groups for the series built from the N_3P_3 core and up to 384 NH, allyl or propargyl groups for the series built from the P=S core (Scheme 19).

3.4.2
Multitri- and Tetrafunctionalization

Multitrifunctionalization is first attempted by treating **10-[G$_1$]–10-[G$_4$]** or **12-[G$_1$]–12-[G$_7$]** with allylamine in the presence of triethylamine at 0 °C. Quantitative and selective monosubstitution occurs, giving rise to dendrimers possessing up to 192 sets of three functional groups on the surface: NH and allyl groups, P-Cl bonds. Analogous reactions are conducted with the same starting dendrimers and propargylamine [20] (Scheme 20). Remarkably, while attempts to disubstitute dendrimers **10-[G$_1$]–10-[G$_4$]** with bisallylamine failed (see above), the monosubstituted compounds **36-[G$_1$]–36-[G$_7$]** are still reactive towards propargylamine leading to dendrimers **39-[G$_1$]–39-[G$_7$]** incorporating both bisallylamino and propargyl amino groups on the same phosphorus atom. Dendrimers **39-[G$_1$]–39-[G$_7$]** are also formed from compounds **37-[G$_1$]–37-[G$_7$]** and allylamine.

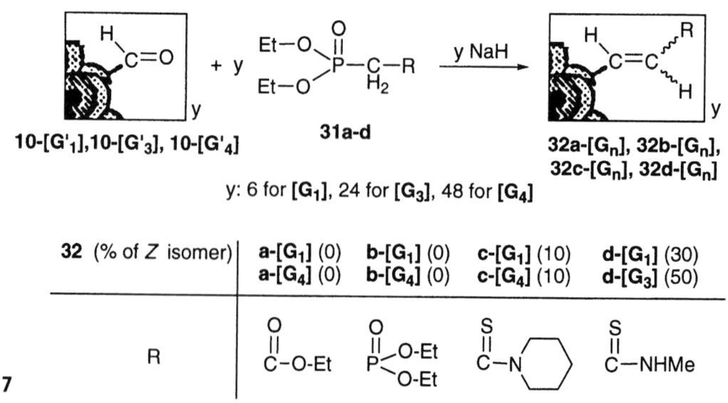

Scheme 17

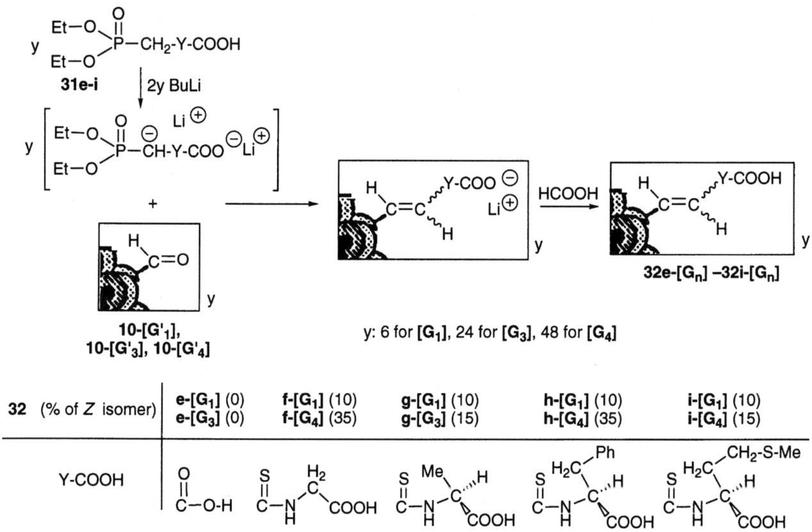

Scheme 18

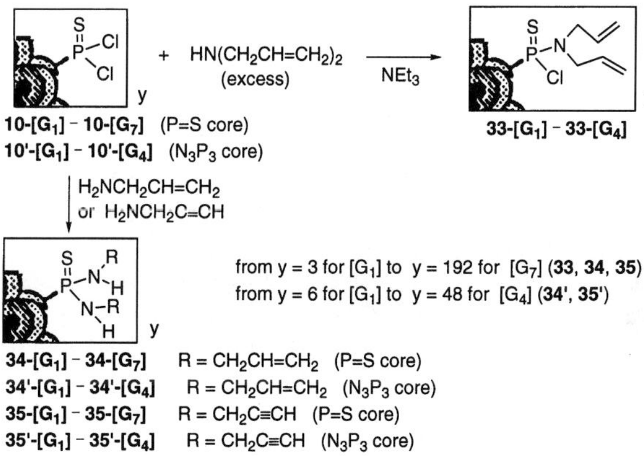

Scheme 19

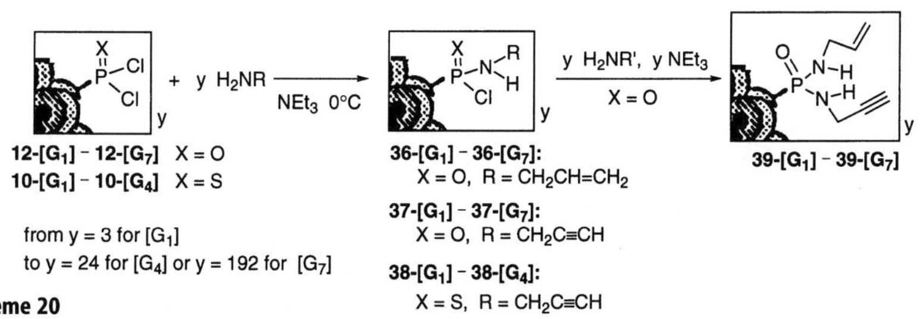

Scheme 20

The remaining P-Cl bonds of the monosubstituted compounds **36-[G_1]–36-[G_3]**, **37-[G_1]–37-[G_3]** are substituted with the sodium salt of 4-hydroxybenzaldehyde, affording dendrimers in which the surfaces are covered by NH, aldehyde and allyl or propargyl groups. These latter macromolecules are good models for the preparation of the first multiply "tetrafunctionalized" dendrimers. This can be examplified by two types of reaction involving terminal aldehyde groups: reaction with hydrazine and a Wittig reaction with $Ph_3P{=}CH\text{-}CN$ (Scheme 21).

Another specific monosubstitution which does not occur on phosphorus but on the substituents linked to phosphorus consists of a Mannich type reaction between Ph_2PCH_2OH and dendrimers bearing NH groups in α position relative to the terminal phosphorus atoms (Scheme 22). This reaction only occurs on one of the two NH chain ends, and no disubstitution takes place here [21].

All these experiments clearly confirm the unique specificity of the substitution reactions of phosphorus-containing dendrimers. The presence of phosphorus not only allows the construction of these macromolecules to be monitored without difficulty, but also permits the formation in high yield of a large number of dendrimers with multiple di-, tri- and tetrafunctionalizations. These results should open new perspectives in dendrimer chemistry. For example the attachment of several active substances showing different properties can be envisaged as well as the grafting of dendrimers to a second polymer by means of a suitable functionality leaving other functional groups available for further reactions.

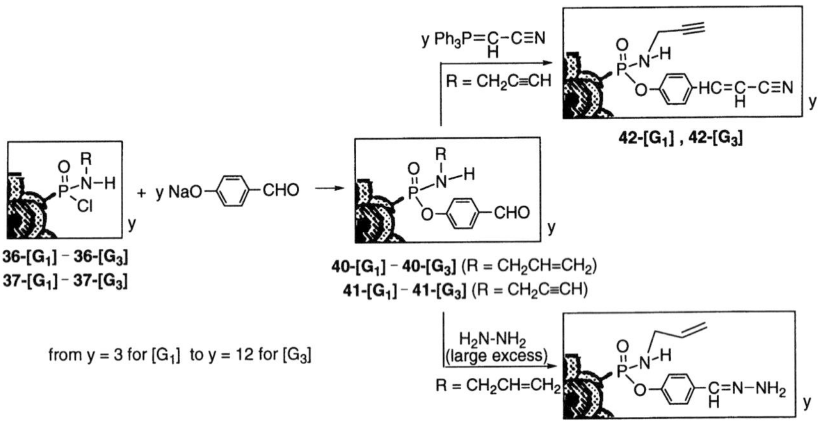

Scheme 21

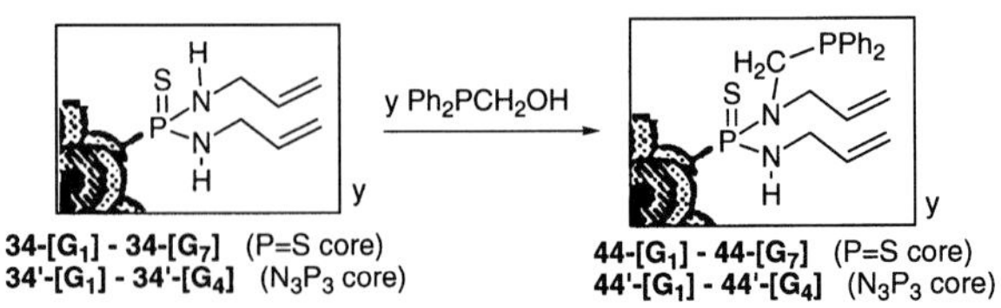

from y = 3 for [G_1] to y = 192 for [G_7] (**34,44**)
or from y = 6 for [G_1] to y = 48 for [G_4] (**34',44'**)

Scheme 22

3.5 Complexation

The search for new catalytic systems has stimulated the design of new ligands in organometallic chemistry over the past few years. In this respect the recent development of dendrimeric molecules has attracted a considerable interest as potential ligands. Examples of complexes of groups 6–11 have been recently prepared [3, 22–28] and some catalytic applications have been studied such as the Kharash addition of polyhalogeno alkanes to carbon-carbon double bonds [27]. The expected advantage of dendrimeric ligands concerns first the catalyst recovery which could be a valuable alternative to stabilization at the surface of a polymer or to biphasic reactions. Furthermore, the stabilization of reactive complexes at the surface of dendrimers could allow regioselective enhancements as a result of the blocking of conformations due to the steric crowding induced by the dendrimer.

It was demonstrated (see above) that steric considerations that dictate the globular shape do not affect the reactivity of surface functionalities. Since it was shown that phosphino groups can be easily grafted on the periphery of these macromolecules it appears interesting to investigate the possibilities of grafting a large number of these groups and studying their complexation properties towards a number of metals. The new complexes might present properties in different areas and of course as catalysts.

3.5.1 *Phosphino Groups as Terminal Chain Ends*

First, efforts were directed towards the preparation of dendrimers containing up to the theoretically predicted 3072 terminal phosphino groups (generation 10). As previously indicated, phosphino groups are introduced via the treatment of aldehyde terminated molecules with methylhydrazine and then with the phosphine Ph_2PCH_2OH. Complexation of dendrimers of generations 1–10 (P=S or P_3N_3 core) with AuCl (tetrahydrothiophene) affords the corresponding dendritic complexes [29, 30]. Some of these complexes (Scheme 23) were imaged by high resolution electron microscopy in order to compare the size of consecutive generations and to see if higher ordered structures are formed.

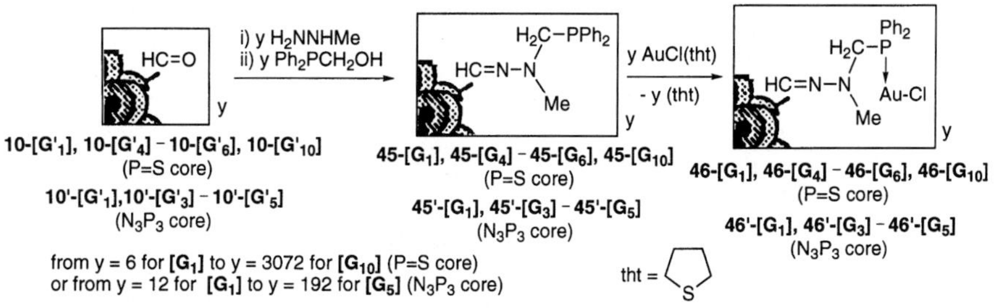

Scheme 23

Isolated spheres with diameters of 60 ± 5 (**46′**-[G_3] generation 3, P_3N_3 core), 75 ± 5 (**46′**-[G_4] generation 4, P_3N_3 core), 90 ± 5 (**46′**-[G_5] generation 5, P_3N_3 core) and 150 ± 5 (**46**-[G_{10}] generation 10, P=S core) Å were observed (Fig. 7). Besides these isolated spheres, aggregates generally adopting a spherical shape were detected with sizes up to 10000 Å.

It can be noted that no marked difference of reactivity towards Au-Cl (tetrahydrothiophene) was observed when moving from dendrimer of generation 1 to dendrimer of generation 10, whatever the nature of the core. Moreover the terminal Au-Cl bonds remain available for further reactions. Indeed an exchange reaction occurs readily between for example a dendrimer of generation 4 and Cp_2ZrMe_2 as alkylating agent [30]. This exchange can easily be followed since it is characterized by the deshielding of the signal due to the Ph_2P-Au-Cl group in ^{31}P NMR and the presence of a doublet for the Au-Me groups in ^{1}H NMR.

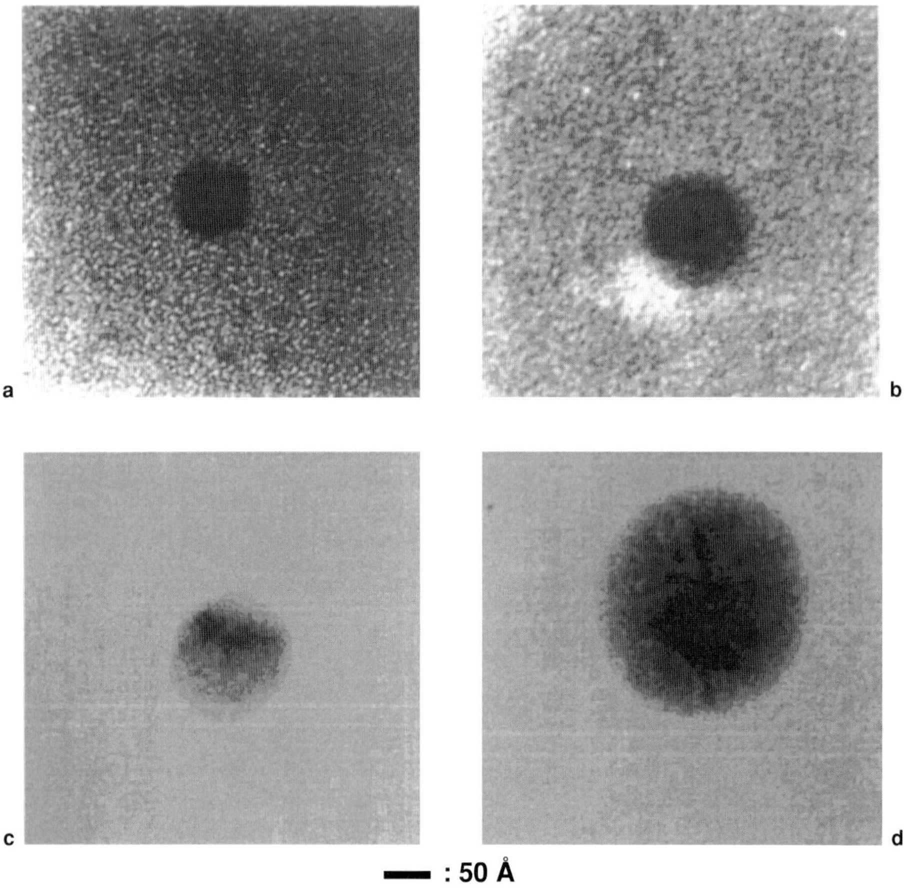

Fig. 7 a–d. High-resolution transmission electron micrograph of: **a** **46′**-[G_3] (N_3P_3 core); **b** **46′**-[G_4] (N_3P_3 core); **c** **46′**-[G_5] (N_3P_3 core); **d** **46**-[G_{10}] (P=S core)

Complexation with $Fe_2(CO)_9$ and $W(CO)_5$ (THF) occurs in a few hours at room temperature to give the corresponding iron (0) and tungsten (0) derivatives respectively (Scheme 24).

Dendrimers possessing terminal phosphino groups but also other functional chain ends such as allylamine present the same classical behavior towards $Fe_2(CO)_9$ and $W(CO)_5(THF)$ [21].

The reaction of Rh(I) derivatives such as $Rh(acac)(CO)_2$ (acac=acetylacetonate) with dendrimers of generation 1, 5 and 6 also proceeds readily at room temperature (Scheme 25). The complexation is unambiguously characterized in all cases by the appearance of a doublet ($^1J_{PRh}$=175 Hz) in the ^{31}P-NMR spectra and corroborated by ^{1}H NMR (two different CH_3 groups for the acac moieties due to the decrease of symmetry of rhodium in complexes). The poor solubility of complexes of generations 5 and 6 precludes their characterization by ^{13}C NMR.

This problem of poor solubility is not encountered starting from another Rh(I) derivative, $[Rh(\gamma\text{-}Cl)(COD)]_2$ at least up to the fourth generation (Scheme 25).

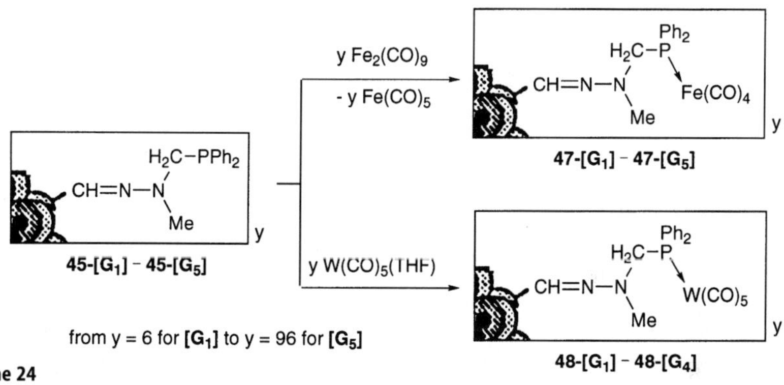

Scheme 24

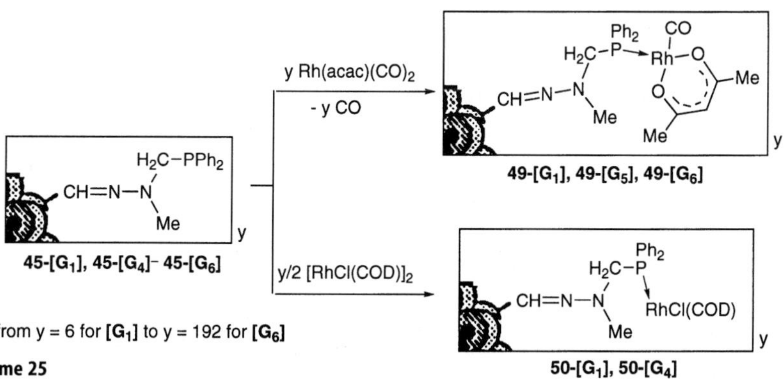

Scheme 25

3.5.2
Diphosphino Groups as Terminal Chain Ends

All the previous experiments consist of the reaction of terminal phosphines with M(0) or M(I) derivatives. Analogous complexations can be envisaged with multidentate dendrimers possessing diphosphino groups on the surface. These ligands can be prepared using the strategy outlined in Scheme 26: addition of hydrazine on terminal aldehyde groups followed by treatment of the resulting polyhydrazones with Ph_2PCH_2OH (2 equiv. per hydrazone moiety).

Dendrimers of generations 1–4 incorporating 6–48 $N(CH_2PPh_2)_2$ terminal groups are thus obtained. The ability of these dendrimers to act as ligands towards $PdCl_2(COD)$, $PdBr_2(COD)$, PdMeCl(COD) was investigated [31] (Schemes 27–29). In all cases diphosphino groups act as bidentate and it has been demonstrated that organometallic chemistry on the dendritic surface can be readily developed. An easy halogen exchange takes place when a dichloromethane solution of **53-[G$_1$]** for example is reacted with KBr while treatment of the same derivative with 12 equiv. of the Grignard reagent BrMgMe leads to the unsymmetrical bis-substituted palladium complex **55-[G$_1$]**: here halogen exchange and alkylation on palladium take place simultaneously (Scheme 27).

On the other hand, monomethylation, exclusively, can be clearly perfomed with Cp_2ZrMe_2.

A third type of reaction, namely CO insertion, is also described. The rate of CO insertion into the Pd-Me bond in several classical diphosphine monomers has been found to be very dependent on the nature of the ligand. Full carbonylation of, for example, dendrimer **56-[G$_1$]** occurs when this compound is pressurized to 1 bar of CO. Experiments with ^{13}CO corroborate the formation of the acetyl complex while insertion of norbornene into the Pd-acetyl bond can be easily performed. Dendrimers of generations 2 and 3 react similarly (Scheme 27).

Complexation of the same terminated diphosphino dendrimers and $PtCl_2(COD)$ leads to complexes incorporating 6, 12 or 24 $PtCl_2$ units on the dendrimer surface (Scheme 28). Substitution on platinum takes place in the presence of Grignard reagent MeMgBr but the final derivative was found to be unstable in the presence of ClMgBr generated in the reaction: thus the complex with PtMeBr end groups is formed, then another one with $PtBr_2$ end groups (Scheme 28).

As terminated phosphino dendrimers, terminal diphosphino dendrimers (generations 1–3) react with Rh(acac) (COD) to give the expected diphosphinorhodium complexes (Scheme 29).

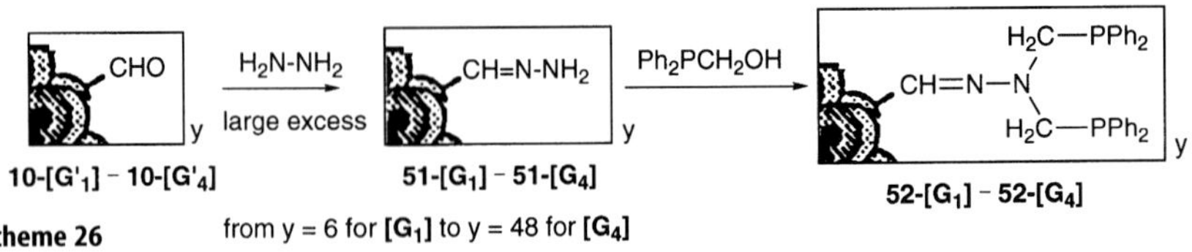

Scheme 26

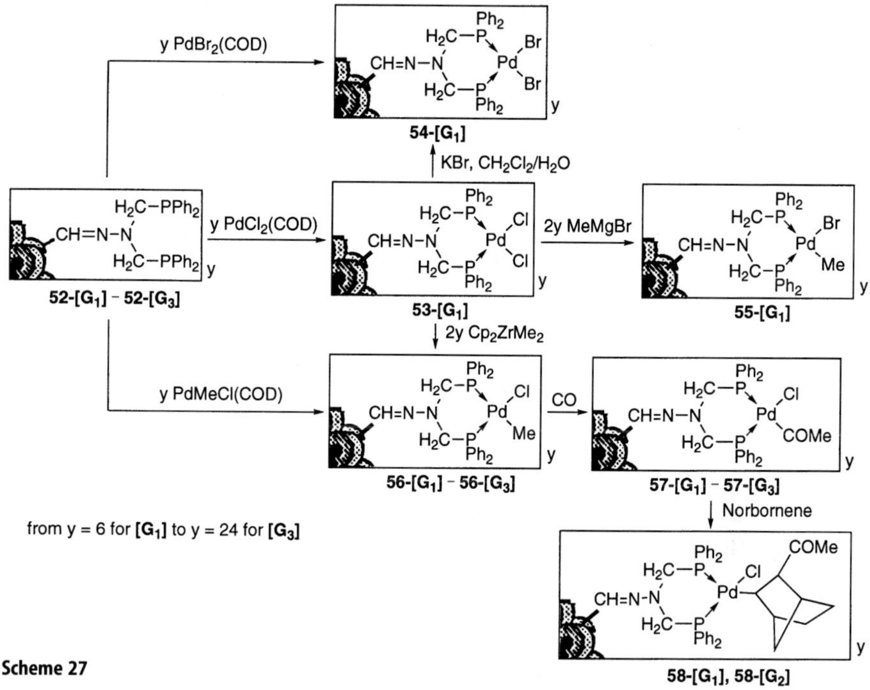

Scheme 27

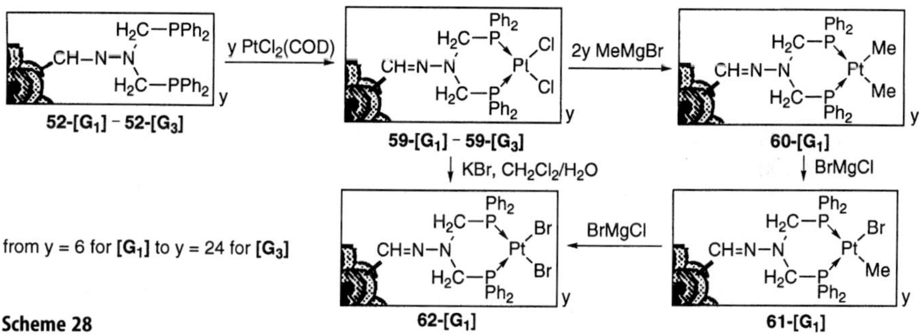

Scheme 28

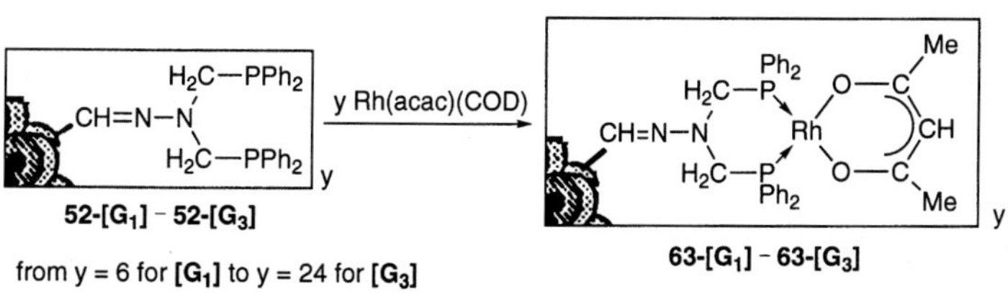

Scheme 29

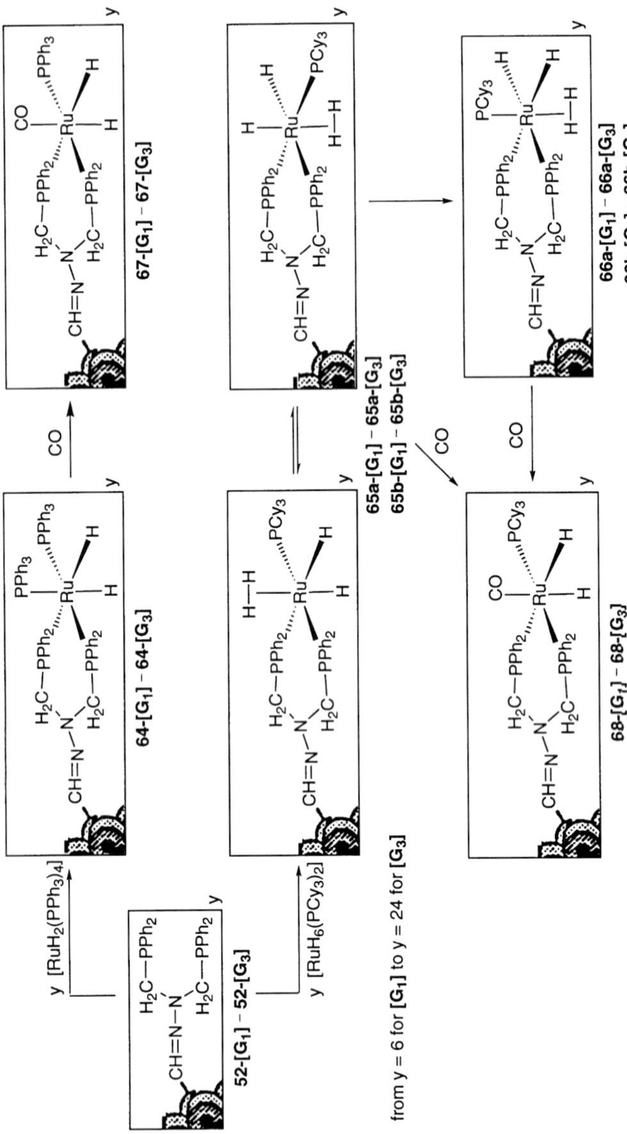
52-[G1] - 52-[G3]
y [RuH2(PPh3)4]
y [RuH6(PCy3)2]
64-[G1] - 64-[G3]
65a-[G1] - 65a-[G3]
65b-[G1] - 65b-[G3]
CO
67-[G1] - 67-[G3]
66a-[G1] - 66a-[G3]
66b-[G1] - 66b-[G3]
68-[G1] - 68-[G3]
from y = 6 for [G1] to y = 24 for [G3]

Scheme 30

The reaction of the ruthenium polyhydrides $RuH_2(PPh_3)_4$ and $RuH_2(H_2)_2(PCy_3)_2$ with dendrimers G_x $(NCH_2PPh_2)_2$ (1st, 2nd and 3rd generation) **52-[G$_1$]–52-[G$_3$]** clearly leads to the new complexes $[G_xN(CH_2PPh_2)_2\text{-}RuH_2(PPh_3)_2]$ and $G_xN(CH_2PPh_2)_2\text{-}RuH_2(H_2)$ (PCy_3) which exist as a mixture of isomers **65a-[G$_1$]–65a-[G$_3$], 65b-[G$_1$]–65b-[G$_3$]** [32] (Scheme 30). The former complexes were shown to accommodate a stretched dihydrogen ligand and to display a high temperature for the minimum of the T_1 relaxation time in agreement with their binding to the dendrimer in solution. The treatment of these complexes with CO leads to new dihydrido carbonyl ruthenium complexes (Scheme 30). The very characteristic spectroscopic properties of these complexes, in particular the ^{31}P and the high field ^1H-NMR spectra, which in mononuclear complexes are very sensitive to subtle changes of their environment, are invariant as a function of the dendrimer generation used as a ligand. The most interesting derivatives contain dihydrogen ligands for which it has been calculated in the case of the third generation a H-H distance (around 0.9 Å, rapid rotation; around 1.15 Å, slow rotation) similar to those calculated for mononuclear ruthenium dihydrogen complexes. It can be noted also that the mixture of isomers **65a-[G$_1$]–65a-[G$_3$], 65b-[G$_1$]–65b-[G$_3$]**, whatever the relative concentration of each species, reacts cleanly with thiols RSH to give a single compound which displays spectroscopic properties similar to $RuH(SR)(PPh_3)_3$, and with silanes and olefines. Furthermore preliminary catalytic tests indicate a good activity of these compounds for ketone hydrogenation.

4
Reactivity Within the Cascade Structure

Rengan and Engel reported the preparation of polycationic phosphorus-containing dendrimers but these ionic compounds are insoluble from the third generation upwards (40 phosphonium centers) [2, 5]. Polycationic dendrimers can also be formed by reacting the $P(S)Cl_2$ chain ends of dendrimers with $H_2N(CH_2)_2NEt_2$ [33] while polyanionic species are also isolated via a Horner Wadsworth Emmons reaction involving terminal aldehyde groups (see above). These last two types of charged polymers are water soluble.

These methods allow the incorporation of charges either at each generation during the construction of the dendrimer or only on the surface taking advantage of the high reactivity of terminal groups.

In order to be able to introduce selectively charges in different layers of the dendrimers, a new methodology of construction of dendrimers was recently proposed (Schemes 31 and 32) [34]. This synthesis offers the possibility to introduce, when and where required, very reactive P=N-P(S) fragments, in the framework of the dendrimer. Some of the neutral molecules bearing up to 322 P=N-P(S) units are represented in Scheme 33. From these derivatives the respective polycationic dendrimers are readily prepared by selective alkylation of the thiophosphoryl groups of P=N-P(S) units with methyltrifluoromethane sulfonate. Once again the reaction can easily be monitored by ^{31}P $\{^1H\}$-NMR spectroscopy since the two doublets attributed to the P=N-P(S) neutral fragments are replaced by two new doublets with different $^2J_{PP}$ values due to the new

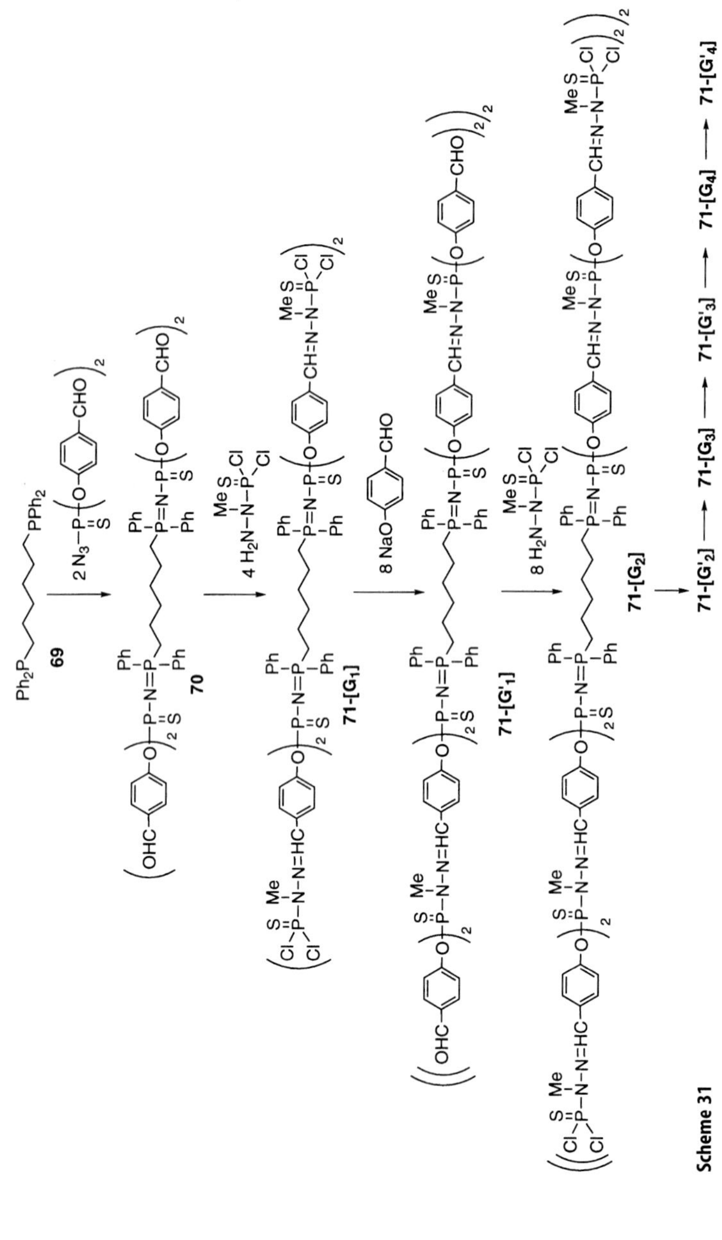

Scheme 31

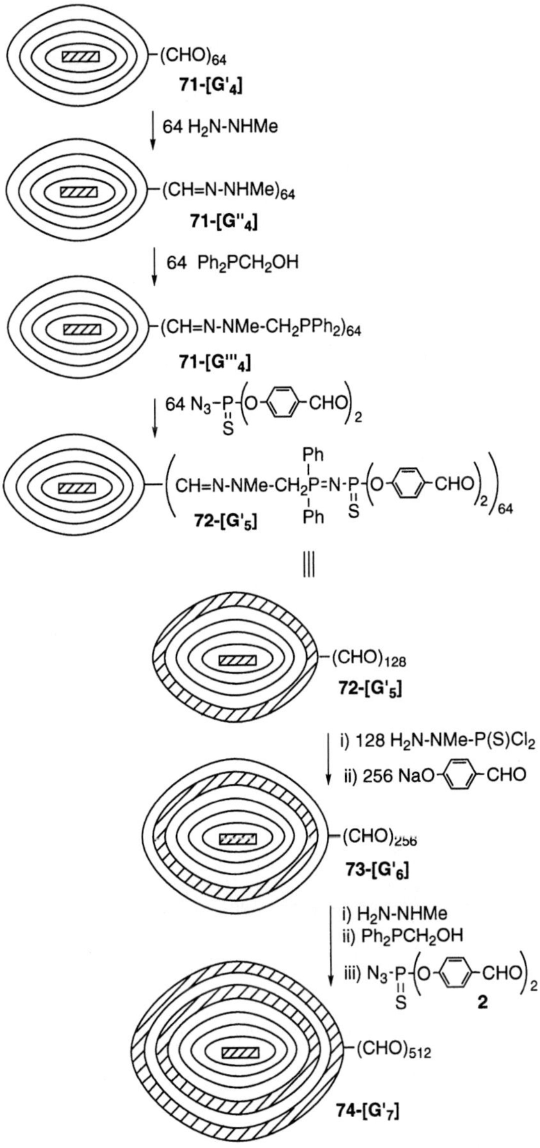

Scheme 32

$P{=}N^{+}{=}P(SMe)$ units. ^{31}P NMR indicates that the other P=S bonds of the macromolecule are not alkylated.

Remarkably, alkylation of the core of a dendrimer of generation 6 (two P=N-P(S) fragments on the core, 64 for the fifth generation, and an upper generation incorporating 256 terminal aldehyde groups) can be performed. This clearly demonstrates that the core of this sixth generation dendrimer is available for certain reactions.

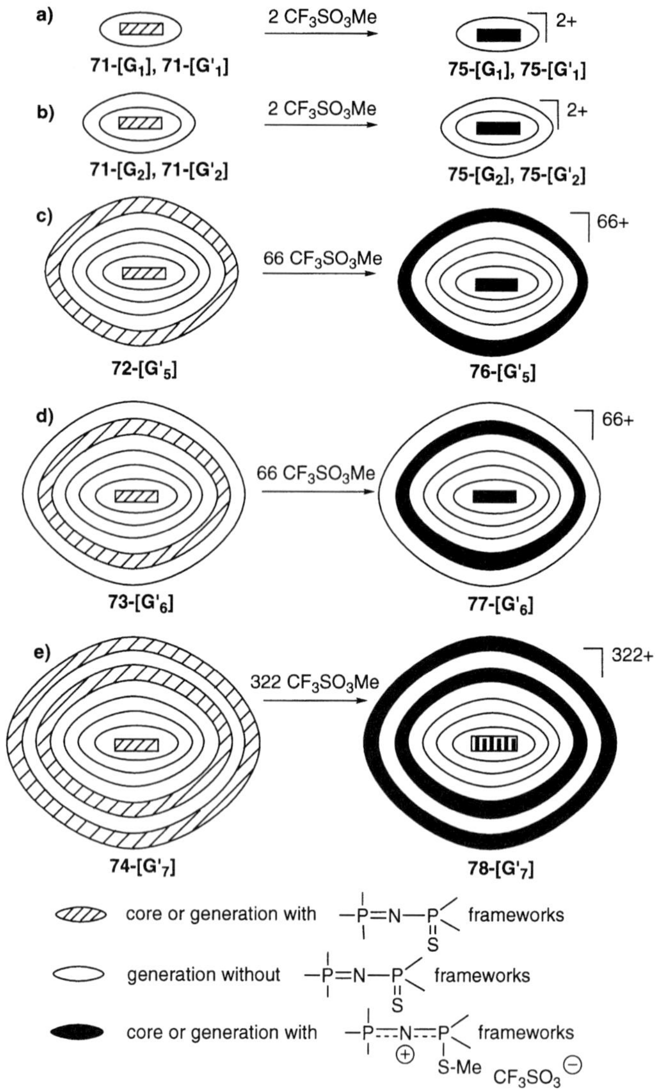

Scheme 33

Indeed, up to 322 charges can be chemoselectively incorporated on the core, within the cascade structure (core and internal generations) or even within the cascade structure and on the surface. The polycationic species remain fairly soluble in several solvents.

Such a methodology is also useful for the chemoselective functionalization of internal voids of dendrimers. This can be accomplished for example by adding 2 equiv. of allyl, propargyl or phosphonate trifluoromethane sulfonate on the dendrimer of generation 1 **71-[G$_1$]** (Scheme 34). Functionalization occurs on the sulfur atom of the two P=N-P(S) units with the quantitative formation of the

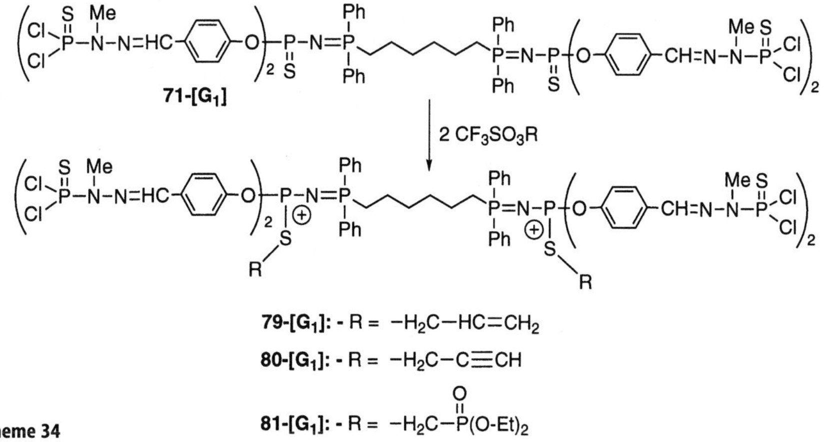

Scheme 34

dicationic species **79-[G$_1$]**, **80-[G$_1$]** and **81-[G$_1$]**. X-ray crystallography studies confirm the structure of the bis allyl derivative (Fig. 8). This is the largest dendrimer characterized by X-ray, even larger than all the dendrimers of generation 2 whose structures were also determined by X-ray.

Selective functionalization with allyl, propargyl and phosphonate triflates occurs with a dendrimer of generation 4 **82-[G$_4$]** possessing eighteen P=N-P(S) units (Scheme 35). The reaction proceeds instantaneously when allyl or propargyl triflates are used and gives quantitatively the polycationic dendrimers **83-[G$_4$]** or **84-[G$_4$]** incorporating eighteen allyl or propargyl fragments in the internal cavities. The same reaction with the phosphonate triflate takes two weeks at 60 °C in CH_2Cl_2 to go to completion [35]. Therefore this is a general methodology allowing one to introduce simultaneously charges and functional groups when and where required into the internal voids.

Remarkably, it is possible from these polycationic dendrimers to create new reactive centers into the framework of the molecules, these centers being the starting points for the development of a macromolecular chemistry into the cavities. Indeed, dendrimer **86-[G$_3$]** (prepared as shown in Scheme 36) possessing six internal P=N-P(S) fragments within the cascade structure can be treated with methyltrifluoromethane sulfonate; the resulting hexacationic dendrimer **87-[G$_3$]** is submitted to react with (tris dimethylamino)phosphine: a clean transfer of the SMe units from the P=N-P(SMe) groups to $P(NMe_2)_3$ takes place with the formation of six internal and very reactive P=N-P moieties incorporated in the skeleton of the new dendrimer **88-[G$_3$]** [36].

At this stage two strategies (one of them outlined in Scheme 37) can be developed to start the construction of dendrimers within the cascade structure of **88-[G$_3$]**. Each of them allows the synthesis of six internal dendrimers of generation 4 into the internal voids of **88-[G$_3$]**. ^{31}P NMR constitutes an extraordinary and unique tool for monitoring the construction of these controlled polydendritic structures (see Fig. 9 for illustration). Chemical shifts of phosphorus groups are different from one generation to another and the intensities of signals are of

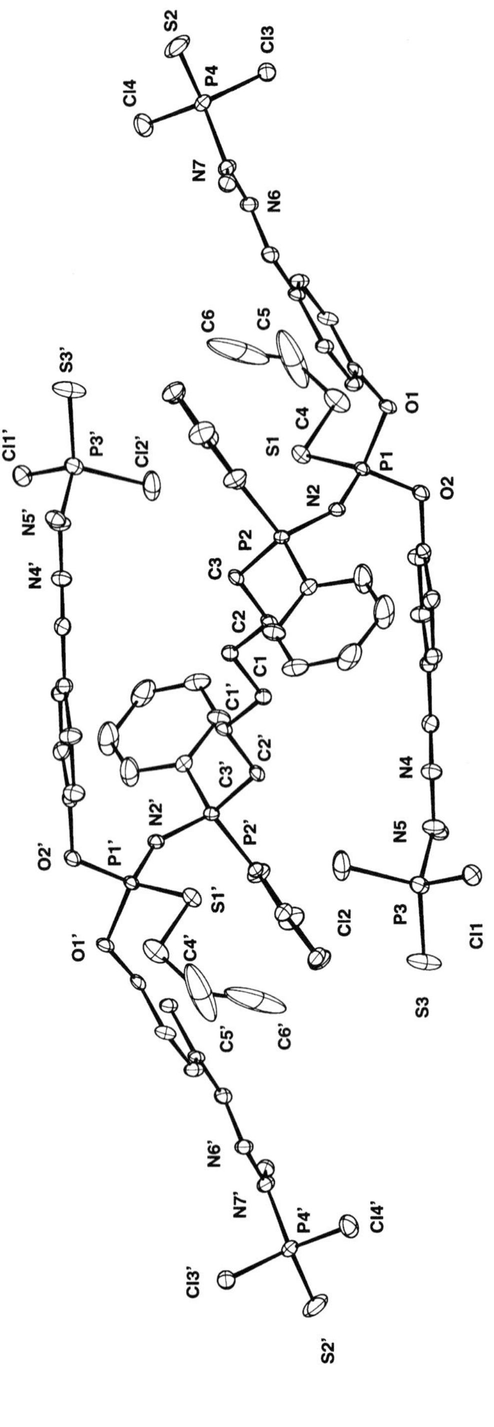

Fig. 8. ORTEP view of 79-[G_1]

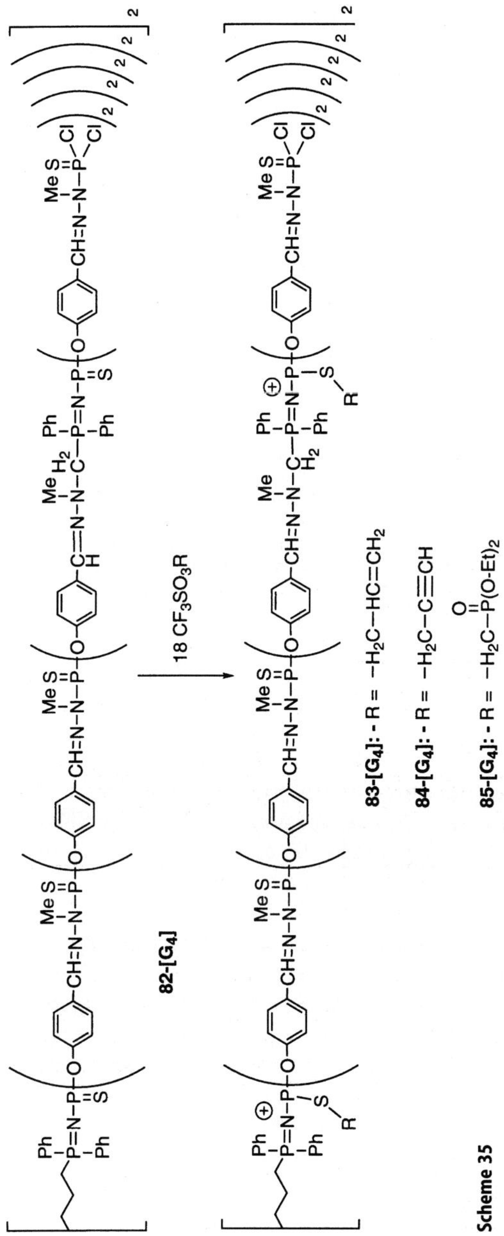
82-[G4]
18 CF3SO3R
83-[G4]: - R = −H2C−HC=CH2
84-[G4]: - R = −H2C−C≡CH
85-[G4]: - R = −H2C−P(O)(O-Et)2

Scheme 35

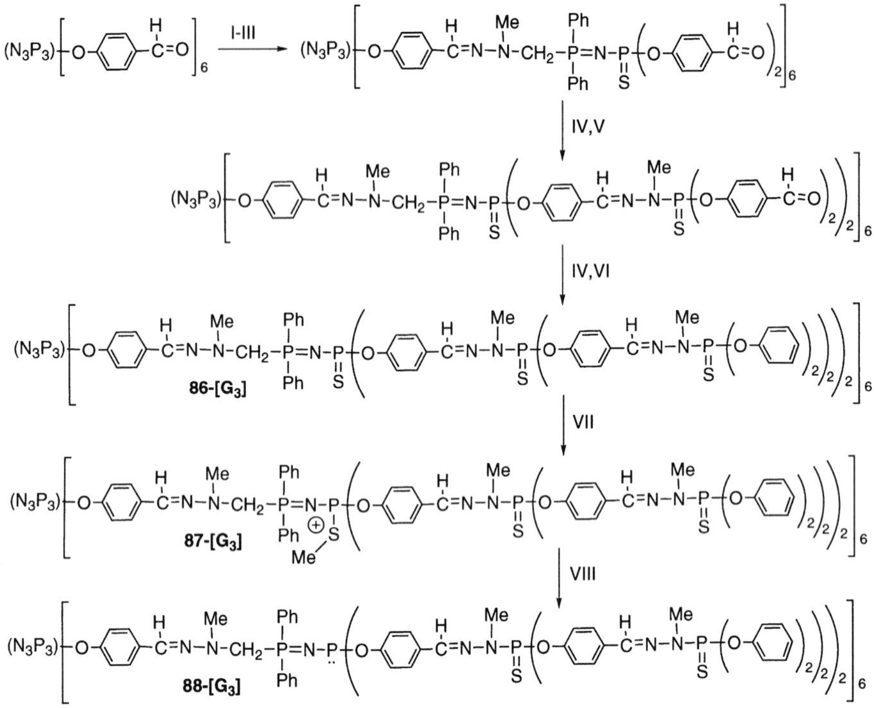

I) H_2NNHMe II) Ph_2PCH_2OH III) $N_3P(S)(OC_6H_4CHO)_2$ IV) $Cl_2P(S)NMeNH_2$
V) $NaOC_6H_4CHO$ VI) $NaOC_6H_5$ VII) CF_3SO_3Me VIII) $P(NMe_2)_3$

Scheme 36

course very different. Indeed, N-N-P-O fragments give a singlet, P=N-P(S) linkages two doublets ($^2J_{PP}$=31 Hz) and the P=N-P=N-P(S) unit a doublet of doublets for the central phosphorus atom ($^2J_{PP}$=22–23 and 58–59 Hz) and a doublet for each of the two other phosphorus atoms.

This work demonstrates that functionalization of the internal cavities of various dendrimers can be done via a post modification of the skeleton. Various functional groups can be selectively introduced: aminophosphite, aldehyde, hydrazone, dichlorophosphane sulfide. Therefore all the chemistry reported on the surface of dendrimers can be now envisaged to be done into the cavities and it is demonstrated for the first time that a macromolecular chemistry can be performed into the internal voids of a dendrimer.

5 Dipole Moments

Dipole moments for dioxan solutions of both CHO or P-Cl terminated dendrimers were measured at 25 °C. Remarkably, values of dipole moment vs generation increase exponentially from 8.43 (**10-[G$_1$]**) to 258 (**10-[G$_{10}$]**) and from 8.27 (**10-[G$'_1$]**) to 328 (**10-[G$'_{11}$]**) debyes (Fig. 10a), the highest dipole moment values

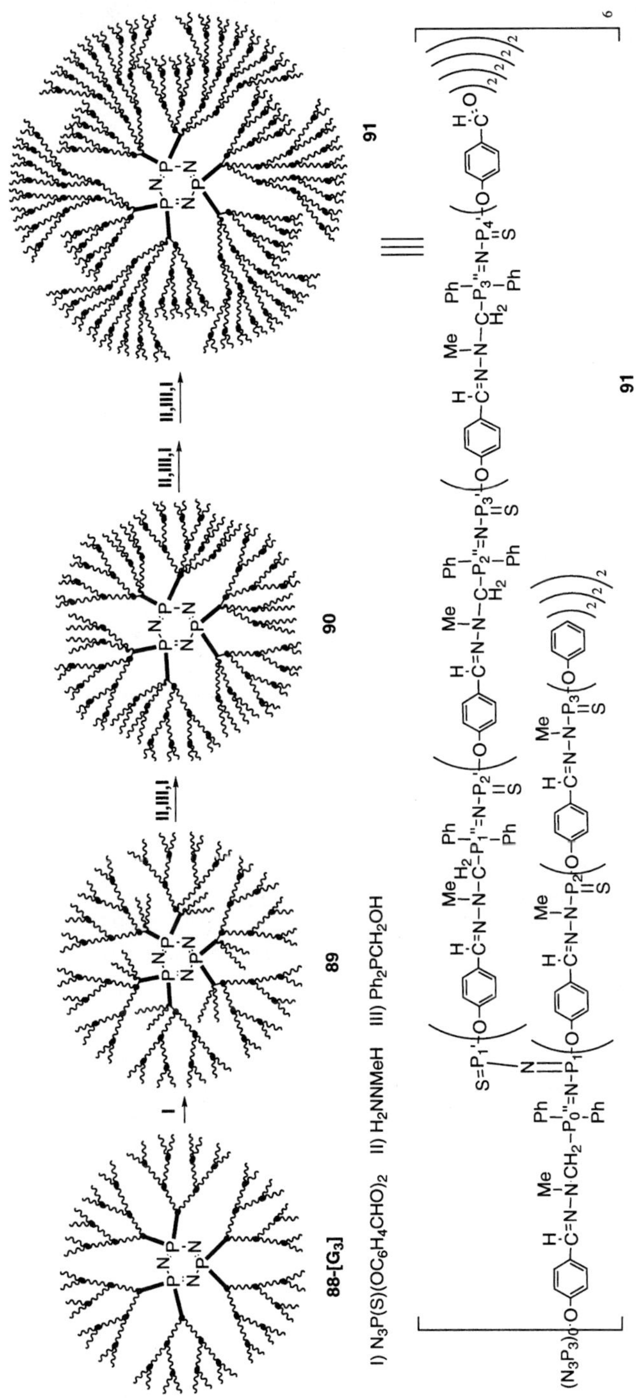

Scheme 37

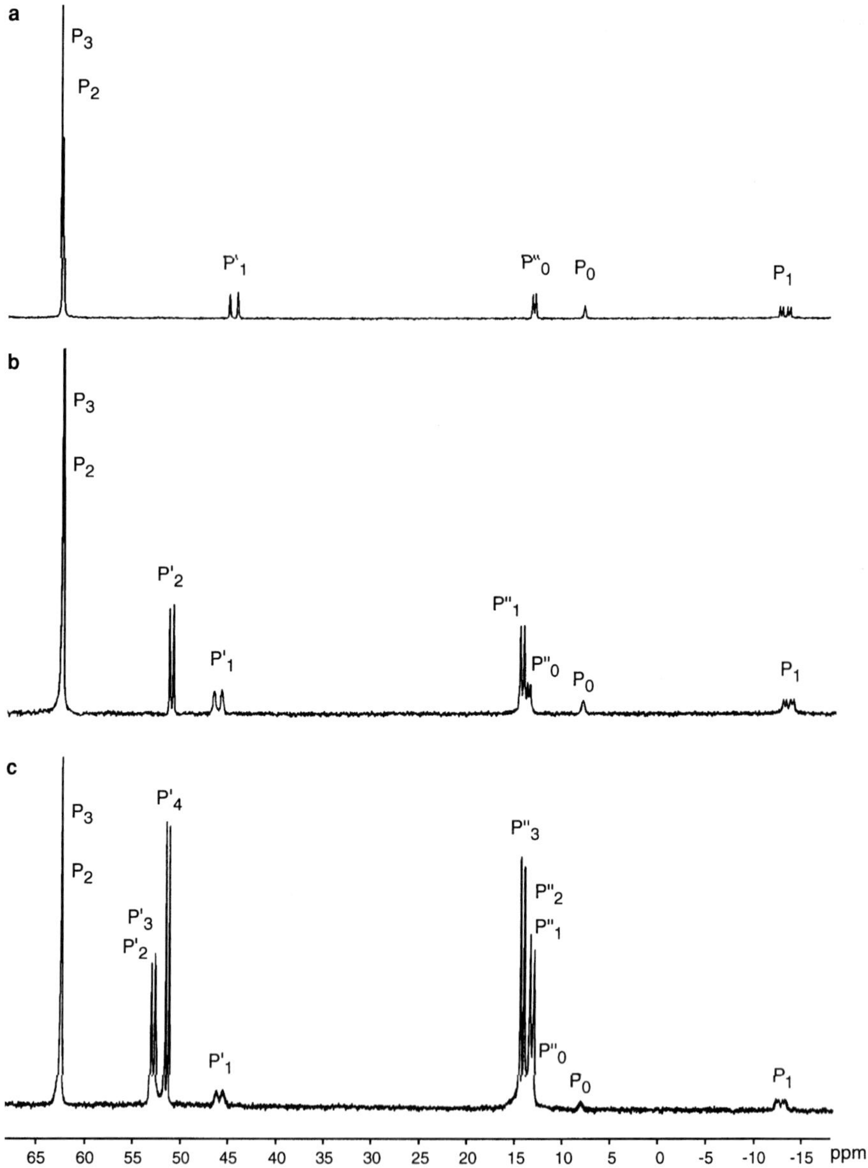

Fig. 9a–c. ^{31}P-NMR spectra of the polydendritic macromolecules: **a** compound **89** – central dendrimer P_0, P_0'', P_1, P_2, P_3 – internal dendron P_1'; **b** compound **90** – central dendrimer P_0, P_0'', P_1, P_2, P_3 – internal dendrimers P_1', P_1'', P_2'; **c** compound **91** – central dendrimer P_0, P_0'', P_1, P_2, P_3 – internal dendrimers P_1', P_1'', P_2', P_2'', P_3', P_3'', P_4'

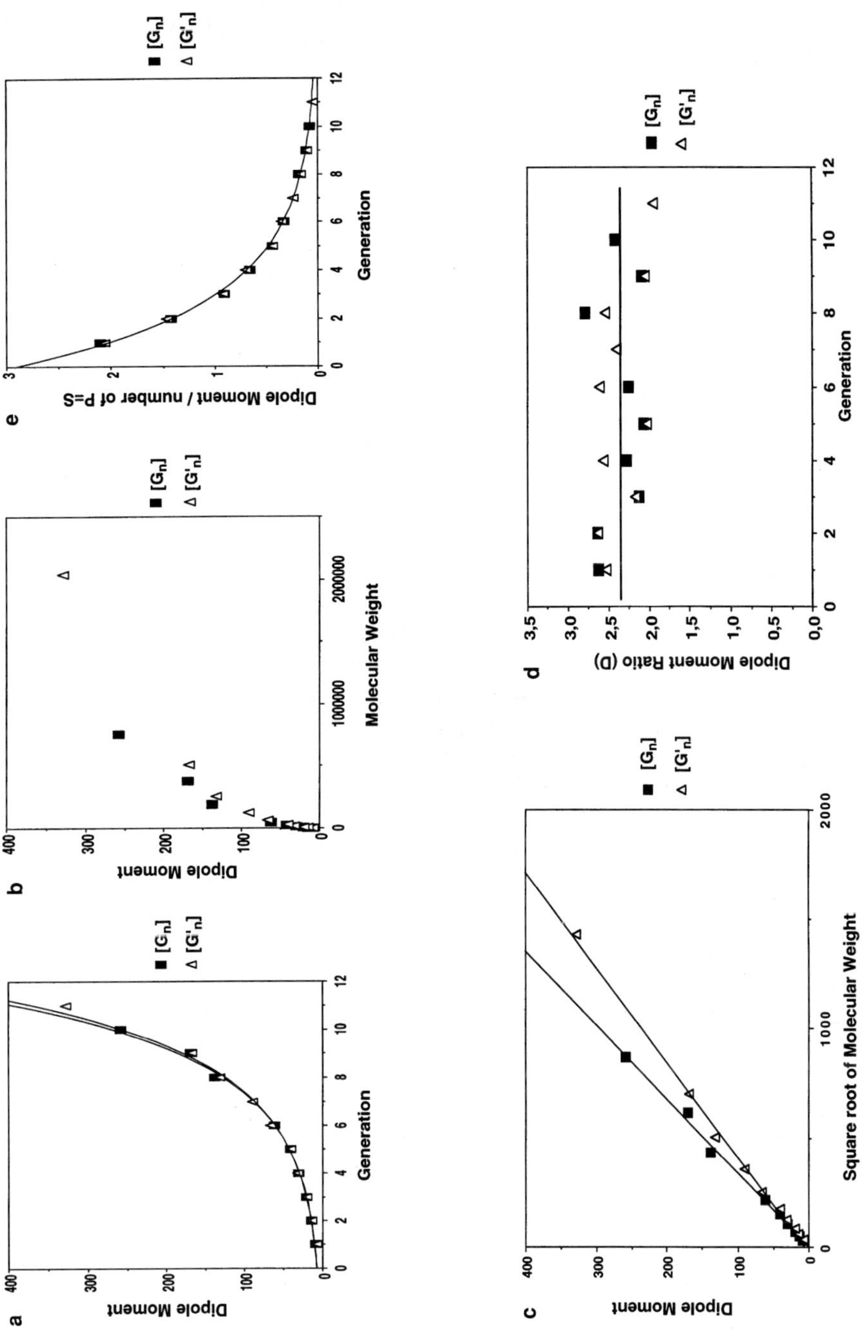

Fig. 10. **a** Dipole moment vs generation for dendrimers **10-[G$_1$]–10-[G$_{10}$]** and **10-[G$'_1$]–10-[G$'_{11}$]**. **b** Dipole moment vs molecular weight for dendrimers **10-[G$_1$]–10-[G$_{10}$]** and **10-[G$'_1$]–10-[G$'_{11}$]**. **c** Dipole moment vs square root of molecular weight for dendrimers **10-[G$_1$]–10-[G$_{10}$]** and **10-[G$'_1$]–10-[G$'_{11}$]**. **d** Dipole moment ratio vs generation for dendrimers **10-[G$_1$]–10-[G$_{10}$]** and **10-[G$'_1$]–10-[G$'_{11}$]**. **e** Dipole moment/number of P=S vs generation for dendrimers **10-[G$_1$]–10-[G$_{10}$]** and **10-[G$'_1$]–10-[G$'_{11}$]**

reported up to now for dendritic structures [9b]. Dipole moments are not dependent of the nature of chain ends since close values are obtained for CHO and P-Cl terminated dendrimers of a given generation. The dipole moment of these phosphorus-containing dendrimers vs molecular weight does not increase in a linear fashion (Fig. 10b). However, the representation of the dipole moments in function of the square root of molecular weights is a straight line (Fig. 10c). This is in perfect agreement with previous calculations on statistical mean dipole moment of polymers. Moreover, the dipole moment ratio D ($D=\mu^2/N(\mu_0)_2$ where μ is the dipole moment of the dendrimer, N the number of monomer units and μ_0 the dipole moment of the monomer unit) vs molecular weight is a constant (D=2.3). This unusually large value indicates considerable local correlation between neighboring dipoles as might be expected (Fig. 10d).

Remarkably too, the contribution of each polar unit, the P=S group, to the global dipole moment decreases exponentially from 2.1 (**10-[G_1]**) or 2.07 (**10-[G'_1]**) to 0.08 (**10-[G_{10}]**) or 0.05 (**10-[G'_{11}]**) debyes (Fig. 10e). Therefore a strong compensation (98.5% for **10-[G_{11}]**!) takes place. One can postulate that this strong compensation might be due to steric hindrance restricting the orientation of individual dipole vectors: as a result the contribution of each polar unit is largely attenuated. Nevertheless obtaining rather sharp resonances in solution spectra in ^{31}P NMR demonstrates that conformational flexibility which determines the correlation times of the nuclei is still high even at the **10-[G_{10}]**–**10-[G_{12}]** levels; otherwise no well resolved signals would be obtained. This can explain why the contribution of each polar unit, although weak, is not equal to zero and why that even at high generations a maximum of μ is not observed in spite of the globular shape of these dendrimers.

6 Conclusion

Even if only a few methods of synthesis of phosphorus-containing dendrimers are reported up to date, it is clear that these macromolecules now constitute a well known class of polymers whose studies appear to be an emerging field of research.

The contribution of phosphorus in the area of dendrimer chemistry is essential for many reasons.

- Phosphorus dendrimers are among the largest dendrimers reported till now (generation 12, molecular weight higher than 3000000).
- A variety of phosphorus cores can be used. These cores can be di-, tri-, tetra- or hexafunctionalized, enhancing the possibilities to prepare dendrimers of various forms at least for the lower generations ("cauliflower", "bow-tie", bolaform, ball shaped).
- Phosphorus at the core can be tri-, tetra- or pentacoordinated and therefore a number of reactions on the core can be envisaged.
- $P(X)Cl_2$ (X=S,O) terminated units allow the grafting of various other functional groups and the preparation of multiplurifunctionalized dendrimers, i.e. dendrimers possessing a large number of sets of two, three or four different functionalities on the surface.

- Phosphorus also permits the postmodification of the backbone of dendrimers. Indeed charges, and also various functional groups, can be selectively introduced into the internal voids. Such a transformation can be performed where and when required.
- The facile functionalization of the cavities allow the development of a macromolecular chemistry within the cascade structure of dendrimers. As an example six dendrimers of generation 4 were built into the internal voids of a dendrimer of generation 3.
- ^{31}P NMR appears to be a useful tool and the method of choice to follow rigorously the construction of dendrimers since up to generation 6 the signal of the phosphorus atom of the core can be detected; lack of substitution on the surface at one or several of the terminal functional groups from generations 1–6 would thus be observed. Moreover each generation gives a different chemical shift. Furthermore substitution reactions on the surface generally result in a shielding or deshielding effect (depending on the type of substitution) of the signal due to the phosphorus atoms of the top generation n and a slight deshielding effect for the phosphorus atoms of generation n–1.
- The easy grafting of up to 3000 phosphino groups or up to 96 diphosphino groups on the surface allows the development of an organometallic chemistry and the use of the corresponding complexes in catalysis. Indeed Au, Pt, Pd, Ru, Rh, Fe, W derivatives can be anchored at the periphery.
- Even if attention has not been focussed on the physical properties of phosphorus-containing dendrimers in this review, one can mention the high dipole moment values observed for these polymers: up to 328 debyes for generation 11.

In summary, phosphorus-containing dendrimers have their own specificity: applications of these derivatives in different areas would be numerous in the near future. Some of these applications are in progress.

Acknowledgement. The authors would like to thank all their co-workers whose names appear in the references.

7 References

1. Buhleier E, Welmer W, Vögtle F (1978) Synthesis 78:155
2. Rengan K, Engel R (1990) J Chem Soc Chem Commun 1084
3. (a) Lange P, Schier A, Schmidbaur H (1996) Inorg Chem 35:637; (b) Lange P, Schier A, Schmidbaur H (1995) Inorg Chim Acta 235:263
4. (a) Miedaner A, Curtis CJ, Barkley RM, DuBois DL (1994) Inorg Chem 33:5482; (b) Herring AM, Steffey BD, Miedaner A, Wander SA, DuBois DL (1995) Inorg Chem 34:1100
5. (a) Rengan K, Engel R (1990) J Chem Soc Perkin Trans I 987; (b) Engel R (1993) Phosphorus Sulfur and Silicon 77:221; (c) Engel R, Rengan K, Chan CS (1993) Heteroatom Chem 4:181
6. Sournies F, Crasnier F, Graffeuil M, Faucher JP, Lahana R, Labarre MC, Labarre JF (1995) Angew Chem Int Ed Engl 34:578
7. Hudson RHE, Damha MJ (1993) J Am Chem Soc 115:2119
8. (a) Launay A, Caminade AM, Lahana R, Majoral JP (1994) Angew Chem Int Ed Engl 33:1589; (b) Caminade AM, Majoral JP, Slany M French Patent 9,506,281

9. (a) Launay N, Caminade AM, Majoral JP (1995) J Am Chem Soc 117:3282; (b) Lartigue ML, Fayet JP, Donnadieu B, Galliot C, Caminade AM, Majoral JP (1997) Macromolecules 30:7335
10. Lartigue ML, Launay N, Donnadieu B, Caminade AM, Majoral JP (to be published)
11. Launay N, Caminade AM, Majoral JP (1997) J Organomet Chem 529:51
12. Galliot C, Prévôté D, Caminade AM, Majoral JP (1995) J Am Chem Soc 117:5470
13. Newkome GR, Lin X, Weis CD (1991) Tetrahedron Asym 1:957
14. (a) Jansen JFGA, Peerlings HWI, de Brabander-van den Berg EMM, Meijer EW (1995) Angew Chem Int Ed Engl 34:1206; (b) Peerlings HWJ, Jansen JFGA, de Brabander-van den Berg EMM, Meijer EW (1995) Polymer Mater Sci 73:342; (c) Jansen JFGA, Meijer EW, de Brabander-van den Berg EMM (1996) Macromol Symp 102:27
15. Issberner J, Böhme M, Grimmes S, Nieger M, Paulus W, Vögtle F (1996) Tetrahedron Asym 7:2223
16 Lartigue ML, Caminade AM, Majoral JP (1997) Tetrahedron Asym 8:2697
17. (a) Prévôté D, Caminade AM, Majoral JP (unpublished results); (b) Prévôté D, Caminade AM, Majoral JP (1997) J Org Chem 62:4834
18. Prévôté D, LeRoy-Gouvennec S, Caminade AM, Masson S, Majoral JP (1997) Synthesis 1199
19. Launay N, Slany M, Caminade AM, Majoral JP (1996) J Org Chem 61:3799
20. Lartigue ML, Slany M, Caminade AM, Majoral JP (1996) Chem Eur J 2:1417
21. Slany M, Caminade AM, Majoral JP (1996) Tetrahedron Lett 37:9053
22. (a) Campagna S, Giannetto A, Serroni S, Denti G, Trusso S, Mallamace F, Micali N (1995) J Am Chem Soc 117:1754; (b) Campagna S, Denti G, Serroni S, Juris A, Venturi M, Ricevuto V, Balzani V (1995) Chem Eur J 1:211; (c) Newkome GR, Cardullo F, Constable EC, Moorefield CN, Thompson AMWC (1993) J Chem Soc Chem Commun 925; (d) Liao YH, Moss JR (1995) Organometallics 14:2130; (e) Haga MA, Ali MM, Akakawa R (1996) Angew Chem Int Ed Engl 35:76
23. (a) Achar S, Puddephatt RJ (1994) Angew Chem Int Ed Engl 33:847; (b) Achar S, Puddephatt RJ (1994) J Chem Soc Chem Commun 1895
24. (a) Miedaner A, Curtis CJ, Barkley RM, DuBois DL (1994) Inorg Chem 33:5482; (b) Heering AM, Steffey BD, Miedaner A, Wander SA, DuBois DL (1995) Inorg Chem 34:1100; (c) Huck WTS, van Veggel FCJM, Kropman BL, Blank DHA, Keim EG, Smithers MMA, Reinhoudt DN (1995) J Am Chem Soc 117:8293; (d) Huck WTS, van Veggel FCJM, Reinhoudt DN (1996) Angew Chem Int Ed Engl 1213:35
25. (a) Astruc D (1991) Top Curr Chem 160:47; (b) Fillaut JL, Linares J, Astruc D (1994) Angew Chem Int Ed 30:2460 and references therein; (c) Alonso B, Cuadrado I, Moran M, Losada J (1994) J Chem Soc Chem Commun 2575
26. a) Moors R, Vögtle F (1993) Chem Ber 126:2133; (b) Newkome GR, Moorefield CN (1994) Macromol Symp 77:63
27. Knapen JWJ, van der Made AW, de Wilde JC, van Leeuven PWNM, Wijkens P, Grove DM, van Koten G (1994) Nature 372:659
28. Ottaviani MF, Bossmann S, Turro NJ, Tomalia DA (1994) J Am Chem Soc 116:661
29. Slany M, Bardaji M, Casanove MJ, Caminade AM, Majoral JP, Chaudret B (1995) J Am Chem Soc 117:9764
30. (a) Slany M, Bardaji M, Caminade AM, Chaudret B, Majoral JP (1997) Inorg Chem 36:1939; (b) Bardaji M, Slany M, Lartigue ML, Caminade AM, Chaudret B, Majoral JP (1997) Main Group Chem 2:133
31. Bardaji M, Kustos M, Caminade AM, Chaudret B, Majoral JP (1997) Organometallics 16:403
32. Bardaji M, Caminade AM, Majoral JP, Chaudret B (1997) Organometallics 16:3489
33. Loup C, Caminade AM, Majoral JP, Meunier B (unpublished results)
34. Larré C, Caminade AM, Majoral JP (1997) Angew Chem Int Ed 36:595
35. Larré C, Caminade AM, Majoral JP (1998) J Am Chem Soc (in press)
36. Galliot C, Larré C, Caminade AM, Majoral JP (1997) Science 277:1981

Chiral Dendrimers

Dieter Seebach* · P. Beat Rheiner · Guy Greiveldinger · Thomas Butz · Holger Sellner

Laboratorium für Organische Chemie der Eidgenössischen Technischen Hochschule, ETH Zentrum, Universitätstrasse 16, CH-8092 Zürich
* *E-mail: seebach@org.chem.ethz.ch*

The synthesis of chiral dendrimers from various building blocks, their – difficult – structure determinations, and their – potential – use in physiological applications, in bioassays, and in enantioselective catalysis are reviewed.

Keywords: Chiral dendrimers, dendrimers, asymmetric synthesis, asymmetric catalysis.

1 Introduction

From the atomic to the macroscopic level chirality is a characteristic feature of biological systems and plays an important role in the interplay of structure and function. Originating from small chiral precursors complex macromolecules such as proteins or DNA have developed during evolution. On a supramolecular level chirality is expressed in molecular organization, e.g. in the secondary and tertiary structure of proteins, in membranes, cells or tissues. On a macroscopic level, it appears in the chirality of our hands or in the asymmetric arrangement of our organs, or in the helicity of snail shells. Nature usually displays a preference for one sense of chirality over the other. This leads to specific interactions called *chiral recognition*.

For organic chemists large chiral molecules offer not only a synthetic challenge but also access to a large spectrum of new molecules. The introduction of each new stereogenic center increases the possible number of different molecules by a factor of two. Therefore ten stereogenic centers already lead to 2^{10} or 1024 possible different stereoisomers. The – seemingly minor – change of configuration on a single stereogenic center may be sufficient to change the shape and the function of a whole macromolecule. The influence of chiral units on supramolecular structures and possible applications of such molecules in areas such as catalysis, (bio)sensor research, optical devices etc. are thus worth studying.

Chiral dendrimers are a class of compounds which offer the possibility to investigate the impact of chirality in macromolecular systems. Their specific properties are based on their well defined highly ordered structures with nanoscopic dimension (in this report we refer to *dendrimers* if the molecule has a core with at least three branches attached and a defined structure; otherwise we will use the term *dendritic compound*).

Chiral dendrimers should differ from achiral ones by showing the following properties: (a) their overall shape could be chiral and not spherical; (b) the arrangement of the functional groups on their surface could be chiral; (c) chiral substructures within a chiral dendrimer should be detectable by optical measurements (cf. helices in peptides). Interactions between chiral dendrimers and other chiral molecules should also be detectable by optical or kinetic measurements and interactions with small molecules should be enantioselective (chiral recognition). Functional groups in the cavity region which are able to form non-covalent bonds (cf. amino or amide groups, metal complexes, charge-transfer components etc.) may form active sites for catalytic reactions (enzyme models).

As previously reported [1] there are various possibilities of rendering a dendrimer chiral, see Fig. 1.

Here we review the development in and the manifold contributions to the area of chiral dendrimers [2] during the past ten years. The next section (Sect. 2) focuses on the synthesis and the properties of chiral dendrimers which are built of essentially unaltered natural building blocks. Their composition often resembles natural macromolecular systems and therefore opens possibilities for biological and medicinal applications. A variety of chiral dendrimers has been synthesized, containing amino acids, carbohydrates, nucleic acids and other natural chiral components. In part three we will focus on chiral dendrimers which have been synthesized mainly with the aim to investigate the influence of stereogenic centers on the entire structure of the molecule. Either by asymmetric synthesis or by modifying molecules from the pool of chiral building blocks dendritic structures of up to the fourth generation have been obtained. In part four first applications of chiral dendrimers in asymmetric catalysis are described. Several dendrimers bearing different catalytically active sites have been prepared for improving catalytic activity in homogeneous and heterogeneous catalysis.

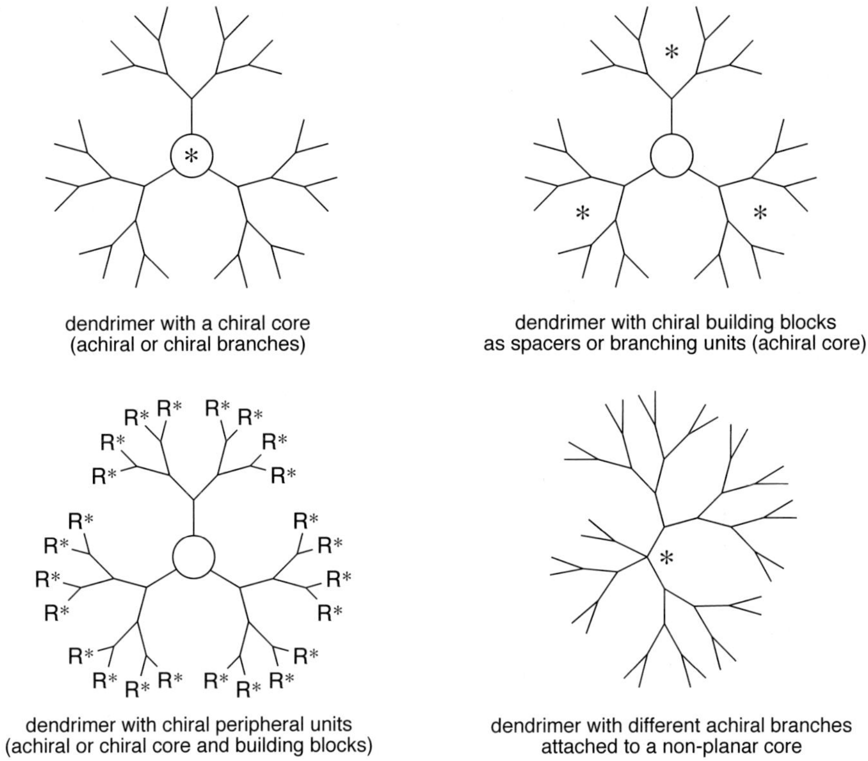

Fig. 1. Introduction of one or more than one stereogenic elements (center, axis, plane or helix) leads to different types of chiral dendrimers

2
Chiral Dendrimers and Dendritic Compounds Containing Unaltered Natural Building Blocks

2.1
Dendritic Compounds Built of or Containing Amino Acids

Amino acids and peptides have played a key role in research on chiral dendrimers. The formation of amide bonds has been an extremely well-studied reaction of the past decades and there is a rich palette of suitable coupling reagents. The highly developed protecting group methodology for amino and carboxy functionalities makes amino acids versatile building blocks for the construction of chiral systems. It is therefore not surprising that, already in 1981, Denkewalter et al. reported for the first time on the synthesis of dendritic poly(α,ε-L-lysines) [3–5]. Starting with the reaction between benzhydrylamine and the monomeric building block *N,N'*-bis(*tert*-butoxycarbonyl)-L-lysine nitrophenyl ester, repetition of a simple coupling and deprotection sequence led to lysine polypeptides containing more than 1000 terminal groups at the 10th generation. As an example, a 2nd-generation dendron (**1**) with four terminal groups, is shown in Fig. 2. Although the dendritic structure has not been exactly proven, the physical characterization by Aharoni et al. [6,7] demonstrated that this compound was monodisperse. Due to their protecting groups, the Denkewalter dendritic lysine derivatives have a hydrophobic surface. Chapman et al. used poly(ethylene glycol, PEG) as a polymer support for the synthesis of similar dendra. Starting from soluble PEG, hydraamphiphiles of up to the 8th generation (polymeric surfactants, see **2** in Fig. 2) have been obtained [8, 9]. By deprotection with trifluoroacetic acid the hydrophilic NH_2-groups can be set free to change the character of the dendrimers completely. By saponification of the glycine ester bond, the PEG chains have been readily cleaved from the dendritic part of the molecule.

Polypeptide dendrimers of 1st and 2nd generation have also been synthesized by Mitchell et al. [10]. They used L-glutamic acid, protected by benzyloxycarbonyl groups and activated as bis(succinyl)ester, as branching units. In contrast to Denkewalter, they chose a convergent growth strategy by treating the

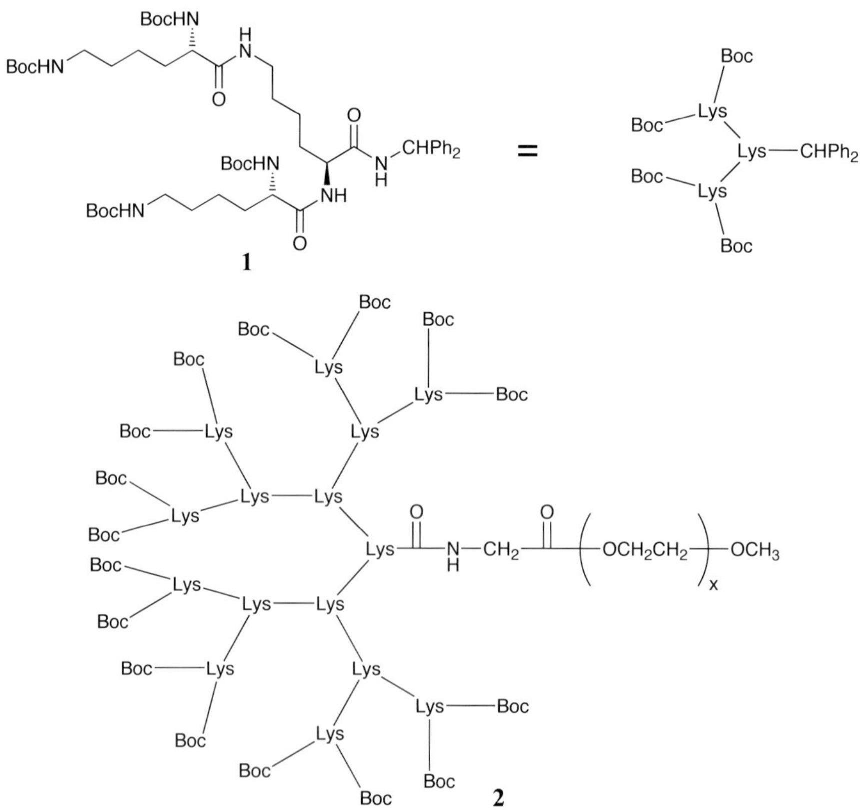

1

2

Fig. 2. Lysine-based dendron 1 of 2nd generation and a similar dendron 2 of 4th generation with PEG as soluble polymer support [8, 9]

diester with L-diethyl-glutamate and subsequently deprotecting the *N*-termini of the resulting 1st-generation dendron. The compounds obtained are monodisperse and have been fully characterized by MS, NMR and elemental analysis. Unfortunately no CD or ORD spectra have been reported.

Other types of branched peptide dendrimers, known as *m*ultiple *a*ntigen *p*eptides (MAPs), have been synthesized to mimic proteins for applications, for instance as synthetic vaccines, serodiagnostics, peptide inhibitors and intracellular delivery vehicles. Since this concept has been recently described in detail elsewhere [11], only the conceptual framework will be briefly presented here. Tam and coworkers have developed a dendritic core based on lysine units for the construction of MAPs [12–15] (Fig. 3). Carrying antigens at their periphery these MAPs have been designed to increase antigenicity and immunogenicity of peptides.

One common approach to the preparation of such dendritic peptides is stepwise solid-phase synthesis, which allows to reach the desired branching level [13, 16]. Thc selected peptide antigen is then added, again stepwise, to the resin-bound lysine core matrix to create the MAP dendron. However, this has not turned out to be a suitable procedure for obtaining dendritic macromolecular products with high purity. Modular approaches were also ineffective due to the poor solubility of the protected segments and the sluggish coupling rates. Tam et al. have therefore developed a more efficient approach for forming dendritic compounds, using unprotected peptides as building blocks and chemoselective "ligation" methods (which they refer to as orthogonal coupling). Examples of this approach include conjugation through thioalkylation, thiol-disulfide exchange, thioester and oxime formation, thiazolidine and oxazolidine ring formation, hydrazone, reverse proteolysis, and fragment coupling ("domain ligation"). Fig. 4 illustrates the synthesis of such a MAP derivative with three different ligation methods. The tetravalent lysine-based core peptide **5** carrying glyoxyl end groups has been directly coupled with a model peptide (VA20), which is derived from the surface protein of feline immunodeficiency virus and

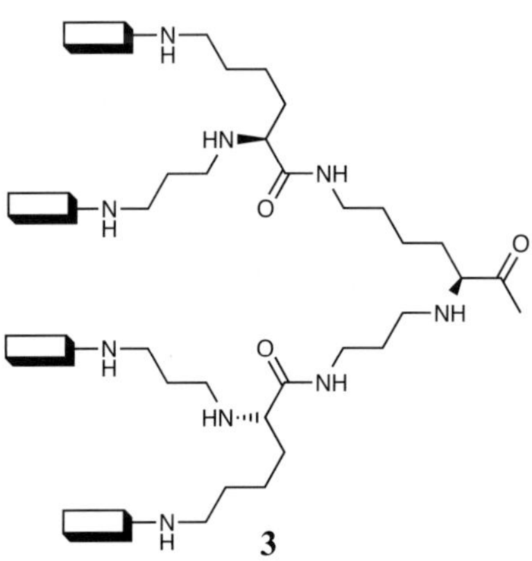

3

"Symmetrical" MAP Core

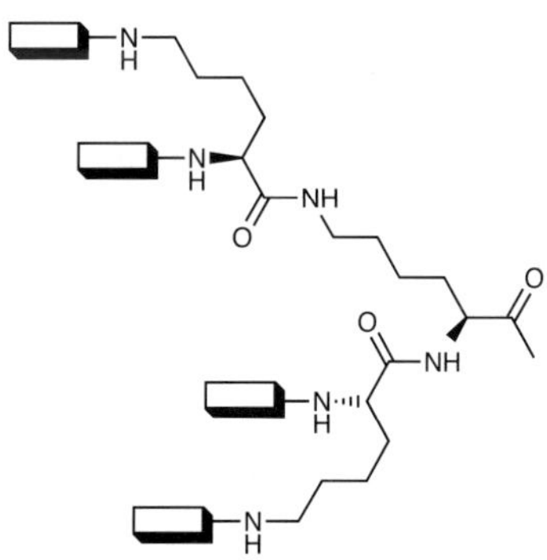

4

Unsymmetrical MAP Core

Fig. 3. Different MAPs based on lysine units. The open bars represent antigens [11]

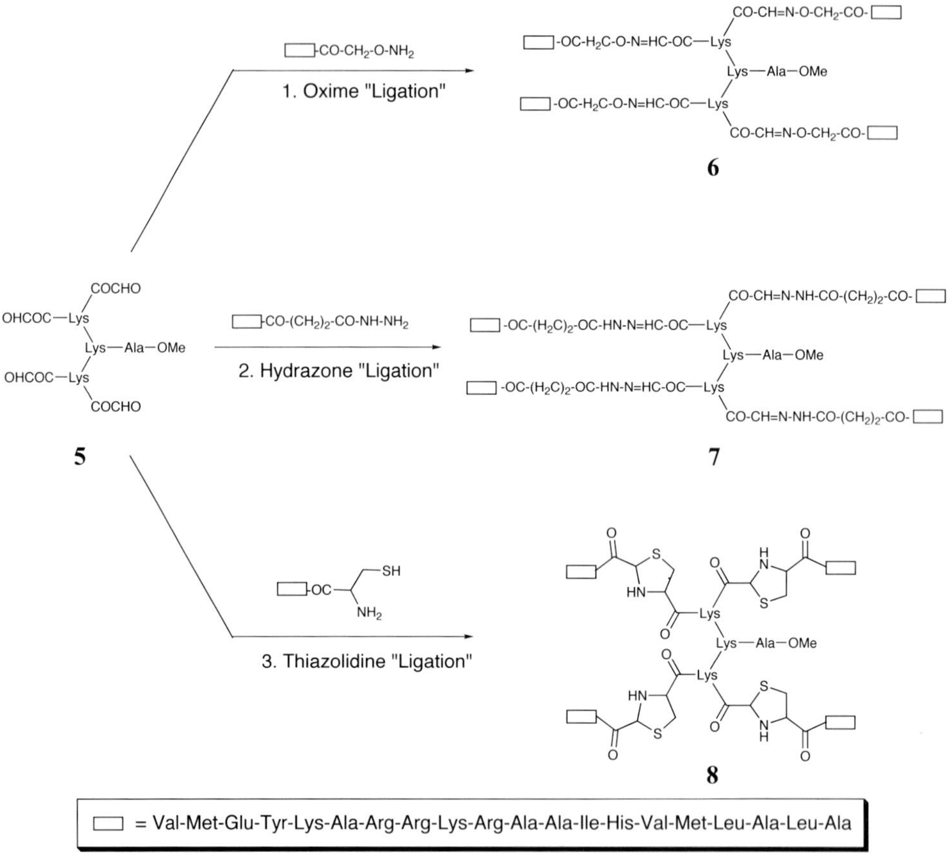

Fig. 4. Lysine-based core peptide **5** coupled with the model peptide VA 20 (represented by open boxes) by various "ligation" methods [17]

which consists of 20 amino acid residues. For the smooth coupling with the aldehyde groups of dendron **5**, VA20 has been elongated with (aminooxy)acetyl (→**6**), monohydrazide succinyl (→**7**), or cysteinyl groups (→**8**) [17]. The circular dichroism (CD) spectra indicate a stronger helical structure of the dendritic peptides compared with the peptide monomer VA20. Cooperative interactions between the branches of the helical peptides might result in a more rigid overall structure.

Recently, Tam et al. extended their orthogonal "ligation" concept, using the thermodynamically driven formation of a thiazolidine ring for the synthesis of dendritic compounds that carry *cyclic* peptides at the surface, which were designated *m*ultiple *c*yclic *a*ntigen *p*eptides (McAPs) [18, 19].

The first example of a "grafted" dendrimer carrying non-racemic amino acid moieties at the surface has been reported in 1991 by Newkome et al. [20]. A four-directional core molecule, which they prepared from pentaerythrol [21], has been elongated with the branching units *tris*[carboxyethoxymethyl]amino-

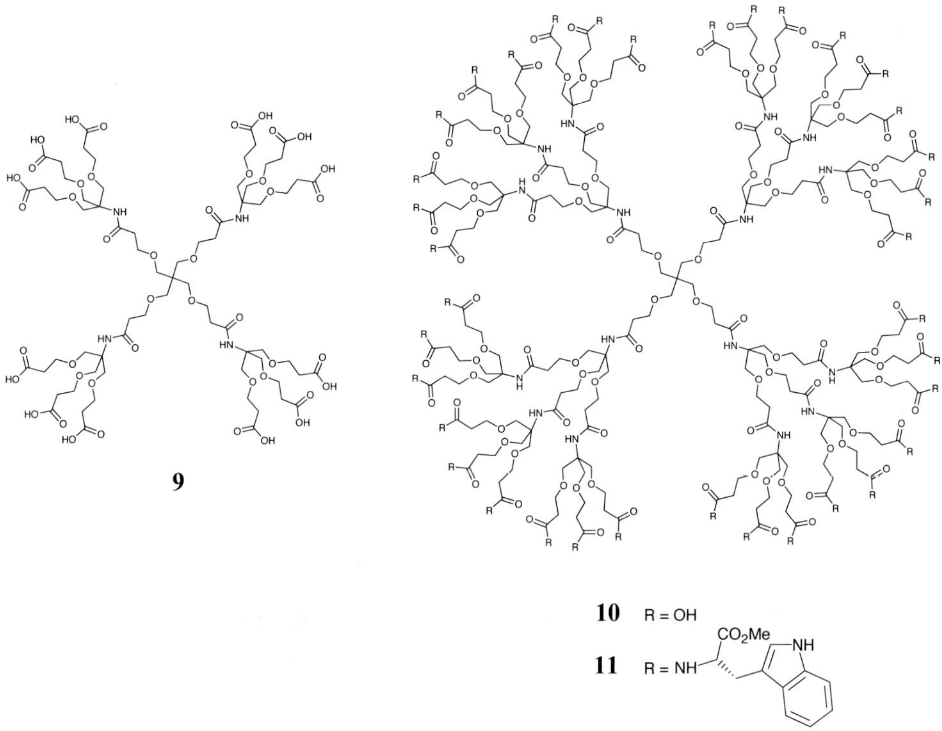

Fig. 5. Dodeca acid **9** and 2nd-generation dendrimers **10** and **11** as examples of "grafted" dendrimers [20]

methane, using standard DCC peptide coupling conditions, to give, after hydrolysis, the dodeca acid **9** (Fig. 5). Similarly, the 2nd-generation dendrimer **10** has been obtained, which could be modified by treatment with tryptophane methyl ester to give dendrimer **11**.

The molar ellipticity of these dendrimers was found to increase proportional to the number of chiral end groups. This is to be expected, in the absence of interactions between the terminal tryptophane moieties. No higher-generation dendrimers of this type have been reported. Other amino-acid-containing chiral dendrimers have been described by Meijer et al. who attached various amino acid derivatives to the periphery of poly(propylene imine) dendrimers (see Sect. 3) and more recently by Liskamp et al. (modification of polyamide dendra) [22] and Ritter et al. (synthesis of "grafted" polymerizable dendrimers containing L-aspartic acid components) [23].

2.2 Glycodendrimers

The important role saccharides play in biology, especially in recognition processes, led researchers to work on the development of multivalent so-called "neoglycoconjugates"; the numerous biological roles of cell surface oligosaccharides

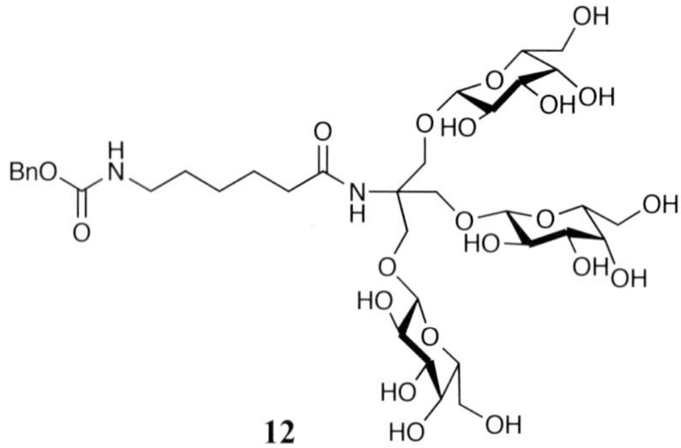

12

Fig. 6. Simple triply branched molecule with carbohydrates at the periphery (carbohydrate "cluster") [28]

have been reviewed in detail [24, 25]. Nevertheless it is worth to mention that, because of the weak nature of carbohydrate-protein interactions, only the large number and the multiplicity of these interactions led to measurable effects [26]. In addition it was found that only a few terminal sugar moieties are necessary to bind to the receptors. The linear increase of sugar density in these neoglycoconjugates led to a logarithmic growth of the binding affinity called "cluster-effect" [27].

Due to these observations, simple glycomimetics presenting the desired sugar moieties at the surface, should replace complex natural oligosaccharides which are only accessible through laborious syntheses. This has been demonstrated to be the case by Y.C. Lee et al. who synthesized the minimal "cluster" **12** (Fig. 6) [28] which already showed a "cluster-effect". During the last decade, various glycopolymers with *O*-, *S*-, and *C*-α-sialosides have been prepared, using copolymerization or "grafting"-methods. More complex structures contained e.g. sialyl *Lewis*^{x} (S*Le*^{x}) or a related 3′-sulfo-*Lewis*^{x} analog. The work in this area has been reviewed extensively [29].

Glycodendrimers [29–32] are supposed to be suitable molecules to fill the gap existing between these high molecular weight polydisperse glycoconjugates (glycopolymers) and small clusters. Due to the possibility to control the size, the molecular weight and the shape, a myriad of glycodendrimers can be envisioned and, in fact, synthesized. This novel class of glycoconjugates can be divided into two subgroups: bi- or tri-directional dendritic compounds and spherical dendrimers that can be built up either by a divergent or a convergent approach.

In 1993, the first synthesis of such neoglycoconjugates was published by Roy et al. [33, 34]. Doubly-branching polylysine dendra of up to 4 generations were prepared by solid-phase synthesis (Wang resin). The terminal amino groups were transformed into electrophilic *N*-chloroacetyl groups by treatment with *N*-chloroacetylglycylglycine hydroxybenzotriazolester to give the corresponding chlorides (cf. **13**, Fig. 7). They were functionalized with peracetylated glycosyl derivatives bearing thiol groups and, in a final step, they were cleaved from the support with trifluoroacetic acid. Thiolated glycosides were chosen in order to obtain dendritic glycosyl compounds that would be resistant to the action of glycohydrolases. After deprotection, dendritic glycosyl compounds

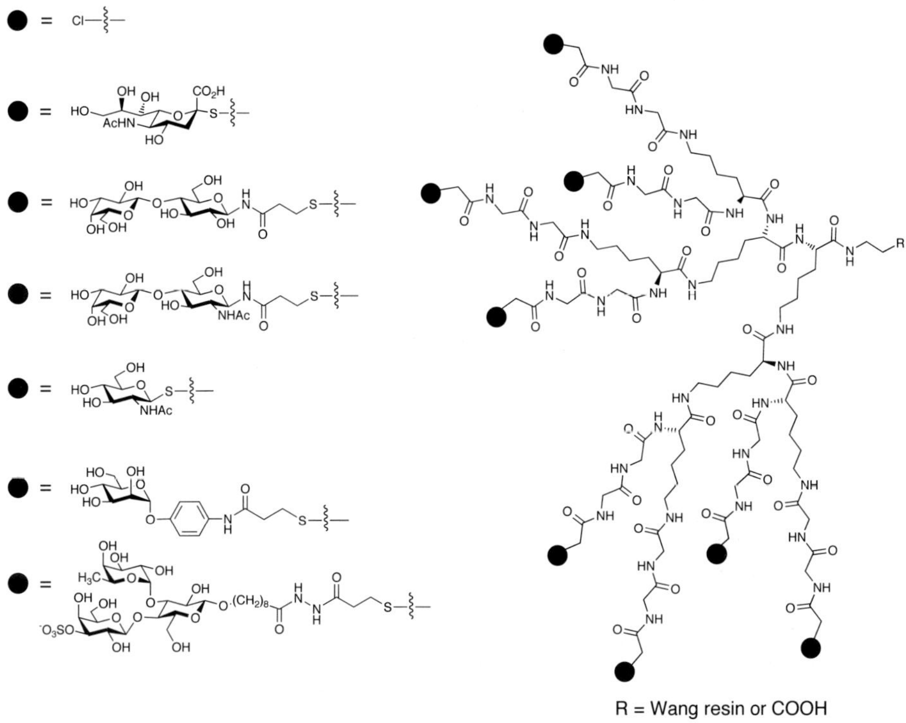

Fig. 7. Doubly-branching polylysine dendra functionalized with a chloroalkyl group and with various glycosides [33–39]

were obtained, carrying α-thiosialosides **14** [35], β-D-lactosides **15**, *N*-acetyllactosamines **16** [36], 1-thio β-D-*N*-acetylglycosamines **17** [37], α-D-mannosides **18** [38], or 3′-sulfo *Lewis*^{x}-glycosides **19** [39] (Fig. 7). These products were fully characterized, mainly by ^{1}H-NMR and mass spectrometry.

An alternative synthesis of compound **16**, using an enzymatic approach and starting from peracetylated 1-*S*-acetyl-1-thio-β-D-*N*-acetylglycosamine, has also been realized [37]. All the lysine dendra have been tested in *e*nzyme-*l*inked-*l*ectin-*a*ssays (ELLA) and showed strong biological activities. Especially compound **19**, containing the 3′-sulfo-*Lewis*^{x}-epitope, an active analogue of S*Le*^{x}, was found to be very active, showing an IC_{50}-value towards L-selectin of ca. 1 μM, 625 times higher than the corresponding monovalent compound [39]. Roy et al. also prepared triply branching dendra such as **20** that were synthesized with gallic acid as core and tetraethyleneglycol amine spacers using the above mentioned functionalization with thiolated glycosyl derivatives [40] (Fig. 8). Similar compounds built from phosphotriester backbones have also been synthesized [41]; they were tested as inhibitors of *vicia villosa* binding to asialoglycophorin. The results indicated a 3 to 10-fold enhanced affinity, thus supporting the "cluster-effect".

Okada et al. were the first to synthesize so-called "sugar-balls" [42]. They functionalized commercially available 3rd- and 4th-generation PAMAM-den-

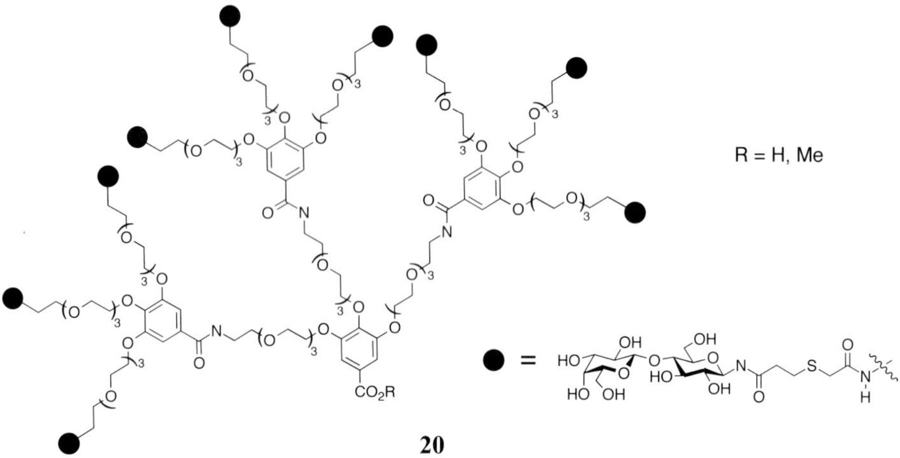

20

Fig. 8. Triply branching dendron functionalized at the surface with a glycosyl derivative [40]

drimers, bearing 24 or 48 amino groups at the surface, with disaccharide lactones of lactose (*O*-β-D-galactopyranosyl-(1,4)-D-glucono-1,5-lactone) via amide bond formation (**21**, Fig. 9). The resulting dendrimers showed strong interactions with the lectin *concanavalin A*.

Recently, the Okada group described a new class of polymerization systems [43]: oligoglycopeptide-type sugar-balls were obtained by a "radial growth polymerization" (RGP) of α-amino acid *N*-carboxyanhydrides with PAMAM dendrimers of different generations.

Using peracetylated glycosyl isothiocyanates of β-D-glucose, α-D-mannose, β-D-galactose, β-cellobiose, and β-lactose, Lindhorst and Kieburg synthesized tetra-, hexa- and octa-valent PAMAM dendrimer derivatives. Using these glycosyl isothiocyanates, dendrimers, bearing a wide variety of sugars at the surface, were readily obtained [44]. Very recently, the same group published the synthesis of small dendrimers, applying the same coupling strategy that does not rely on the use of protecting groups [45].

In a more recent paper, Roy et al. reported on the synthesis and biological properties of mannosylated Starburst™ dendrimers [46]. In addition to the presence of good biological properties in ligand- and inhibitor-tests, these dendrimers were shown to constitute novel biochromatography materials of high affinity for the rapid and easy isolation and purification of carbohydrate-binding proteins from crude mixtures.

Stoddart et al., in a collaboration with Meijer, synthesized "sugar-balls" by modification of poly(propylene imine) dendrimers [47]. An attachment of carboxyl derivatives of D-galactose and D-lactose to the amino surface groups has been achieved by means of amide bond formation, using the *N*-hydroxysuccinimide coupling procedure. The acetate protecting groups, that are still necessary to avoid undesired reactions in the coupling step, have been deprotected under standard Zemplen deacylation conditions, followed by treatment with an aqueous NaOH solution. The interpretation of the ^{13}C-NMR spectra allowed the

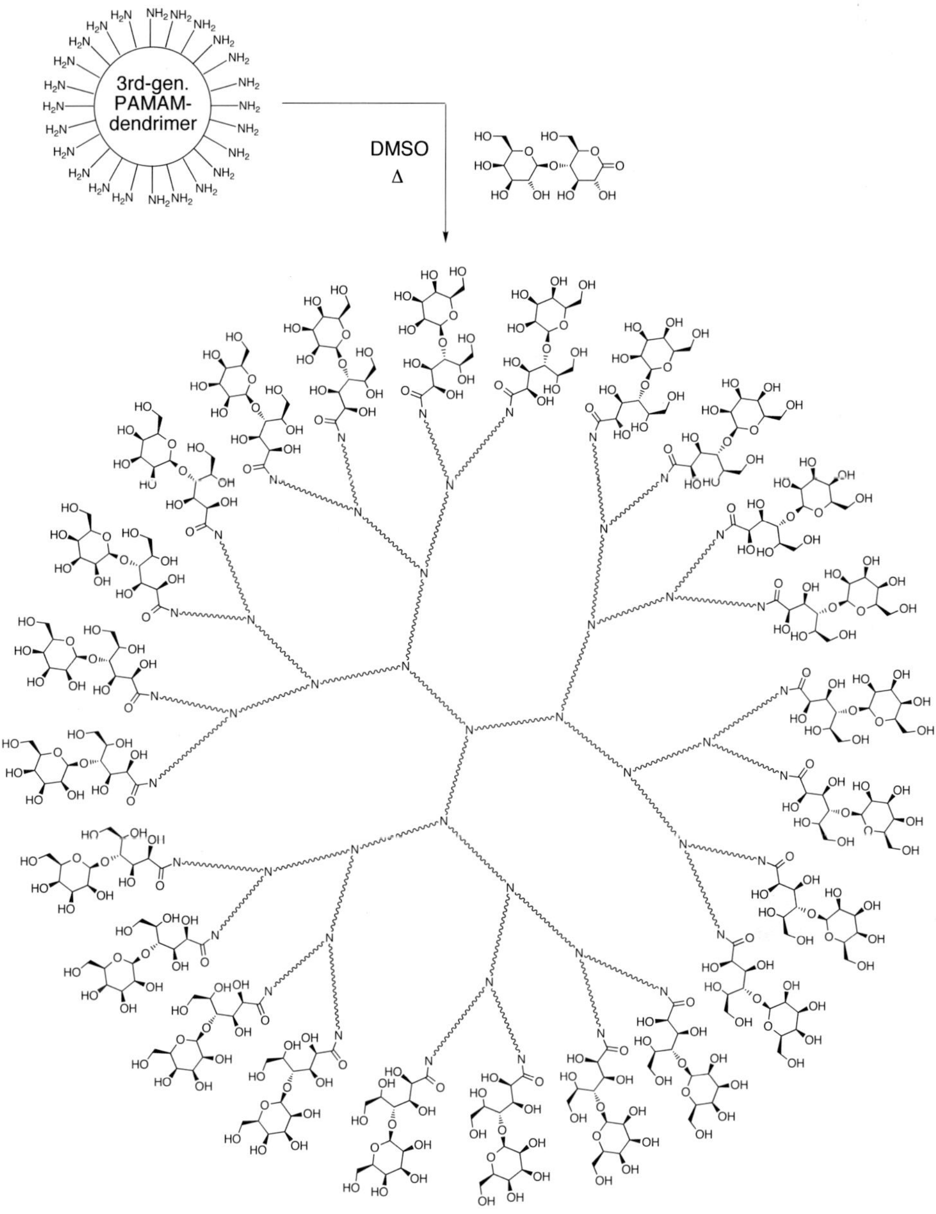

21

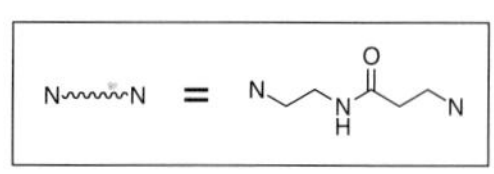

Fig. 9. 3rd-Generation PAMAM-dendrimer functionalized with disaccharide end groups [42]

authors to conclude that the resulting glycodendrimers do not contain any serious defects as a result of the chemical manipulations and that the degree of functionalization by the small saccharides seems to be very high. The statistical defects present in these structures are a result of the divergent nature of the synthesis of poly(propylene imine) dendrimers. Until now no details about biological properties have been reported.

A convergent approach leading to monodisperse glycodendrimers was used by Stoddart et al. [48]. Their strategy involved the synthesis of a triglycosylated derivative of tris(hydroxymethyl)methylamine (TRIS), the introduction of a glycine-derived spacer and 3,3′-iminodipropionic-acid-derived branching units on to the TRIS derivative by amide bond formation, the subsequent coupling of these dendrons with a trifunctional 1,3,5-benzenetricarboxylic acid derivative, used as a core, and finally the deprotection of the saccharide units. An example of an 18-mer (**22**), carrying 18 saccharide units at the periphery is shown in Fig. 10. In this case, the isolated compounds were shown to be monodisperse, an advantage of the convergent approach.

More recently, the above described route was extended by Stoddart et al. to even larger structures [49]. However, due to steric hindrance during the cou-

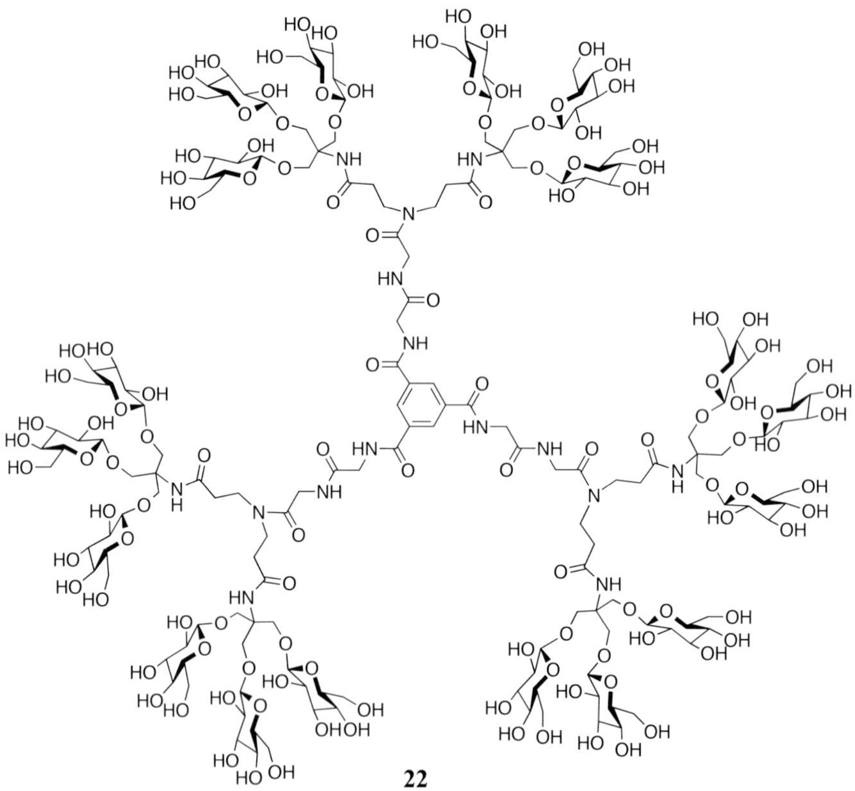

22

Fig. 10. Glycodendrimer with triglycosylated TRIS end groups [48] pling of the large dendrons to the core (a limitation of the convergent approach), the strategy had to be sligthly modified. The biggest molecules contained up to 36 saccharide units with a calculated molecular weight of 14965 Da. The reduced activities of the focal point in dendritic wedges may probably be circumvented by using non-protected carbohydrate moieties [50]. The first biological evaluations of representatives of this class of glycodendrimers furnished promising results [51]. The same group also started a project to synthesize "fully-carbohydrate"-derived dendrimers [52].

In summary, glycodendrimers with a wide variety of shapes, core molecules and carbohydrate residues are now available. A combination of both, the convergent and the divergent approach, seems to be the best strategy to build up these dendritic structures. As far as there exist no X-ray structures of biologically important glycoproteins and lectins, the biological testing of these new classes of neoglycoconjugates can help to obtain structural information about carbohydrate ligands and *c*arbohydrate *r*ecognition *d*omains (CRDs) of lectins. The glycodendrimers are envisaged as potentially useful therapeutic agents in the prevention of bacterial and viral infections and could also find application in the development of cancer drugs. In order to reach these important goals, the toxicological and immunochemical properties of the compounds have yet to be evaluated.

2.3 Chiral Dendrimers Based on Nucleic Acids

Less attention than to dendrimers containing amino acids or carbohydrates has been paid to dendrimers containing nucleic acids. Hudson and Damha reported on the synthesis of a 3rd-generation dendron containing 87 nucleotides (**23**, Fig. 11) [53]. On the surface of controlled-pore glass they synthesized, by a convergent procedure, oligonucleotides from thymidine. By coupling of two adjacent polymer-bound nucleotide chains with a tetrazole-activated adenosine 2′,3′-bis(phosphoramidite) derivative the 1st generation was obtained. Further elongation and coupling steps led, after cleavage from the polymer support, to the 87-mer **23**. Besides this synthetic work, commercial applications (for pharmaceuticals [54] or for signal amplification) and labeling in DNA blots [55] have so far been described.

The latter is an interesting example of self-organizing chiral dendrimers. The construction of the dendrimer is based on the natural property of nucleic acids to recognize and specifically bind to complementary sequences. Pairwise hybridization of two designed DNA strands results in the formation of large "monomers" which have four single stranded "arms" and a double stranded "waist" (**24**, Fig. 12).

The surface of each layer has two types of single stranded arms (e.g. one $3' \rightarrow 5'$and one $5' \rightarrow 3'$ strand) which can bind to other monomers to render 1st- and 2nd-generation "dendrimers" **25** and **26**. Therefore the molecular scaffold grows exponentially with each sequential layer of hybridization. If an oligonucleotide contains a sequence complementary to those at the surface of these networks it should be hybridized. The remaining free sequences from the other type of arms then bind in a standard nucleic acid blot (after they are bound to

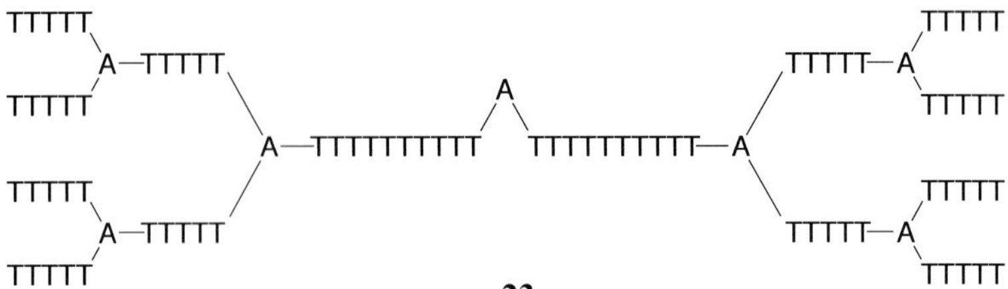

23

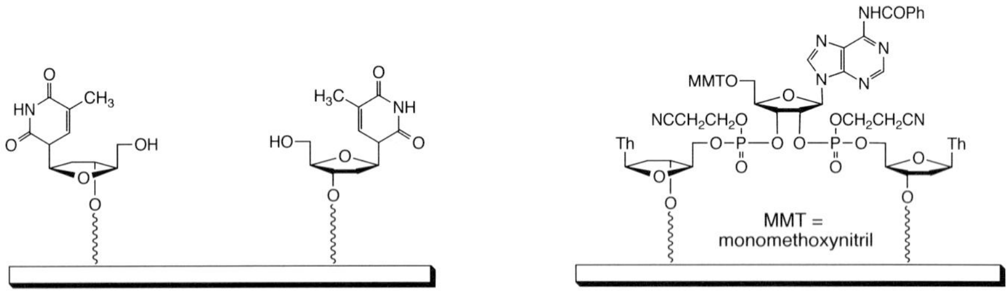

Fig. 11. Dendron **23** of 3rd generation containing 87 nucleotides. Two key intermediates in the synthesis of dendra such as **23**. The box represents long-chain alkylamine controlled-pore glass [53]

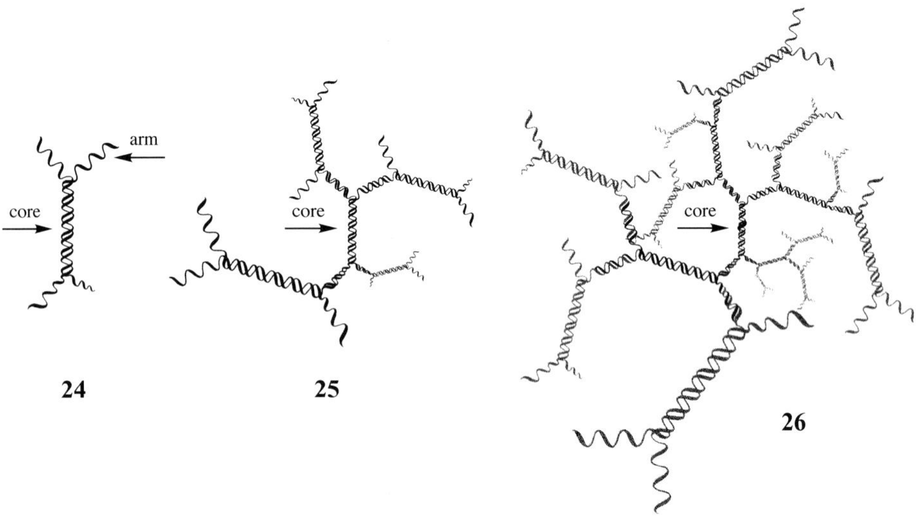

Fig. 12. DNA "core" and DNA dendrimers of "generations" 1 to 3 [55]

the target) to hundreds of so-called "label molecules". The signals of blots that were probed using oligonucleotides with dendrimers showed amplification of over 100-fold when compared to identical blots probed with the specific oligonucleotide alone [55].

2.4
Chiral Dendrimers Containing Oligo-(3-Hydroxybutanoate) Units

In our group, several dendrimers based on (*R*)-3-hydroxybutanoic acid (HB) have been prepared [56–58]. The dendrimers were synthesized by the convergent strategy. Trimesic acid has been used as core unit and the benzyl esters of the dimer and the tetramer of HB as elongation units. In such a way dendrimers of 1st and 2nd generation (**27**–**30**) have been constructed (Fig. 13). Since poly(*R*)-3-hydroxybutanoic acid (PHB) is known to be biodegradable [59, 60] the stability of the dendrimers **27**–**30** was tested in the presence of PHB-de-

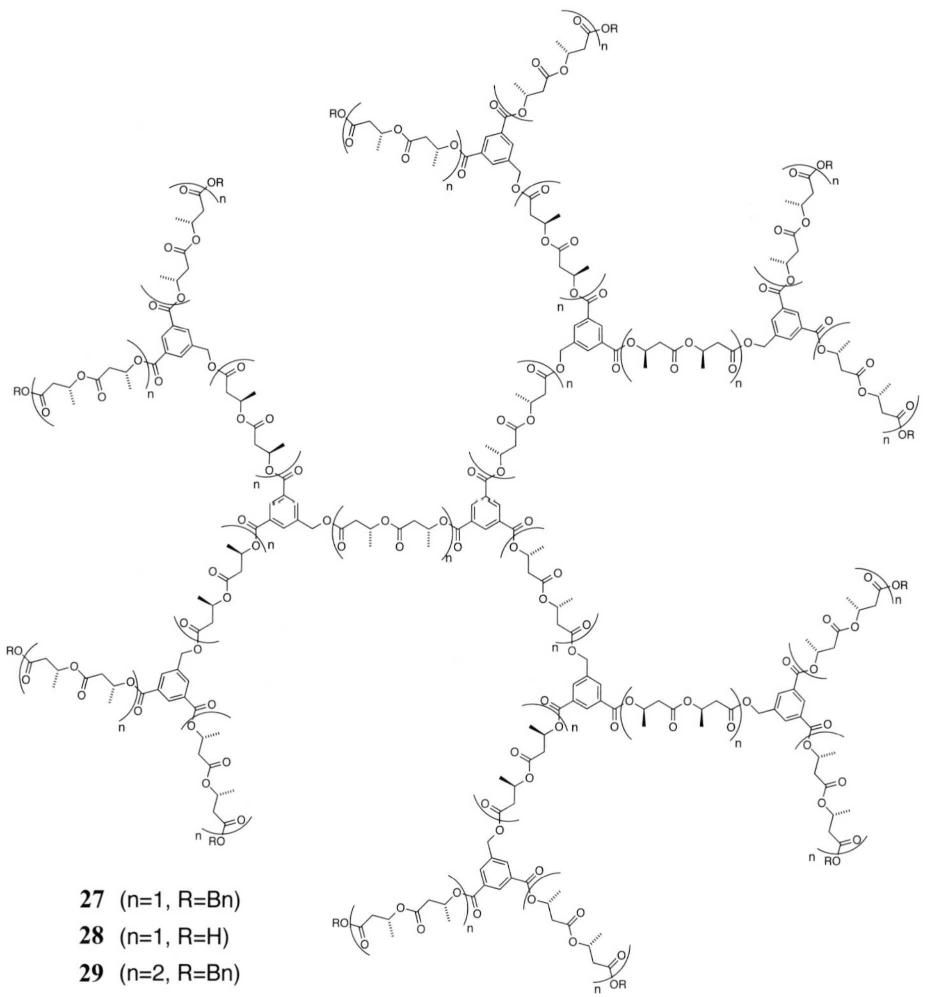

27 (n=1, R=Bn)
28 (n=1, R=H)
29 (n=2, R=Bn)
30 (n=2, R=H)

Fig. 13. Chiral dendrimers of 2nd generation from trimesic and (*R*)-3-hydroxy-butanoic acid [56, 57] polymerase. It was shown that the benzyl-protected dendrimer with dimeric HB-elongation units was not degraded by the depolymerase whereas the free acid was a surprisingly good substrate for the enzyme, even though the simple dimeric HB is not [58]. All deprotected dendrimers with tetrameric HB-elongation units were degraded very well. The rate of the first degradation step was about one hundred times faster than the degradation of dendrimers with dimeric HB-elongation units. An esterase, a lipase and a protease were shown to be able to degrade the dendritic compounds as well [56].

3
Dendrimers Containing Synthetic Chiral Building Blocks

Because of their high molecular weight and their defined structure, dendrimers offer themselves for studying the "expression of chirality" on a macromolecular level. The construction of configurationally uniform macromolecules is otherwise a complex task but can be achieved more easily with dendrimers because of repetitive synthesis from identical (chiral) building blocks. Comparison of optical rotation values and circular dichroism (CD) spectra should demonstrate what influence there is of the chiral building blocks on the structure of the whole dendrimer.

The stereogenic centers of chiral dendrimers synthesized so far are either generated by asymmetric synthesis, or they are derived from molecules of the pool of chiral building blocks. The only investigation on chiral dendrimers, consisting of achiral building blocks exclusively, was published by Meijer et al., who synthesized dendrimers such as **31** [61] (Fig. 14). This compound ows its chiral-

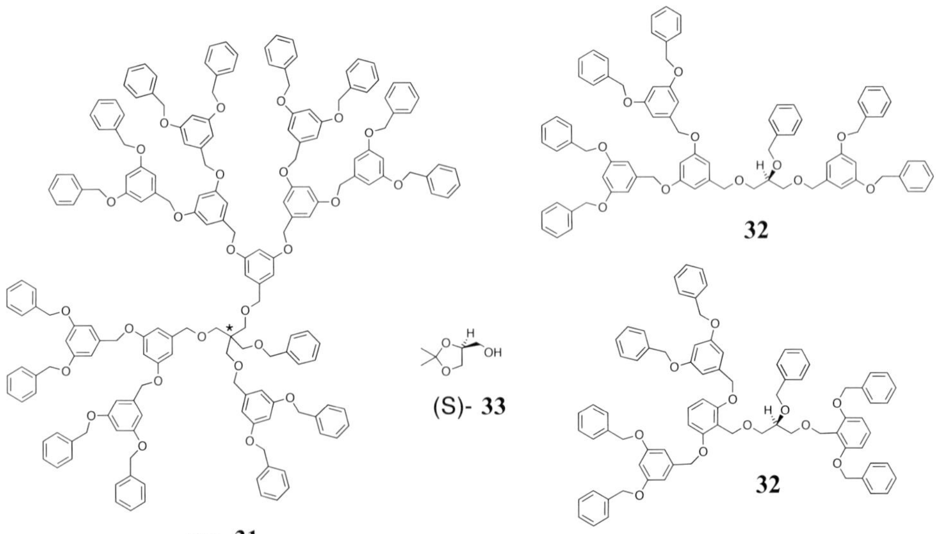

Fig. 14. Chiral dendrimers **31** (prepared as racemic mixture) with a core chirality center. Compounds **32** and **34** are derived from (*S*)-solketal **33** as enantiopure precursor [61, 64, 66]

ity to four achiral Fréchet-type [62] branches of different generations, which are attached to a pentaerythrol core [63]. Unfortunately the enantiomers of the racemic compound **31** could not be separated. By another route the chiral compound **32** was synthesized, starting from the chiral glycerol derivative (*S*)-solketal **33** [64]. Even though all intermediate products of the steps leading to **32** were optically active, the final product **32** was not (**32** was thus designated as being "cryptochiral" [65], a terrible term, considering that chirality is the geometrical property of an entity which is non-superimposable with its mirror image, and that an optically active compound must consist of chiral molecules, but that a compound consisting of chiral molecules needs not be optically active !).

The conformational flexibility and the lack of difference of the electronic properties of the polyether branches in **32** have been forwarded to explain this zero rotation. Therefore a similar dendrimer **34** has been prepared which carries a more sterically demanding branch, leading to a more rigid structure [66]: interestingly, this dendrimer indeed exhibited a very small but measurable optical activity, which underlines the thesis that "nanoscopic chirality" depends on the rigidity of the investigated structure.

In their studies on so-called dendritic boxes [67–69] Meijer et al. also modified achiral poly(propylene imine) dendrimers with several protected amino acids. The resulting dendrimers of generations 1 to 5 showed no constant optical activity per chiral end group. Rather, a decrease with increasing generation number was observed (Fig. 15). This effect has been most pronounced with the aromatic amino acids (*S*)-tyrosine and (*S*)-phenylalanine, and it is in contrast to results described by Newkome et al. [20] who found a comparable contribution of chiral appendages to the optical rotation in a series of dendrimers of different generations. There is no obvious explanation for the unexpected chiroptical behavior of the dendritic boxes garnished with amino-acid-derived groups, but racemization or "dilution effects" [70] can be excluded. In model studies, the contribution to the optical activity of the *N*-BOC-(*S*)-phenylalanine end groups has been shown to be very sensitive to the local environment and, therefore, also to the solvent. From ^{13}C-NMR relaxation-time measurements [67] a "solid-phase" behavior of the peripheral groups on going to higher generations is indicated. Thus, the local environment of the *N*-BOC-(*S*)-phenylalanine end groups changes when they are getting closer to each other. According to the authors' interpretation, the end groups in a dense packing seem to adopt "frozen-in" conformations with contributions to the optical rotation that internally cancel each other to give a resulting average of almost zero. Evidence for these conformational changes has also been derived from UV/VIS spectroscopic data [70].

When *t*-butoxy methoxy benzyl acetate groups are attached to the same dendrimers, which are similar in shape to but less dependent on the solvent than the phenylalanine moieties, a roughly constant optical rotation per end group is obtained for all generations [2]. In this case the contributions to optical activity of the end groups seem to be additive and insensitive to differences in packing.

In another experiment, alkyl chains have been introduced as spacers between the surface NH_2 groups of the dendrimer and the *N*-BOC-(*S*)-phenylalanine groups. In this case, too, the optical activity per end group remained constant for both, the dendrimer of the 1st (with four end groups) and of the 5th generation

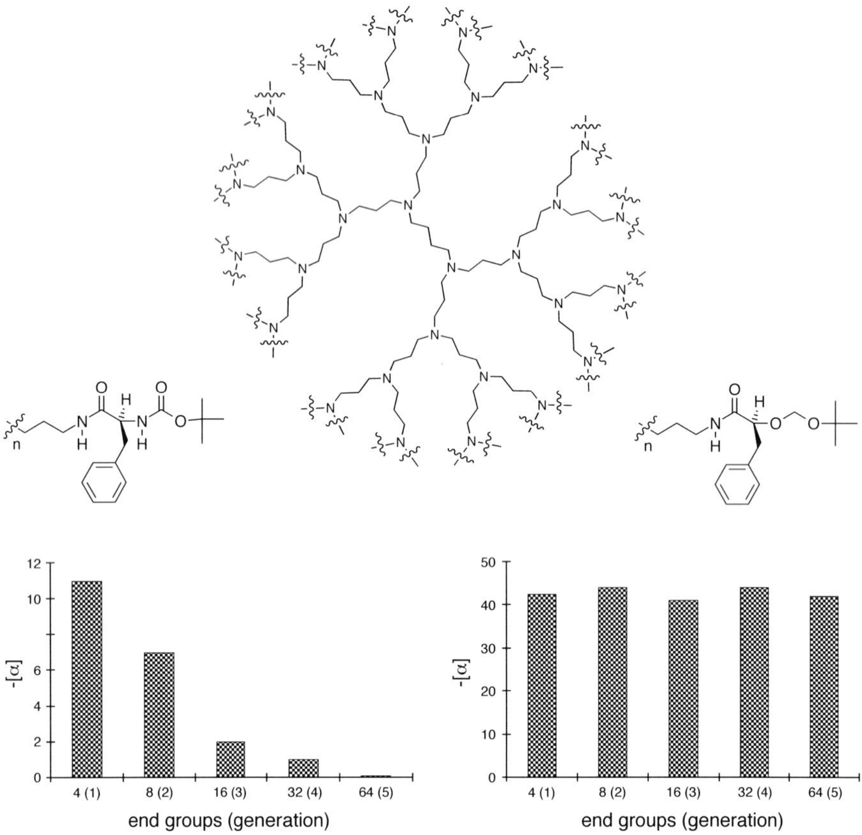

Fig. 15. Optical activities of poly(propylene imine) dendrimers, functionalized at the periphery with protected phenylalanine or *t*-butoxy methoxy benzyl acetate groups, depend on the number of end groups [2]

(with 64 end groups). Obviously, the end groups can now freely adopt their preferred conformation. After all, an *N*-BOC-(*S*)-phenylalanine group may be considered a suitable terminal group to determine the (local) density of a dendrimer surface.

Vögtle et al. have prepared chiral poly(imine) dendrimers of various generations by condensation of non-racemic 5-formyl-4-hydroxy[2.2]paracyclophane moieties with poly(amine) dendrimers [71]. They have found that the optical activity of these dendrimers was nearly constant with increasing generation number.

The first dendrimers with stereogenic centers generated by asymmetric reactions rather than being derived from natural chiral building blocks were prepared by Sharpless et al. [72]. With the goal "to find reliable strategies for the efficient construction of dendrimers and to introduce chiral cavities into these compounds", they synthesized chiral 1,2-diols as branching units. Starting from *para*- or *meta*-(chloromethyl)-phenyl substituted acetonides (from enantio-

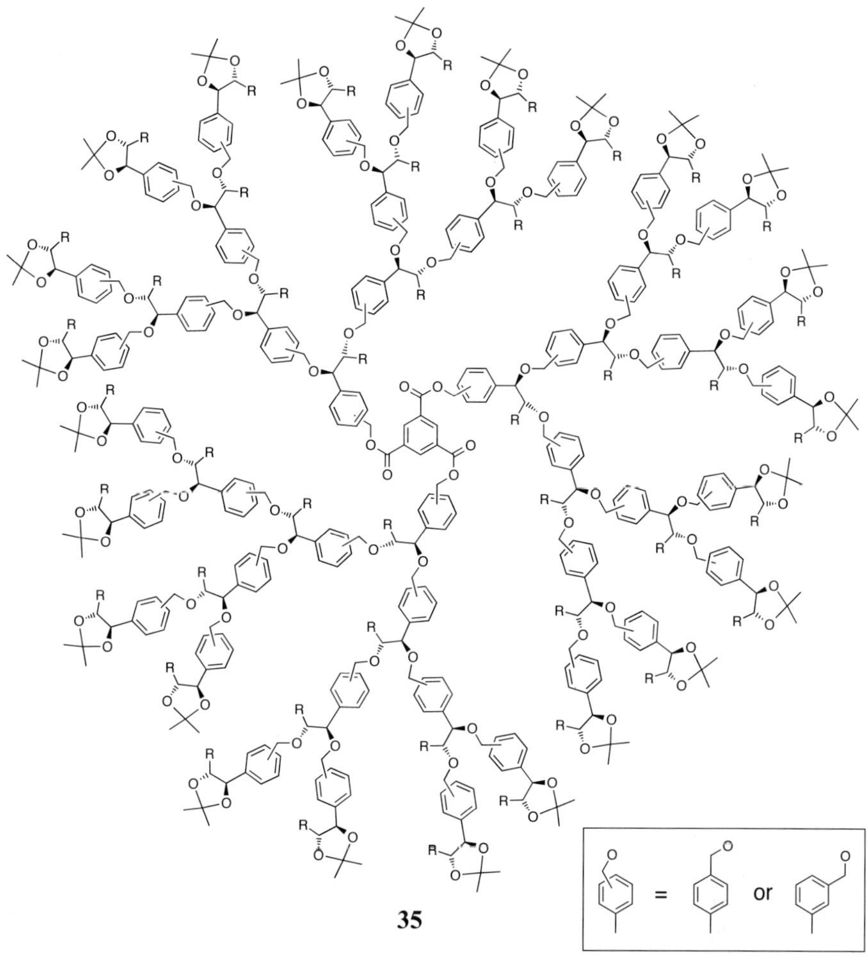

35

Fig. 16. Chiral 4th-generation dendrimers from *para-* or *meta-*(chloromethyl)-phenyl substituted acetonides and a 1,3,5-benzenetricarboxylic acid center piece [72]

selective dihydroxylations) they were able to synthesize chiral polyether dendrimers up to the 4th generation (**35**, Fig. 16), following a convergent "double exponential dendrimer growth" approach [73], which implies direct coupling of four 2nd-generation dendra with a 2nd-generation dendron, skipping the third generation.

A similar approach, employing the same asymmetric dihydroxylation reaction to synthesize chiral doubly branching monomers **36**–**39** has been described by McGrath et al. (Fig. 17) [74]. In their reactivity, monomers **36** and **37** are similar to Fréchet-type [62] branching units since each possesses two phenolic and one benzylic hydroxy group. With these building blocks dendrimers of up to the 3rd generation (**40**) have been synthesized, using, again, the achiral 1,3,5-benzenetricarbonyl moiety as core [75]. Now, OH groups on the stereogenic centers

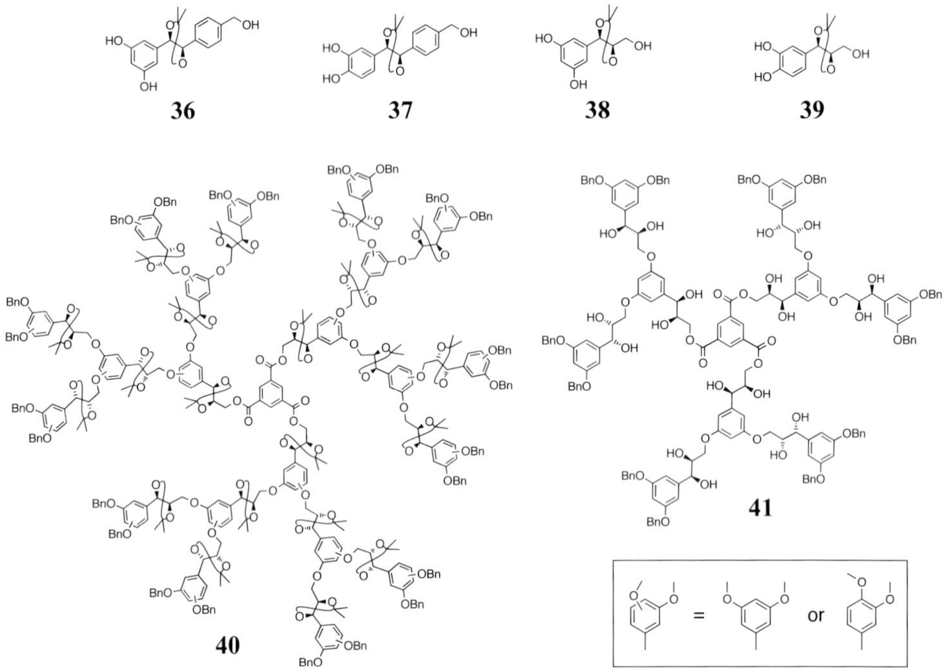

36 37 38 39 40 41

Fig. 17. Chiral acetonide building blocks **36**–**39**, protected 3rd-generation dendrimer **40** and deprotected 2nd-generation dendrimer **41**, derived from enantioselectively dihydroxylated styrenes and cinnamic alcohols [74–76]

in the inner part of the dendrimers are not the branching points used to combine the monomer units by etherification; rather, they are introduced in the acetonide-protected form, and thus, after deprotection, they are ready for reactions and interactions with each other or with other compounds: the free hydroxy groups can give rise to either *intramolecular* interactions between two branches or to *intermolecular* interactions between the dendrimer and other molecules (host-guest chemistry). As an example, the 2nd-generation chiral dendrimer **41** is shown in Fig. 17 [76]. Of course, the solubility of dendrimer **41** with 18 hydroxy groups in non-polar organic solvents is very low. As a possible application, these dendrimers, which should have the ability to anchor metal complexes, could be used for asymmetric synthesis. On the other hand, their hydroxyester functionality is expected to render them sensitive to H^+ or Lewis acid catalyzed transesterifications (the OH to CO distances of 5 and 6 atoms are ideal for acyl shifts !).

McGrath et al. have also thoroughly studied the chiroptical properties of dendrimers such as **40**. They compared the optical activities of the series of 1st-, 2nd- and 3rd-generation compounds of type **40**, considering the molar rotation per chiral unit ($[\Phi]_D/n$) [75]. A big difference of the values was found between the generations which could possibly indicate chiral conformations inside the dendrimers, that enhance the optical rotation values per unit when

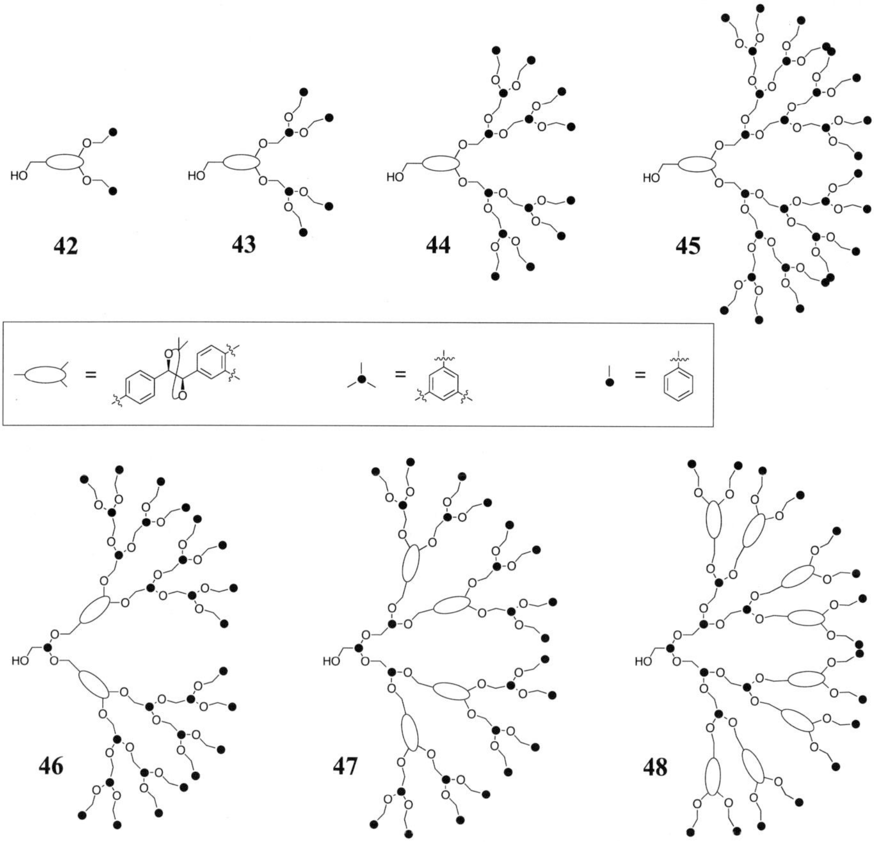

Fig. 18. Series of dendra **42**–**45** of 1st to 4th generation containing a chiral branching unit in the first shell and 4th-generation dendra 45–48 containing a chiral branching unit in different shells [77]

going to larger molecules. But as it was found in the case of the "fully chiral" dendrimers prepared by Seebach et al. (see later in this chapter) they also discovered that slight changes in constitution are causing these effects. By adding the molar rotation values of three model compounds (representing the core, the interior and the peripheral units) the calculated molar rotations agreed with the observed values (with deviations of less than 15%). To find out how the location of individual chiral units affects the chiroptical properties of single branches, they prepared two series of up to 4th-generation dendra (Fig. 18) [77]. In one series they attached Fréchet-type [62] benzyl aryl ether branches of 0th to 3rd generation to a chiral acetonide-protected hydrobenzoin unit (→**42**–**45**). No increase of the molar optical activity of these dendra was observed with increasing size of the achiral branches.

In a next series of experiments McGrath et al. synthesized and compared some 4th-generation dendra **45**–**48**, where the chiral unit(s), when placed in the

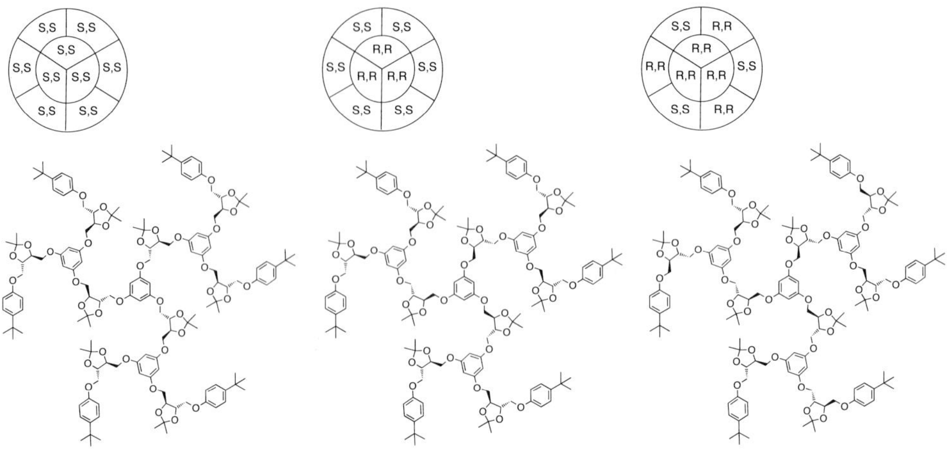

49 **50** **51**

Fig. 19. A series of 2nd-generation dendrimers with chiral spacers derived from (*R,R*)- or (*S,S*)-tartaric acid. Note that the (*R,R*)-threitol-acetonide building blocks in **49**–**51** are derived from (*S,S*)-tartaric acid, and vice versa [78–80]

interior shells (**46** and **47**), have a larger influence on the molar optical activity per chiral unit than when placed at the periphery or in the center. However the authors are not yet fully convinced that this is not also a constitutional effect and further studies in this area are under way.

Other studies to investigate the relationship between the "stereo-spatial" [78] properties of chiral building blocks and the overall chiroptical properties of the entire dendrimer, they are part of, were performed by Chow et al. [78–80]. They synthesized different chiral 2nd-generation dendrimers by the convergent approach, using enantiopure threitol derivatives as spacers between achiral 1,3,5-benzenetriol (phloroglucine) derived branching units (Fig. 19). By combining spacers derived from (*R,R*)- or (*S,S*)-tartaric acid they were able to construct what they call "homo dendrimer" **49** [80] which contains only spacers of (*S,S*)-configuration and "layer-block dendrimers" **50** and **51** with combinations of spacers of opposite chirality sense. These dendrimers could, similarly to those of McGrath et al. [76], be hydrolyzed to (chemically stable) polyhydroxylated compounds.

The molar rotation of the dendrimers **49**–**51** is proportional to the excess of (*R,R*)- or (*S,S*)-threitol units. This means that the chiroptical effects of threitol building blocks of opposite chirality cancel out each other. For the "homo dendrimers" an average positive molar optical rotation value of 146 for each (*R,R*)-threitol unit was calculated whereas a value of –185 resulted for each (*S,S*)-threitol building block. From CD spectra of dendrimers of the types **49**–**51** it could be derived that the chiroptical effect of an (*S,S*)-chiral unit in the outer shell of the dendrimer did not compensate that of a (*R,R*)-chiral unit in the inner shell [81], a result which led the authors to state that the different dendritic layers are "chiroptically slightly different" [80].

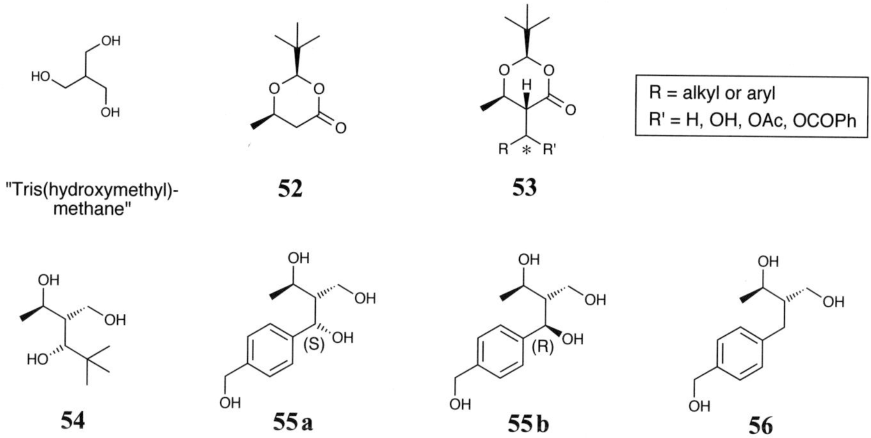

Fig. 20. Enantiopure chiral building blocks **54**–**56** and intermediates **52** and **53** obtained from 3-(*R*)-hydroxy-butanoate [1, 83–88]

At the beginning of investigations on chiral dendrimers in our own group was the question of how to synthesize chiral, non-racemic derivatives of "tris(hydroxymethyl)-methane" [82], which we wanted to use as dendrimer center pieces. We have developed efficient diastereoselective syntheses of such triols [83–85] from (*R*)-3-hydroxybutanoic acid, readily available from the biopolymer PHB [59, 60] (cf. Sect. 2.4). To this end, the acid is converted to the dioxanone **52** [86, 87], from which various alkylation products and different aldol adducts of type **53** were obtained selectively, via the enolate (Fig. 20). These compounds have been reduced to give a variety of enantiopure chiral building blocks for dendrimers, such as the core unit **54**, triply branching units **55a** and **55b** or doubly branching unit **56** [1, 88].

In 1994 we published the first chiral dendrimers built from chiral cores and achiral branches [1, 89], see for instance dendrimer **57** with a core from hydroxy-butanoic acid and diphenyl-acetaldehyde and with twelve nitro-groups at the periphery (Fig. 21). As had already been observed with starburst dendrimers, compound **57** formed stable clathrates with many polar solvent molecules, and it could actually only be isolated and characterized as a complex [2 · (**57**· EtOAc · (8 H_2O))]. Because no enantioselective guest-host complex formation could be found, and since compounds of type **57** were poorly soluble, and could thus not be easily handled, we have moved on and developed other systems to investigate how the chirality of the core might be influencing the structure of achiral dendritic elongation units.

Following the convergent procedure, dendrimers of type **58**, **59** and **60** have been prepared from the chiral core triol **54** and achiral Fréchet-type [62] benzylic branch bromides. In the series of dendrimers with aromatic spacers (**60**) and without spacers (**58**), the optical activity $[\alpha]_D$ decreased on going from the 1st (not shown in Fig. 21) to the 2nd generation, whereas with aliphatic spacers

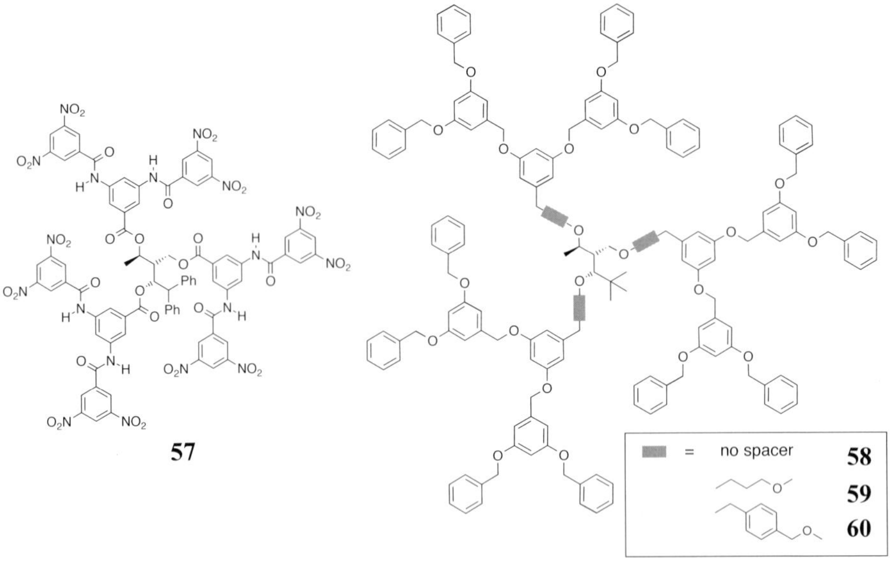

Fig. 21. 1st-Generation dendrimer **57** with nitro groups at the periphery and 2nd-generation dendrimers **58**–**60** used for optical and ^{1}H-NMR measurements [1, 89]

(**59**) hardly any optical activity could be detected at all. Obviously, no chiral substructures (which would contribute to the optical activity) are present in the achiral branches, and there is merely a kind of dilution effect upon the optical activity: for the dendrimers with aromatic spacers the *molar* optical rotation values $[\Phi]_d$ are constant. An anomaly was observed only with the two dendrimers containing no spacer: the *molar* rotation doubled when going from generation 1 to generation 2 (i.e. **58**). Also, the signals of the diastereotopic benzylic H-atoms observed in the ^{1}H-NMR spectra can be taken as a measure for the degree of dissimilarity of their environment: the heterotopic benzylic H-atoms become isochronous when going from the inner to the outer layer of the 2nd-generation dendrimers of type **59** and **60**. Again, the dendrimer **58** without spacers is an exception, in that there were no singlets for any of the sets of analogous benzylic H-atoms.

In order to continue our studies about the influence of chiral building blocks on the overall shape of dendrimers, the synthesis of "fully chiral" dendrimers (with stereogenic centers in the core and in all branching units) became the next goal. To obtain a dense shell even with low-generation dendrimers, triply branching units like **55** (see Fig. 20) with stereogenic centers at every branching point [1, 88, 90] were chosen. The synthetic limit turned out to be reached with the 2nd generation. During the synthesis a remarkable case of diastereoisomer differentiation (i.e. chiral recognition) was observed (Fig. 22).

When branch bromide **61**, consisting of four type-**55a** building blocks of (*S*)-configuration at the benzylic stereogenic centers, was coupled with the chiral triol **54**, dendrimer **63** was obtained in 51% yield after purification. With the

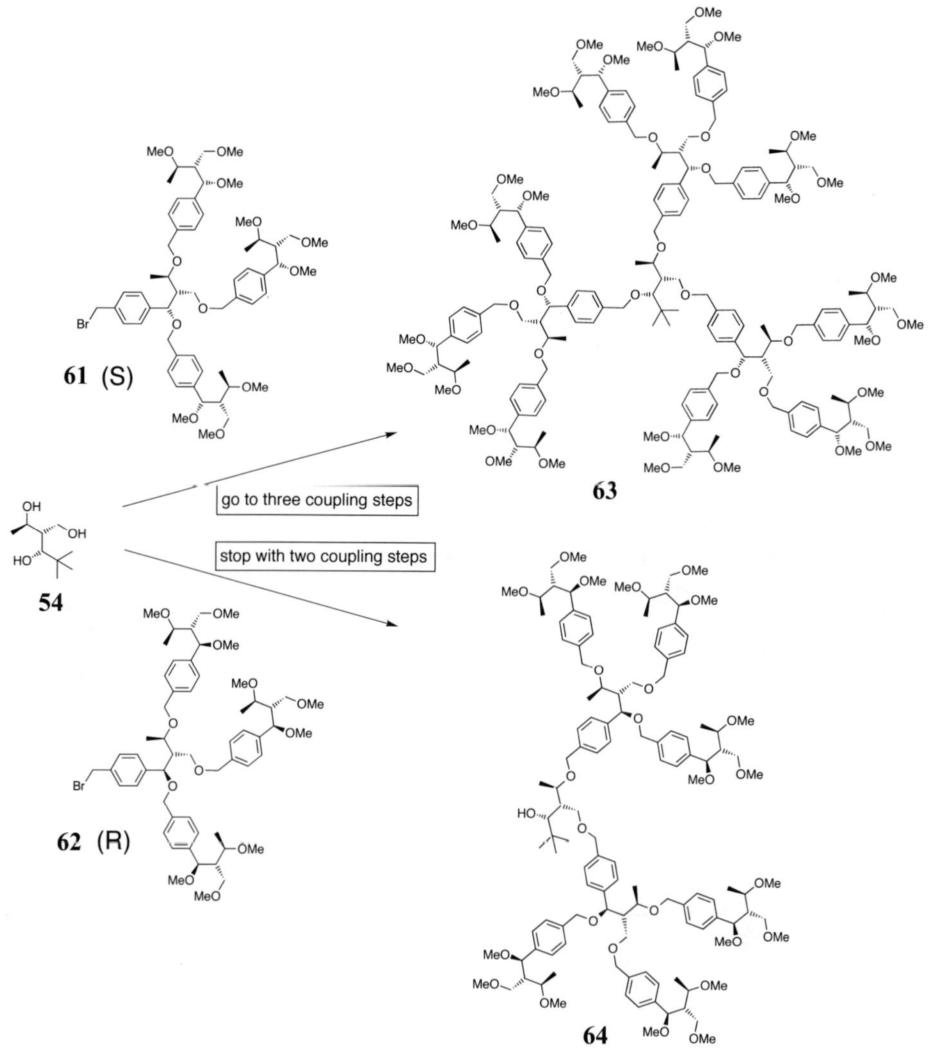

Fig. 22. The influence of different configurations in the dendritic branch bromides **61** and **62** on coupling with the same chiral core **54** [88, 90]

diastereomeric branch bromide **62**, however, that differs from **61** only by having (*R*)-configuration at the four benzylic centers, only two branches could be coupled to give dendritic alcohol **64**; even under forcing conditions, the corresponding dendrimer was not formed to any detectable degree. With the enantiomeric center triol *ent*-**54**, on the other hand, both, the (*S*)- and the (*R*)-branches reacted to give the desired dendrimers (**65, 66** in Fig. 23).

Further work has revealed, that some combinations of diastereoisomeric 2nd-generation branch bromides (**61, 62, 67, 68**) and various cores (**54**, *ent*-**54**, **69**) smoothly reacted to give the desired dendrimers (**63, 65, 66, 74, 75**) whereas

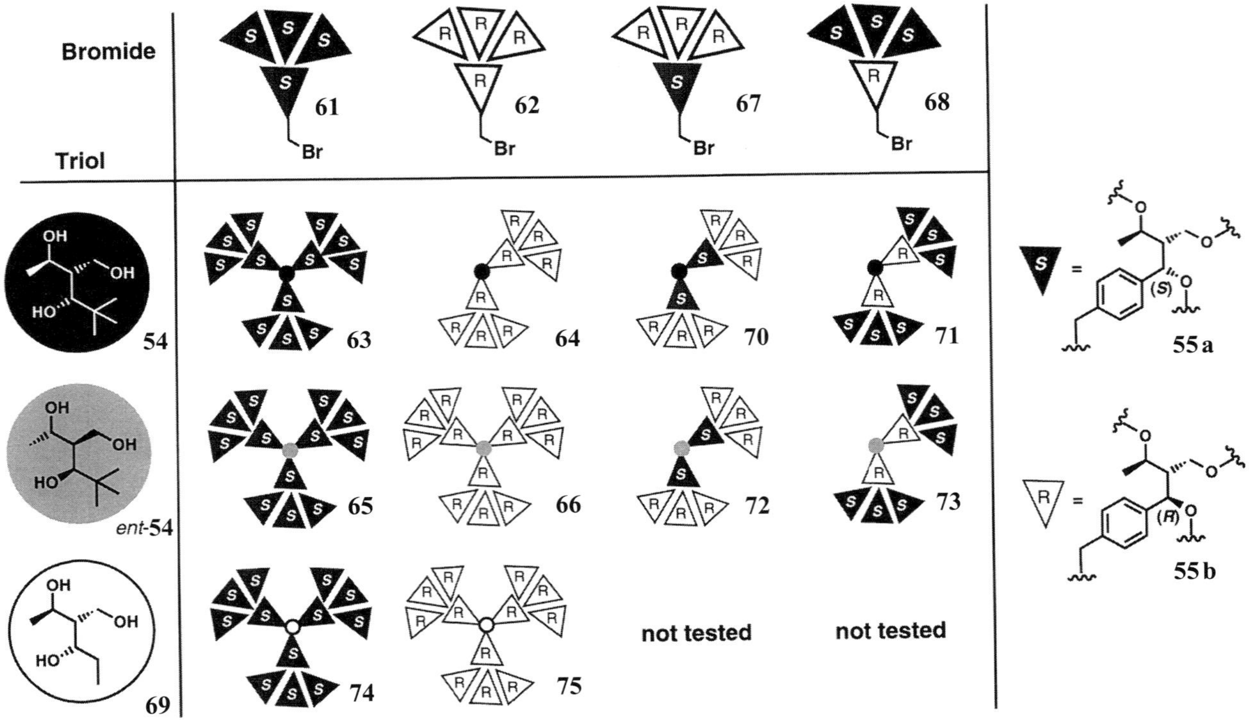

Fig. 23. Formation of various 2nd-generation dendrimers or dendritic compounds using branch building blocks of (*S*)- or (*R*)-configuration at the benzylic centers [90]. Three of the eight combinations tested by us led to dendrimer formation (products **63**, **65**, **66**) while five halted at the stage of the doubly coupled products **64**, **70**–**73**, cf. Fig. 22. Core molecule **69** has less hindered OH groups and is, of course, not an isomer of **54** and *ent*-**54**

others (see **64, 70, 71, 72, 73**) did not (Fig. 23). With a switch of configuration at (some of) the benzylic centers of the branches the conformation of the branch and/or of the doubly coupled dendritic alcohols of type **64** must be altered in such a way that the third etherification may take place or be entirely blocked.

Considering the distance from the core OH groups to the stereogenic center(s) at which there is a configurational difference (17 bonds in the case of **70** vs **63**, of **72** vs **65**, and of **73** vs **66**) this recognition phenomenon might be called "outrageous". The components are able to discriminate between each other and show either a fit or a misfit (coupling or no coupling). Since the numerous single bonds in these dendritic structures have low energy barriers for rotation and hence the molecules have even more numerous conformations separated by shallow energy surfaces we [90] consider it impossible to reproduce and thus rationalize the observed phenomenon, for instance by molecular modelling. The coupling combinations shown in Fig. 23 are but a (minimal) few, considering the fact that of the 10^{11} possible stereoisomers (which exist of a molecule with 39 independent stereogenic centers) we have tested the synthesis of only eight.

Comparison of the optical activity showed that the dendrimer **63** with branches of (*S*)-configuration has a specific rotation and a molecular ellipticity which clearly deviate from the expected values [88, 90]. All other 2nd-generation dendrimers (even those with additional spacers between the branches and the core) have specific rotations that are comparable to those expected by simple addition of appropriate values for their building blocks. The deviation may therefore signal the presence of chiral conformational substructures in the 2nd-generation dendrimer **63**.

Besides MALDI-TOF mass spectroscopy, by which the monodispersity of all the above described dendritic compounds was proven, ^{1}H-NMR spectroscopy was again found to be a most informative characterization method, since most signals from the hydrogens at the different stereogenic centers have unique shifts. The resonances from analogous protons of the peripheral, interior and central units were always well separated and shifted towards lower field on going from outside to inside (for detailed discussion see our recent full paper [90]).

Another possibility to find out more about the structure of these dendrimers was chosen by incorporating fluorine atoms. The use of ^{19}F-NMR spectroscopy offered an additional tool to study the conformation of the dendrimer, especially with the fluorines attached close to the stereogenic centers [91]. Following our previously developed methods [92], fluorine-containing 1st- and 2nd-generation chiral dendrimers such as **76** were synthesized (Fig. 24).

The CF_3 groups not only had interesting effects upon certain synthetic steps on the way to such dendrimers [93], but also could be used as probes: the ^{19}F-NMR spectra which were measured in different solvents and at different temperatures exhibited a clear-cut difference between the inner CF_3 groups and those close to the periphery.

It turned out that with the sterically demanding triply branching building blocks no dendrimers larger than of 2nd generation could be prepared. We therefore switched to dendrimers with analogous doubly branching building units (see **56** in Fig. 20) as our target molecules. The components for the

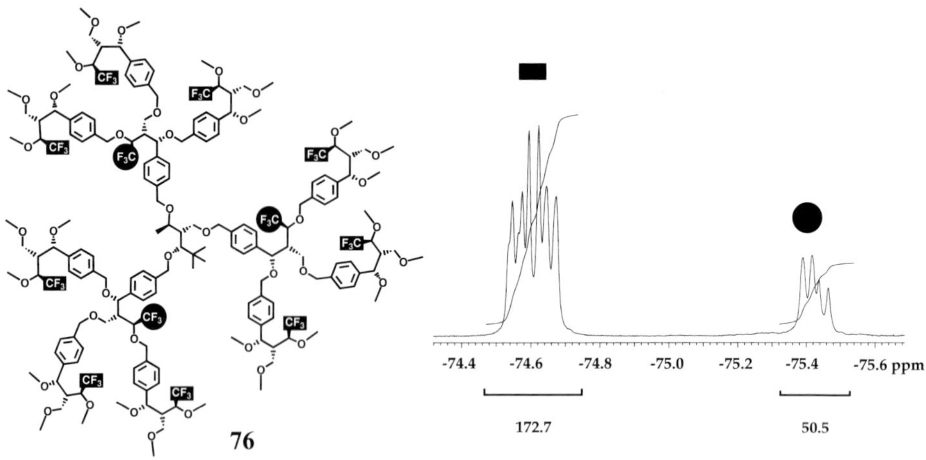

76

Fig. 24. Fluorine-containing 2nd-generation dendrimer **76** and ^{19}F-NMR spectrum showing the different chemical environment in the two dendrimer-layers [91]

branches are available by alkylation of the enolate of **52** with a benzylic bromide. Using a convergent strategy, "fully chiral" branches of up to 5th and dendrimers of up to 4th generation have been synthesized [88, 94, 95]. Dendrimer **77** was built from a triol with an aromatic elongating group and from 45 (24 peripheral, 21 interior) doubly branching units. It contains 93 stereogenic centers (and, thus, is one out of 10^{28} possible stereoisomers) and is monodisperse (Fig. 25).

By ^{13}C-NMR spin-lattice relaxation time measurements, it was shown that the segmental mobility of the peripheral units of all dendrimers of this type (1st up to 4th generation) is higher than that of the interior units [95]. This result is comparable to that obained by similar measurements with achiral dendrimers [96, 97]. By size-exclusion chromatography we tried to estimate the size of these dendrimers. By assuming that the molecules have a spherical shape and by using the intrinsic viscosity values, a diameter of five nanometers has been calculated for dendrimer **77**. Polarimetry showed, that even for dendrimers with so many stereogenic centers, the overall specific rotation $[\alpha]_D$ can be derived as arithmetic sum of increments of suitable model compounds resembling the building blocks. As shown in Fig. 26, compound **78** was used as reference for the elongated core, **79** for the interior and **80** for the peripheral building blocks.

A comparison of the thus calculated with the measured specific rotations of the 0th- to 4th-generation dendrimers of this kind gave a close resemblance, with a curve, approaching asymptotically a limiting value (Fig. 26). It was also shown that the shape of this curve was independent of solvent, concentration and temperature. This was not the case when CD spectra of these dendrimers were compared (Fig. 27): in solvents such as CH_2Cl_2 and *t*-butyl methyl ether a constant rise of the Cotton effect was observed, which correlates with the increasing amount of benzene chromophores in the dendrimers. However, in the

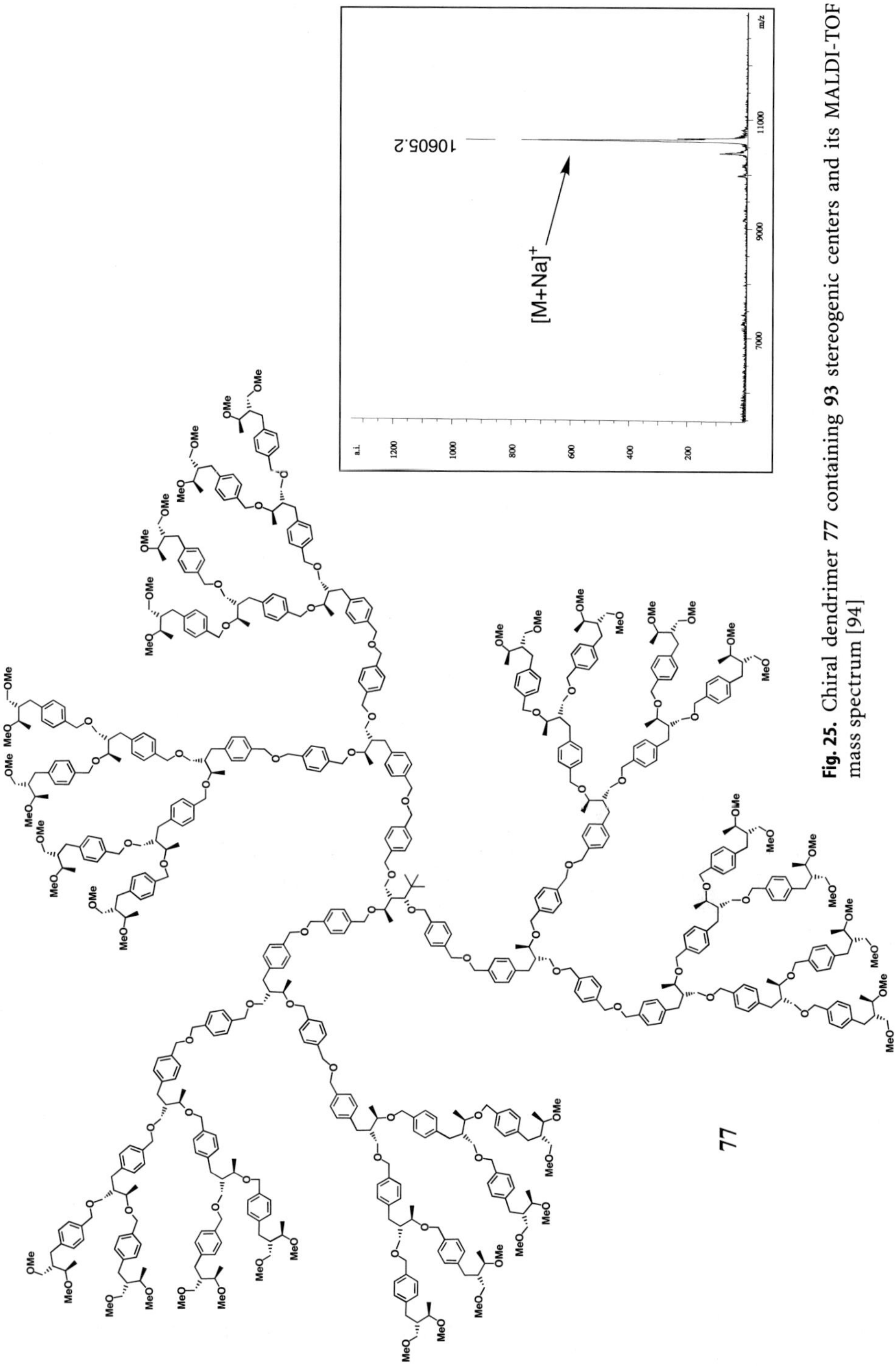

Fig. 25. Chiral dendrimer **77** containing **93** stereogenic centers and its MALDI-TOF mass spectrum [94]

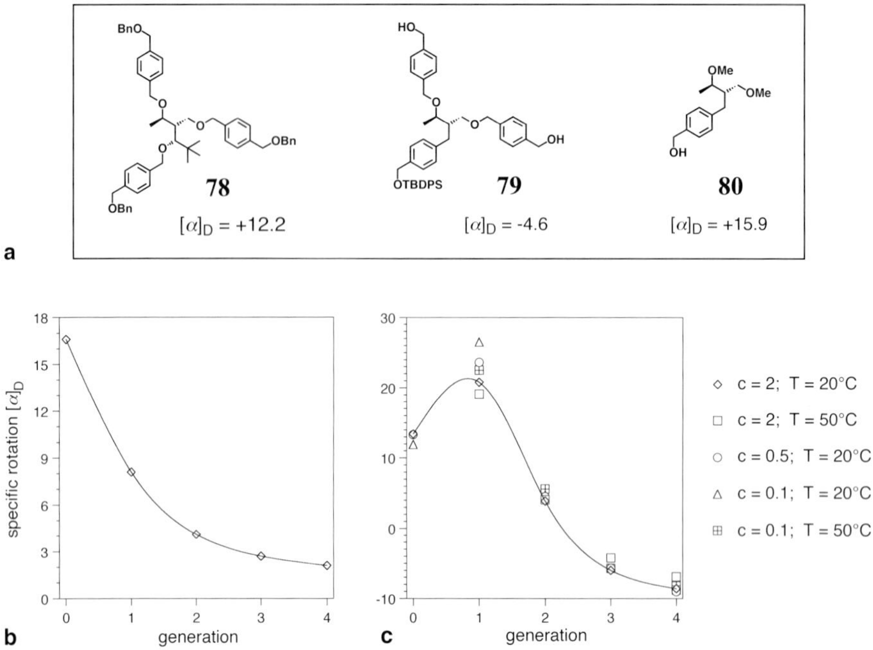

Fig. 26. **a** Specific optical rotations for the model building blocks for "core" **78**, "interior building block" **79** and "peripheral unit" **80** of dendrimers of type 77; **b** calculated average optical rotations for the doubly branching dendrimers of 0th to 4th generation; **c** measured optical rotations in $CHCl_3$ at different temperatures and concentrations [94]

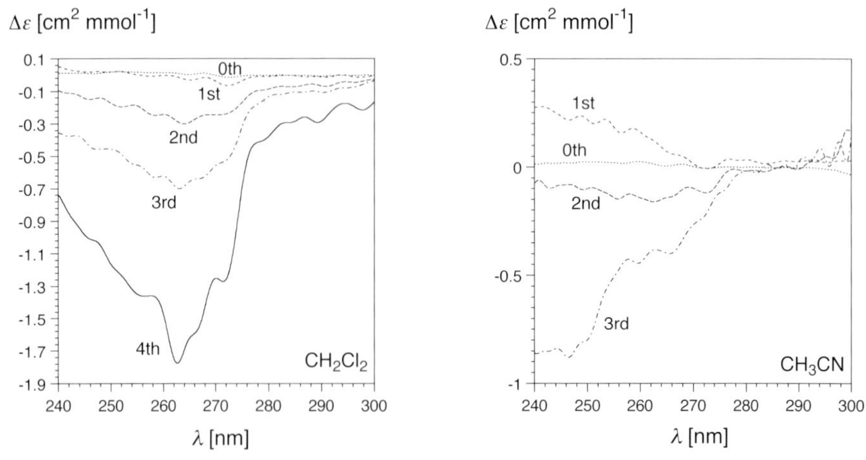

Fig. 27. CD spectra of "fully chiral" dendrimers (0th up to 4th generation of type 77) in CH_2Cl_2 and in CH_3CN [94]

more polar solvent CH_3CN, dendrimer **77** was not soluble and the curves found for lower generation dendrimers changed shape (*and* sign) from generation to generation. To date, this effect could not be rationalized.

In conclusion, we have learned a lot from studying chiral dendrimers, about the behavior of such large chiral molecules and about the contributions of the different building blocks to the whole structure. It remains a great challenge to rationalize the origin of the dramatic diastereoselectivity effects observed in the synthesis of certain chiral dendrimers.

4
Chiral Dendrimers in Catalysis

Due to their regular shape and their controllable surface and size dendrimers show promising properties as carriers for catalytically active sites. Such dendrimers with achiral ligands attached to their periphery were first described by van Koten et al. [98]. They prepared polycarbosilane dendrimers with diamino arylnickel(II) complexes on the surface to catalyze the Kharash addition of polyalkanes to double bonds. Compared to the smaller monomers, commonly used in homogeneous catalysis, dendrimeric molecules should be more easily recovered after the reaction because of their size and their stable shape and constant volume (e.g. by ultrafiltration). In this way, dendrimers combine the advantages of homogeneous catalysis (i.e. fast kinetics and good accessibility to the active sites) and heterogeneous catalysis (i.e. easy recovery) [99].

To catalyze asymmetric transformations, catalytically active sites can be incorporated in different areas of a dendrimer: a) chiral sites at the periphery, b) chiral sites in cavities or at the core, c) achiral sites which are surrounded by chiral branches in the interior of the dendrimer.

The rate of a catalytic reaction depends on the rate of diffusion of both substrates and products to and from the catalytic sites. Therefore it is of outmost importance that the catalytically active sites are freely accessible for reactions. Only dendrimers of *low* generation number can possibly be expected to be suitable carriers for catalytically active sites, especially when these are located in the interior. In high-generation dendrimers with crowded surfaces catalytic activity of an internal site would be prevented. On the other hand, a crowded surface will not only hinder access to an interior ligand site but will also cause steric hindrance between groups attached to it and thus prevent high reactivity of sites at the periphery.

The latter effect has been demonstrated by Meijer et al., who attached chiral aminoalcohols to the peripheral NH_2-groups of poly(propylene imine) dendrimers of different generations [100]. In the enantioselective addition of diethylzinc to benzaldehyde (mediated by these aminoalcohol appendages) both the yields and the enantioselectivities decreased with increasing size of the dendrimer (Fig. 28). The catalyst obtained from the 5th-generation dendrimer carrying 64 aminoalcohol groups at its periphery showed almost no preference for one enantiomer over the other. This behavior coincides with the absence of measurable optical rotation as mentioned in Sect. 3 above. The loss of activity and selectivity was ascribed to multiple interactions on the surface which were

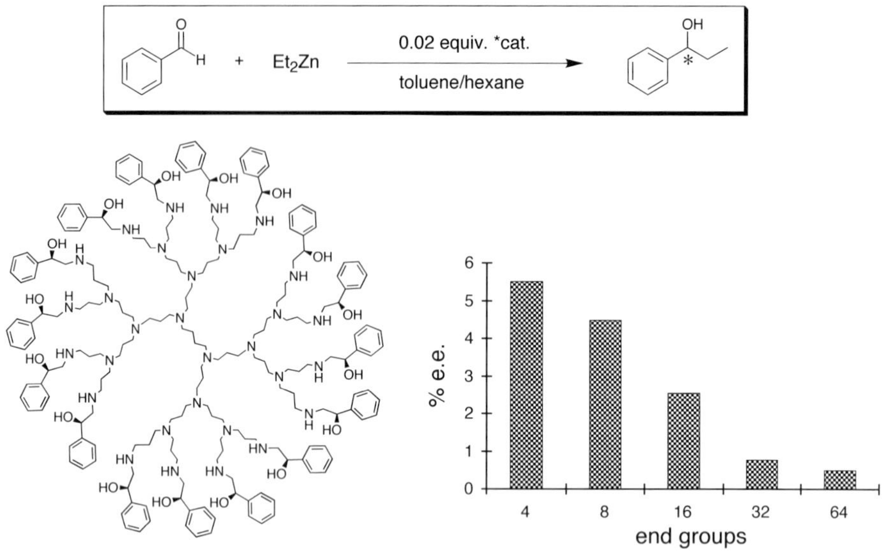

Poly(propylene imine) dendrimer of 3rd generation with 16 end groups

Fig. 28. Dependence of enantioselectivity from the number of generations or endgroups in the enantioselective addition of Et_2Zn to benzaldehyde, using 0.02 equiv. amino alcohol equivalents in each case [100]

proposed to cause steric hindrance and formation of different conformers of the terminal groups, so that opposite selectivities result.

Brunner et al. attached chiral branches to non-chiral catalytically active sites. With the aim to influence the enantioselectivity of transition metal catalyzed reactions they synthesized several dendritically enlarged diphosphines such as **81** [101] (Fig. 29). In situ prepared catalysts from $[Rh(cod)Cl]_2$ and **81** have been tested in the hydrogenation of (α)-*N*-acetamidocinnamic acid. After 20 hours at 20 bar H_2-pressure (Rh/substrate ratio 1:50) the desired product was obtained with an enantiomer ratio of 51:49.

In another reaction dendritic pyridine derivatives such as **82** or **83** were tested as co-catalysts for enantioselective cyclopropanation of styrene with ethyl diazoacetate [102]. Using catalyst **82**, enantiomer ratios of up to 55:45 were obtained. However, with catalyst **83** bearing larger branches yields and selectivities did not increase. The relatively low selectivities were rationalized by the presence of a large number of different conformations that this non-rigid system may adopt.

Bolm et al. attached single achiral Fréchet-type branches of up to the 3rd generation to a chiral pyridyl alcohol, but practically no influence was observed on the selectivity of the catalyzed reaction [103].

In our group, dendrimers carrying the catalytically active part either on the periphery or in the core were investigated. In both cases $\alpha,\alpha,\alpha',\alpha'$-*t*etra*a*ryl-1,3-*d*ioxolane-4,5-*d*imethan*ol*s (TADDOLs) have been employed as ligands in chiral

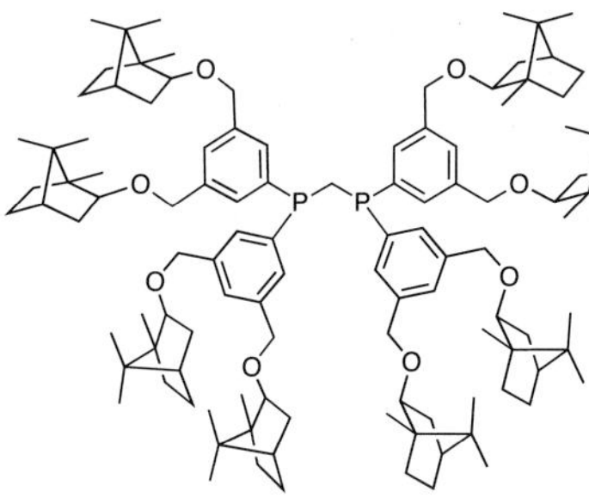

81

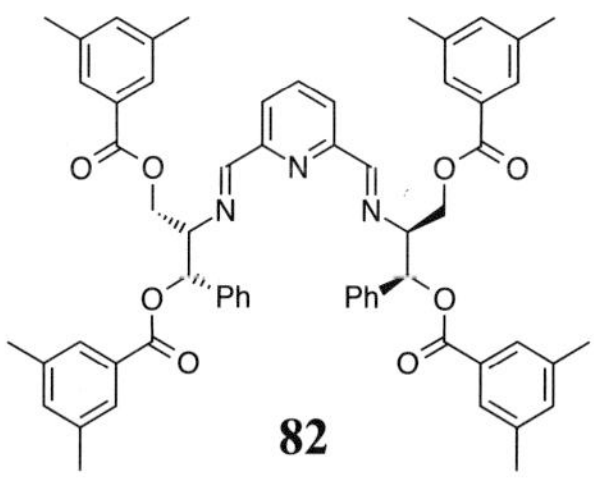

82

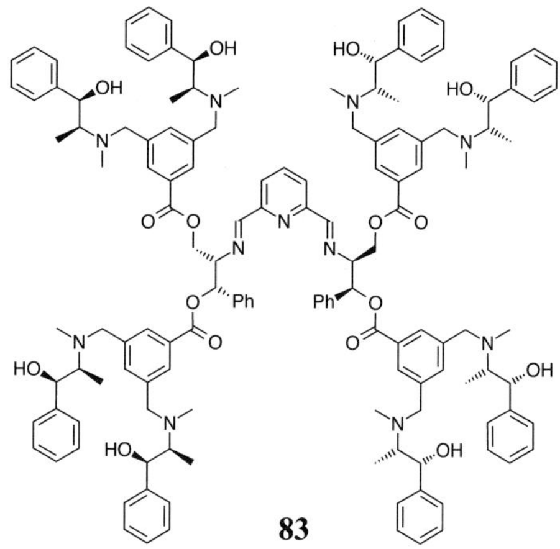

83

Fig. 29. Dendritically enlarged diphosphine **81** and pyridine derivatives **82** and **83** tested for enantioselective catalytic hydrogenations and cyclopropanations [101, 102]

active sites (Fig. 30), since a large variety of catalytic and stoichiometric enantioselective transformations had been shown to be possible with TADDOLate metal complexes [104].

Attached to the periphery of a 1st-generation dendrimer, $Ti(OCHMe_2)_2$ complexes of the six TADDOL moieties in **84** catalyze – in homogeneous solution – the enantioselective addition of diethylzinc to benzaldehyde with about the same selectivity ((*S*):(*R*) 97:3) as do six monomeric TADDOL units [105], but, with a molecular weight of only 3833 Da, dendrimer **84** had to be separated by column chromatography rather than by ultrafiltration methods.

Dendrimers of various sizes with the TADDOL placed in the center have also been prepared. By attaching achiral or chiral branches to the aryl rings of the central TADDOL a series of up to 3rd-generation dendrimers (as an example see **85**) was obtained. By using the titanate of low-generation dendrimers of this type in homogeneous catalysis it was shown that the branches had only a minor influence on the selectivity and rate of the catalyzed reaction. Dendritic branches can be of advantage to change the properties of a catalyst: thus octyl-groups attached to the periphery of dendritic TADDOLs cause their complexes to be very well soluble in apolar media.

To increase efficiency and ease of product separation from reaction mixtures, we also prepared styryl-substituted TADDOL-dendrimers that can act as cross-linkers in styrene suspension polymerizations, and thus lead to beads with intimately incorporated TADDOL sites [106, 107]. Due to the presence of the conformationally flexible dendritic spacers between the chiral ligand and the poly-

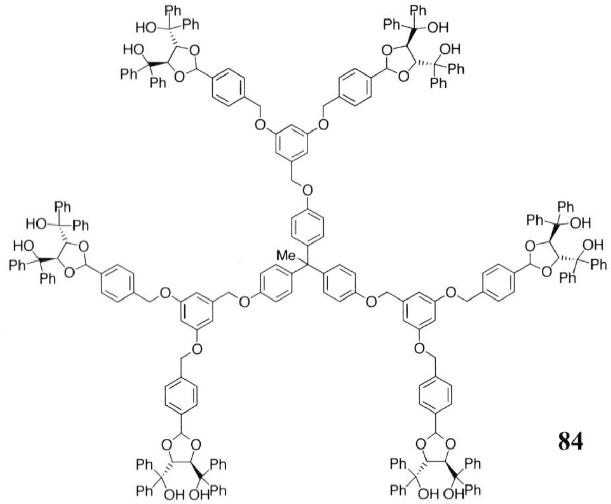

84

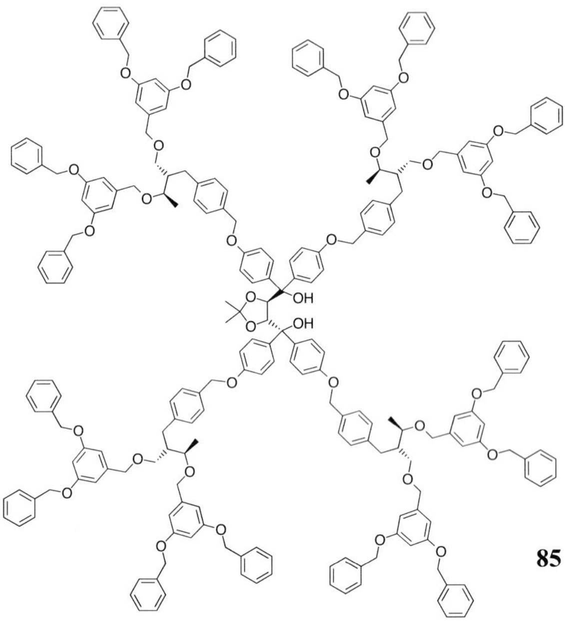

85

Fig. 30. Dendritic TADDOL derivatives carrying the catalytically active site either at the periphery (**84**) or in the center (**85**) [105, 106]

mer backbone(s), such cross-linked insoluble polymers were expected to have readily accessible TADDOL sites and thus provide catalytic activity comparable to that of soluble analogs.

Following a procedure previously employed for simple styryl TADDOLs [105], dendritic styryl-substituted TADDOLs were copolymerized with styrene and loaded with $Ti(OCHMe_2)_4$. The beads thus obtained were tested in the catalysis of the enantioselective addition of Et_2Zn to PhCHO under the usual conditions [105, 108–110] (these first polymer particles which have been prepared with a crosslinking dendrimer do (because of the high degree of crosslinking) not swell in organic solvents which in less crosslinked polymers normally would be considered [111] a sign for poor solvent penetration and thus for poor properties as carriers of polymer-bound active sites or reagents!).

Employing 0.2 equiv. of polymer-bound dendritic Ti-TADDOLates of type **89** (1st and 2nd generation) enantioselectivities up to 98:2 were observed (Fig. 31). This value is comparable to those obtained in heterogeneous reactions using non-dendritic, polymer-bound analogs **88** (er up to 98,5:1,5 [105]) and with the

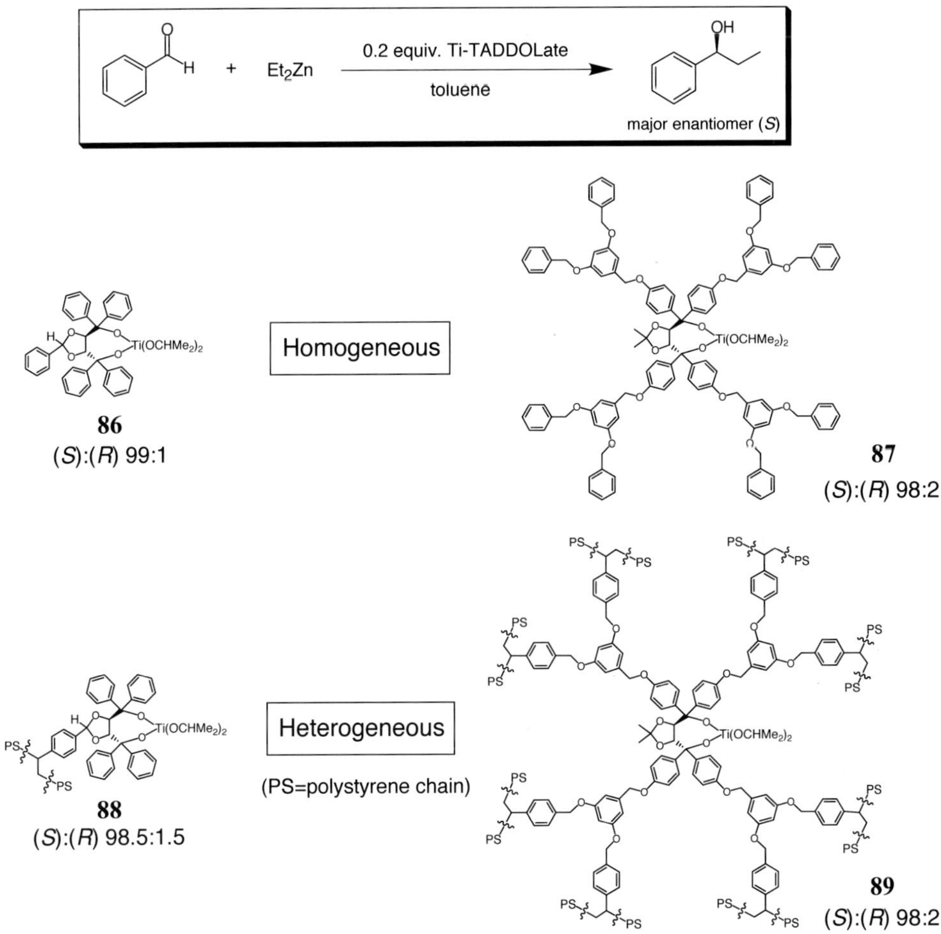

Fig. 31. Selectivity comparison for the enantioselective addition of Et_2Zn to benzaldehyde using different dendritic and non-dendritic homogeneous and heterogeneous Ti-TADDOLates as chiral catalysts [107, 110]. (*S*):(*R*) ratios refer to the 1-phenyl-propanol formed

corresponding homogeneous reactions using dendritic TADDOLate **87** (er up to 98:2) or the simple TADDOLate **86** (er up to 99:1 [108]). For the dendrimer-crosslinked polymers the selectivity increased from the 0th to the 1st generation, whereas with the 2nd generation no further improvement of the enantioselectivity was observed.

A comparison of the rates showed that the polymer-bound Ti-TADDOLate **88** and the dendritic polymer **89** catalyze the Et_2Zn-to-PhCHO addition at a similar fast rate as the monomeric TADDOLate **86** and the dendritic TADDOLate **87** in homogeneous solution [107, 112]. Further experiments also with other ligands are being carried out in our laboratories.

Along another line of work in our group (*S*,*S*)-1,4-bis(dimethylamino)-2,3-dimethoxy butane (DDB), which had been used as cosolvent in asymmetric synthesis [113], was tested as a core moiety for a dendritic amine catalyst. The conformationally flexible DDB-core, which has been synthesized in five steps from diethyl tartrate was coupled with different branches to give dendritically expanded diamines **90**–**92** (molecular weight 3800 Da) [114] (Fig. 32).

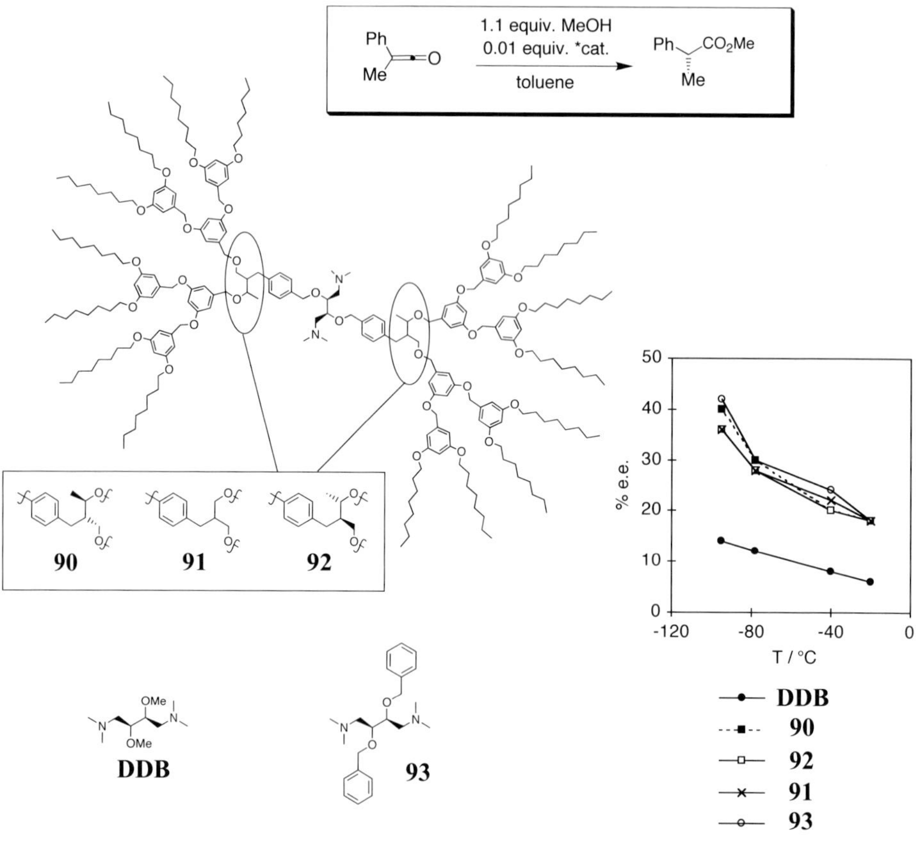

Fig. 32. Dendritically enlarged diamines **90**–**92** and benzylated ligand **93** (DDB analogs) in the enantioselective addition of methanol to ketene using 0.01 equiv. catalyst [114]

These dendritic compounds have been tested in several base-catalyzed reactions, e.g. the addition of ketene to chloral or the Michael-addition of thiophenol to cyclohexenone. In these cases the dendritically expanded DDB provided only minor differences in enantioselectivity, as compared to the monomeric DDB. However, when compound **90** was used to catalyze the enantioselective addition of methanol to methyl phenyl ketene, the selectivity, with which (*R*)-methyl-2-propionate was formed, increased by a factor of three. Also the diastereoisomer **92**, which differs from **90** by containing a branch building block of opposite chirality, and the dendritically enlarged ligand **91** with achiral branches catalyze the reaction more selectively than DDB. This particular ketene reaction includes an enantioselective protonation, i. e. the catalyst is directly involved in the enantio-differentiating step. Since the stereogenic centers of the branching units are remote from the reaction center(s) it follows that the entire dendritic wedge or parts thereof influence the stereochemical outcome of the reaction. The experimental results show that the stereogenic centers of the branching units, which are remote from the reaction center(s) seem to have only minor influence on the reaction. Therefore it was proposed that the size of the whole dendritic branch or parts thereof influence the stereochemical outcome of the reaction. Indeed, when employing non-dendritic dibenzyl ether **93**, an effect on the selectivity was found which is similar to that generated by dendritic elongation.

5 References

1. Seebach D, Lapierre J-M, Skobridis K, Greiveldinger G (1994) Angew Chem 106:457; Int Ed Engl 33:440
2. For a recent review see: Peerlings HWI,. Meijer EW (1997) Chem Eur J 3:1563
3. Denkewalter RG, Kole JF, Lukasavage WJ (1981) US Pat 4,289,872
4. Denkewalter RG, Kole JF, Lukasavage WJ (1982) US Pat 4,360,646
5. Denkewalter RG, Kole JF, Lukasavage WJ (1983) US Pat 4,410,688
6. Aharoni SM, Crosby III CR, Walsh EK (1982) Macromolecules 15:1093
7. Aharoni SM, Murthym NS (1983) Polym Commun 24:132
8. Chapman TM, Hillyer GL, Mahan EJ, Shaffer KA (1994) J Am Chem Soc 116:11195
9. Chapman TM, Mahan EJ (1997) Polym Mater Sci Eng 77:103
10. Twyman LJ, Beezer AE. Mitchell JC (1994) Tetrahedron Lett 35(25):4423
11. Tam JP (1995) In: Gutte B (ed) Peptides: Synthesis, Structures and Applications. Academic Press, San Diego
12. Chang KJA, Pugh W, Blanchard SG, McDermed J, Tam JP (1988) Proc Natl Acad Sci USA 85:4929
13. Tam JP (1988) Proc Natl Acad Sci USA 85:5409
14. Posnett DN, McGrath H, Tam JP (1988) J Biol Chem 263:1719
15. Rao C, Tam JP (1994) J Am Chem Soc 116:6975
16. Tam JP, Lu YA (1989) Proc Natl Acad Sci USA 86:3879
17. Jun S, Tam JP (1995) J Am Chem Soc 117:3894
18. Spetzler JC, Tam JP (1996) Peptide Res 9:290
19. Zhang L, Tam JP (1997) J Am Chem Soc 119:2363
20. Newkome GR, Lin X, Weis CD (1991) Tetrahedron: Asymmetry 2(10):957
21. Newkome GR, Liu X (1991) Macromolecules 24:1443
22. Mulders SJE, Brouwer AJ, Liskamp RMJ (1997) Tetrahedron Lett 38(17):3085
23. Niggemann M, Ritter H (1996) Acta Polymer 47:351

24. Varki A (1993) Glycobiology 3:97
25. Duckamer K, Taylor ME (1993) Ann Rev Cell Biol 9:237
26. Bundle DR, Young NM (1992) Curr Opin Struct Biol 2:666
27. Lee RT, Lee YC (1994) In: Neoglycoconjugates: Preparation and Applications. Academic Press, San Diego, p 23
28. Lee RT, Lee YC (1987) Glycoconjugate J 4:317
29. Roy R (1997) Top Curr Chem 187:241
30. Roy R (1996) Polymer News 21:226
31. Lindhorst TK (1996) Nachr Chem Tech Lab 44:1073
32. Jayamaran N, Nepogodiev SA, Stoddart JF (1997) Chem Eur J 3:1193
33. Roy R, Zanini D, Meunier SJ, Romanovska A (1993) J Am Chem Soc Chem Commun 1869
34. Roy R, Zanini D, Meunier SJ, Romanovska A (1994) ACS Symposium Series 560:104
35. Zanini D, Roy R (1997) J Am Chem Soc 119:2089
36. Zanini D, Park WKC, Roy R (1995) Tetrahedron Lett 36:7383
37. Zanini D, Roy R (1997) Bioconjugate Chem 8:187
38. Pagel D, Zanini D, Roy R (1996) Bioorg Med Chem 4:1949
39. Roy R, Park WKC, Zanini D, Foxall C, Srivastana OP (1997) Carbohydr Lett 2:259
40. Roy R, Park WKC, Wu Q, Wang SN (1995) Tetrahedron 36:4377
41. Park WKC, Kratzer B, Zanini D, Wu Q, Meunier SJ, Roy R (1995) Glycoconjugate J 12:456
42. Aoi K, Itoh K, Okada M (1995) Macromolecules 28:5391
43. Aoi K, Tsutsumiuchi K, Yamamoto A, Okada M (1997) Tetrahedron 53(45):15415
44. Lindhorst TK, Kieburg C (1996) Angew Chem 108:2083; Int Ed Engl 35:1953
45. Kieburg C, Lindhorst TK (1997) Tetrahedron Lett 38:3885
46. Pagel D, Roy R (1997) Bioconjugate Chem 8:714
47. Ashton PR, Boyd SE, Brown CL, Nepogodiev SA, Meijer EW, Peerlings HWI, Stoddart JF (1997) Chem Eur J 3:3974
48. Ashton PR, Boyd SE, Brown CL, Jayamaran N, Nepogodiev SA, Stoddart JF (1996) Chem Eur J 2:1115
49. Ashton PR, Boyd SE, Brown CL, Jayamaran N, Stoddart JF (1997) Angew Chem 109:756; Int Ed Engl 36:732
50. Jayamaran N, Stoddart JF (1997) Tetrahedron Lett 38:6767
51. Ashton PR, Honnsell EF, Jayamaran N, Nilson TM, Spencer N, Stoddart JF, Young M (submitted) J Am Chem Soc, receipt of a copy of the manuscript prior to publication is gratefully acknowledged
52. Colonna B, Harding VD, Nepogodiev SA, Raymer FM, Spencer N, Stoddart JF (submitted) Chem Eur J, receipt of a copy of the manuscript prior to publication is gratefully acknowledged
53. Hudson RHE, Damha MJ (1993) J Am Chem Soc 115:2119
54. Häner R, Skobridis K (1996) PCT Pat WO 96/19240
55. Polyprobe, company's brochure (1996) US Pats 5,175,270; 5,484,904; 5,487,973
56. Seebach D, Herrmann GF, Lengweiler UD, Bachmann BM, Amrein W (1996) Angew Chem 108:2069; Int Ed Engl 35:2795
57. Seebach D, Herrmann GF, Lengweiler UD, Amrein W (1997) Helv Chim Acta 80:989
58. Bachmann B (1997) PhD thesis no. 12074, ETH Zürich
59. Müller H-M, Seebach D (1993) Angew Chem 105:483; Int Ed Engl 32:477
60. Seebach D, Brunner A, Bachmann BM, Hoffmann T, Kühnle FNM, Lengweiler UD (1995) Schering Lecture 28, Ernst Schering Research Foundation, Berlin
61. Kremers JA, Meijer EW (1994) J Org Chem 59:4262
62. Hawker CJ, Fréchet JMJ (1990) J Am Chem Soc 112:7638
63. Kremers JA, Meijer EW (1995) Reactive & Functional Polymers 26:137
64. Struijk MP, Peerlings HWI, Meijer EW (1996) Polym Prepr 37:497
65. Mislow K, Bickart P (1977) Isr J Chem 15:1
66. Peerlings HWI, Trimbach D, Meijer EW (1997) Polym Mater Sci Eng 77:73
67. Jansen JFGA, de Brabander-van den Berg EMM, Meijer EW (1994) Science 266:1226
68. Jansen JFGA, Meijer EW (1995) J Am Chem Soc 117:4417

69. Janssen RAJ, de Brabander-van den Berg EMM, Meijer EW (1995) Adv Mater 7(6): 561
70. Jansen JFGA, Peerlings HWI, de Brabander-van den Berg EMM, Meijer EW (1995) Angew Chem 107:1321; Int Ed Engl 34:1206
71. Issberner J, Böhme M, Grimme S, Nieger M, Paulus W, Vögtle F (1996) Tetrahedron: Asymmetry 7:2223
72. Chang H-T, Chen C-T, Kondo T, Siuzdak G, Sharpless KB (1996) Angew Chem 108:202; Int Ed Engl 35:182
73. Kawaguchi T, Walker KL, Wilkins CL, Moore JS (1995) J Am Chem Soc 117:2159
74. McElhanon JR, Wu M-J, Escobar M, Chaudhry U, Hu C-L, McGrath DV (1997) J Org Chem 62:908
75. McElhanon J, McGrath DV (1997) Polymer Prepr 38:278
76. McElhanon J, McGrath DV (1997) Polym Mater Sci Eng 77:153
77. Junge DM, McGrath DV (1997) Polym Mater Sci Eng 77:157
78. Chow H-F, Mak CC (1997) J Chem Soc Perkin Trans 1 91
79. Mak CC, Chow H-F (1996) J Chem Soc Chem Commun 1185
80. Chow H-F, Mak CC (1997) Pure Appl Chem 69(3):483
81. Chow H-F, Mak CC (1996) Tetrahedron Lett 33:5935
82. Ehrler J, Seebach D (1990) Liebigs Ann Chem 379
83. Amberg W, Seebach D (1990) Chem Ber 123:2413
84. Seebach D, Lapierre J-M, Jaworek W, Seiler P (1993) Helv Chim Acta 76:459
85. Lapierre J-M, Skobridis K, Seebach D (1993) Helv Chim Acta 76:2419
86. Seebach D, Zimmermann J (1986) Helv Chim Acta 69:1147
87. Seebach D, Imwinkelried R, Stucky G (1986) Angew Chem 98:182; Int Ed Engl 25: 178
88. Murer P, Seebach D (1995) Angew Chem 107:2297; Int Ed Engl 34:2116
89. Seebach D, Lapierre J-M, Greiveldinger G, Skobridis K (1994) Helv Chim Acta 77: 1673
90. Murer PK, Lapierre J-M, Greiveldinger G, Seebach D (1997) Helv Chim Acta 80:1648
91. Greiveldinger G, Seebach D (1997) Polym Mat Sci Eng 77:184
92. Gautschi M, Schweizer WB, Seebach D (1994) Chem Ber 127:565
93. Greiveldinger G (1997) PhD thesis no. 12488, ETH Zürich
94. Murer P, Seebach D (1997) Polym Mat Sci Eng 77:69
95. Murer PK (1996) PhD thesis no. 12001, ETH Zürich
96. Meltzer AD, Tirrell DA, Jones AA, Inglefield PT, Hedstrand DM, Tomalia DA (1992) Macromolecules 25:4541
97. Jansen JFGA, de Brabander-van den Berg EMM, Meijer EW (1994) Science 266: 1226
98. Knapen JWJ, van der Made AW, de Wilde JC, Leeuwen PWNM, Wijkens P, Grove DM, van Koten (1994) Nature 372:659
99. Tomalia DA, Dvornic PR (1994) Nature 372:617
100. Sanders-Hovens MSTH, Jansen JFGA, Vekemans JAJM, Meijer EW (1995) Polym Mater Sci Eng 73:338
101. Brunner H, Fürst J (1994) Tetrahedron 50:4303
102. Brunner H, Altmann S (1995) Chem Ber 127:2285
103. Bolm C, Derrien N, Seger A (1996) Synlett 387
104. Dahinden R, Beck AK, Seebach D (1995) In: Paquette L (ed) Encyclopedia of Reagents for Organic Synthesis, vol. 3. J. Wiley, Chichester, p 2167
105. Seebach D, Marti RE, Hintermann T (1996) Helv Chim Acta 79:1710
106. Rheiner PB, Seebach D (1997) Polym Mater Sci Eng 77:130
107. Rheiner PB, Sellner H, Seebach D (1997) Helv Chim Acta 80:2027
108. Seebach D, Plattner DA, Beck AK, Wang YM, Hunziker D, Petter W (1992) Helv Chim Acta 75:2171
109. Seebach D, Beck AK, Schmidt B, Wang YM (1994) Tetrahedron 50:4363
110. Comina PJ, Beck AK, Seebach D (1998) Org Proc Res Der, submitted

111. Flory PJ (1953) In: Principles of Polymer Chemistry. Cornell University Press, Ithaca, New York, p 577
112. The rate for the simple polymer-bound TADDOLate published in [107] was taken from [110]. Newer results show a similar rate for both polymer-bound catalysts described herein.
113. Seebach D, Kalinowski D, Bastani H-O, Crass O, Daum H, Dörr H, DuPreez NP, Ehrig V, Lauger W, Nüssler C, Oei H-A, Schmidt M (1977) Helv Chim Acta 60:301
114. Butz T, Murer P, Seebach D (1997) Polym Mater Sci Eng 77:132

Dendrimers with Polymeric Core: Towards Nanocylinders

A.-D. Schlüter

Freie Universität Berlin, Institut für Organische Chemie, Takustrasse 3, D-14195 Berlin, Germany, *E-mail: adschlue@chemie.fu-berlin.de*

This chapter draws a comprehensive picture of what has been done in the field of dendrimers with polymeric cores putting emphasis first on synthetic issues and then on experiments investigating the aggregation behavior of these intruiging macromolecules both in the solid state and on surfaces. Additionally, experiments will be described which show that some of these dendrimers can be considered cylindrical molecular objects. The macromolecules treated in this chapter may be considered as either dendrimers with polymeric core or alternatively dendronized polymers (or polymers with appendent dendrons) depending on whether one sees them from the vantage point of an organic or macromolecular chemist.

Keywords: Dendrimers, polymers, molecular objects, scanning force microscopy, macromonomers.

1 Introduction

Practically all dendrimers [1] known today have cores with a few, typically three to six functional groups to which the corresponding number of dendrons (dendritic wedges) are attached. The fact that these dendrons are connected to one another by a small, almost dot-like molecule results in considerable steric congestion in the space around the core. This congestion is a unique structural feature and has one led to view dendrimers, specifically those of high generation, as molecular boxes or containers [2] and as entities which assume a spherical shape to which a "surface" can be assigned. The latter view is still unproven and the term surface may only be applied with care. Since their discovery some 20 years ago, dendritic macromolecules have stimulated an almost explosive research effort and many synthetic, analytical, and application-related issues have been addressed [1]. Even industrially applicable syntheses were developed [3]. During this stormy process research has virtually exclusively focused on dendrimers with small cores, in spite of a US patent entitled "Rod-like Dendrimer" filed by Tomalia and Kirchhoff in 1987. Here the dendrimers are like **1** with polymeric

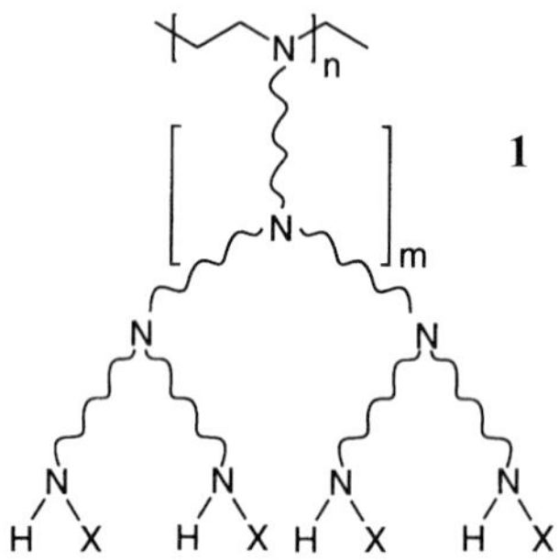

1

a: X = H; **b**: X = CH_2-$CHCO_2^{\ominus}$ $Na^{\oplus}$

cores and proposed as being useful in the production of molecular composites and as crystallinity modifiers for polymeric materials [4]. It took some years more before first (published) steps were undertaken for these dendrimers [5]. It is immediately apparent from looking at van-der-Waals models that they not only complement dendrimers with small cores under structural aspects but as a consequence of the structural differences should also have unique properties. This includes their potential to serve as molecular objects with a defined and pre-determined shape irrespective of the surrounding medium (vide infra). A simple reason for this slow development may be that in the beginning of dendrimer research spherical dendrimers were simply considered a more appropriate and perhaps more important target. As judged by the increasing number of publications, this view is presently undergoing some modification. Another reason may be seen in a reluctance to begin working on a seemingly more complex area than small core dendrimers where chemists are already being confronted with considerable synthetic and analytical difficulty.

This chapter shows that the synthetic and analytical problems associated with the synthesis of the title dendrimers are not outrageously more complex than for small core dendrimers and that the difficulties can be overcome. It draws a comprehensive picture of what has been done in the field of dendrimers with polymeric cores putting emphasis first on synthetic issues and then on experiments investigating the aggregation behavior of these intruiging macromolecules both in the solid state and on surfaces. Finally, experiments will be described which show that some of these dendrimers can be considered cylindrical molecular objects. The macromolecules treated in this chapter may be considered as either dendrimers with polymeric core or alternatively dendronized polymers (or polymers with appendent dendrons) depending on whether one sees them from the vantage point of an organic or macromolecular chemist. Not considered here are polymers which do not carry dendrons at every repeat unit (r.u.) but only at a few ones only, e.g., at both ends [6]. The references to dendrimers with dot-like core are kept at the minimum. The reader is referred to the other chapters of this volume.

2 Syntheses

2.1 General

There are two principally different synthetic routes to dendronized polymers (Scheme 1). In the first, the polymer which becomes the core in the final product serves as starting material. Its anchor groups are used to either divergently or convergently attach a dense sequence of dendrons (Route A). In the second, monomers already carrying dendrons (so-called macromonomers) are subjected to polymerization or polycondensation (Route B). Dendronized polymers formally belong to the class of comb polymers [7]. Strategies and their respective intrinsic problems have both been known for these polymers for a long time. These problems become more serious and limiting the more sterically demanding the substituents (here: dendrons) are. An obviously critical issue associated with route A is to achieve complete coverage of the backbone anchor groups with dendrons. Even if a very efficient coupling chemistry is available, a large dendron excess may be required to drive the reaction to completion. This excess may, in turn, make it difficult to purify the product. If large dendrons are to be attached, steric hindrance comes also into play for two reasons: (a) the kinetic shielding of still unreacted anchor groups on the backbone by dendrons already attached in close proximity and (b) the dendron's conformation which may lead for free energy reasons to a self-shielding of the functional group at the focal point through which the attachment ought to take place. Both factors lead to a decrease in the rate of attachment if they do not suppress the process entirely. Another point of concern is that the slow attachment can favor side reactions which may not just involve themselves or the solvent but also the already attached dendrons.

The problems associated with route B also have something to do with steric hindrance. Here the critical point is the steric demand of both monomer and chain end. Incoming monomer will only be connected to the chain end, if steric hindrance is not too high. Otherwise this process will be slowed down or even rendered impossible. Depending on the kind of polyreaction applied, this may lead to termination of the reactive chain end and/or to side reactions of the monomer, like loss of coupling functionality as in some polycondensations or auto-initiation specifically in radical polymerizations. From this discussion it can be extracted that the basic problems for both routes are: incomplete coverage (route A) and low molecular weight dendronized polymer (route B).

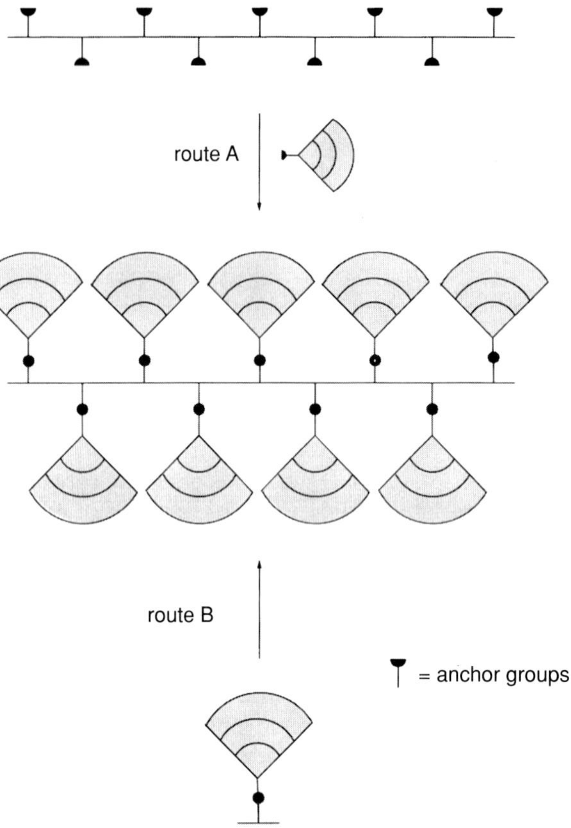

Scheme 1. Routes A and B to dendronized polymers. The dendrons shown are of generation three (G-3). Route A can be done divergently or convergently. Only in the latter case the route is as simple as shown. For the former case, see text. Mixed forms are in principal also possible

Besides synthetic hurdles there are also analytical ones. Dendronized polymers tend to have repeat units (r.u.) with considerable molecular weight. R.u.s with 1 or 2 kDa are no exception. Such high molecular weights sometimes render structural characterization difficult because the proportion of backbone to dendron atoms becomes so unfavorable that NMR spectroscopy reaches its limits. For example, sometimes the degree of attachment (coverage) simply cannot be determined with sufficient accuracy, because the signal intensity of the unreacted anchor groups is too weak for comparison with reference signals in the spectrum. NMR characterization may occasionally be further complicated by large differences in the relaxation times of backbone and dendron nuclei. Thus, NMR integrals are rendered unreliable if not a sufficient puls delay time is applied. Another problem with dendronized polymers is their molecular weight determination. Gel permeation chromatography (GPC) is a quick and easy to use method for a rough estimation of the molecular weight of a polymer [8]. This estimation can only be reasonably used as long as the hydrodynamic volumina of the polymer under consideration and the polymer used for calibration purposes, typically: polystyrene (PS), are comparable. The hydrodynamic volume of dendronized polymers strongly deviates, however, from parent polystyrene and GPC results should be treated with care. Additionally, facile aggregation of these dendrimers is often encountered which leads to further complications. Other methods of molecular weight determination like light scattering have be to used and measures taken to avoid aggregation.

2.2 From Polymeric Cores

The first literary reaction to dendrimer synthesis, one which even has helped lay its foundation, is described in Vögtle's 1978 publication [9]. In this ammonia is used as a trifunctional compound to which the Michael compound, acrylo nitrile, is added in excess. Hydrogenation of the resulting 1:3 adduct furnishes a compound with three primary amino groups, which is then subjected to the same sequence. In today's dendrimer terminology a trifunctional core, namely ammonia, is converted into a hexafunctional one with the result that a first generation dendrimer is born. Tomalia's 1987 patent describes a related approach to dendronized polymers. As core serves poly(ethyleneimine) (PEI) which according to a recent, more detailed publication has a degree of polymerization between $P_n = 100-500$ [10]. To each of these 100–500 anchor groups is added the Michael compound, acrylic ester. The resulting PEI decorated with ester groups at each backbone nitrogen atom is then reacted with an enormous excess of ethylene diamine to convert the esters in aminoethylene amides. The primary amines introduced this way constitute the first two-fold branching points from which the next generations are grown (Scheme 2). This process is repeated several times. Experimental evidence for the dendronized PEI is not yet complete, but there is some interesting support for it from electron micrographs of a fourth generation material whose terminal amino groups (theoretically 16 per side chain) carry the sodium salt of acrylic acid. The view graph shows thin, linear features which are approximately 100–200 nm long and mostly bundled into larger aggregates [10]. From this result it is reasonable to conclude that dendrons divergently grown onto a PEI backbone lead to an elongation of the polymer backbone. Since this approach is a divergent one, all the known advantages and disadvantages associated with this growth technique of course apply [1,11].

One disadvantage of divergent dendrimer synthesis is that structural perfection unavoidably decreases with increasing generation. There is nothing like a 100% reaction in chemistry. In regard to the chapter's target polymers this results, i.e., in a distribution of molecular weight and space demand of the dendrons attached to backbones. In all other approaches to dendronized polymers starting from polymeric cores, convergently synthesized dendrons of varying generations are attached to the backbone. This way the intrinsic problems associated with the divergent route are avoided and the dendrimers obtained are more likely to be structurally perfect. There is, however, also a critical point to mention here, generally not encountered in convergent synthesis of small core

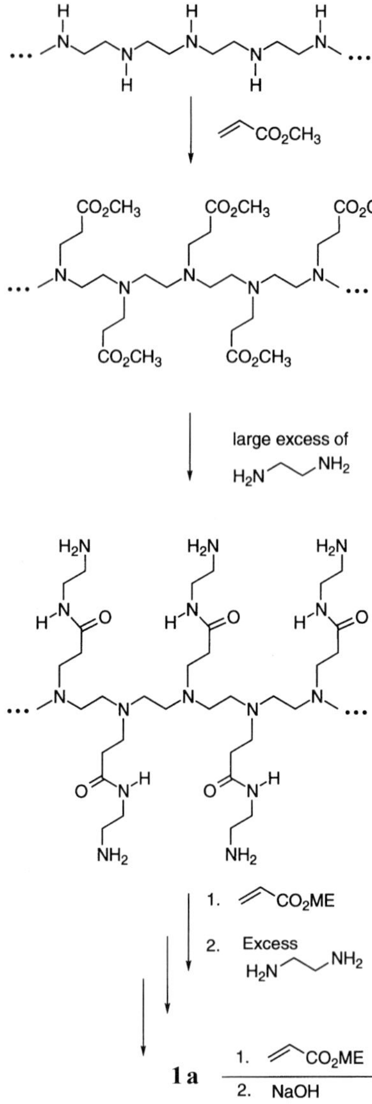

Scheme 2. Tomalia's divergent synthesis of dendronized polyethylene imine (PEI)

dendrimers. If for what ever reason the degree of coverage is not virtually 100%, the molecular weight difference between perfect dendrimer and "side product" with some missing dendrons is considerable for small dendrimers and almost negligible for those with polymeric core. As a consequence, perfect and imperfect product can be seperated for small core dendrimers but not for polymeric core dendrimers. This underlines the importance of the kind of chemistry applied to attach dendrons to a backbone and the strict must for high conversion processes.

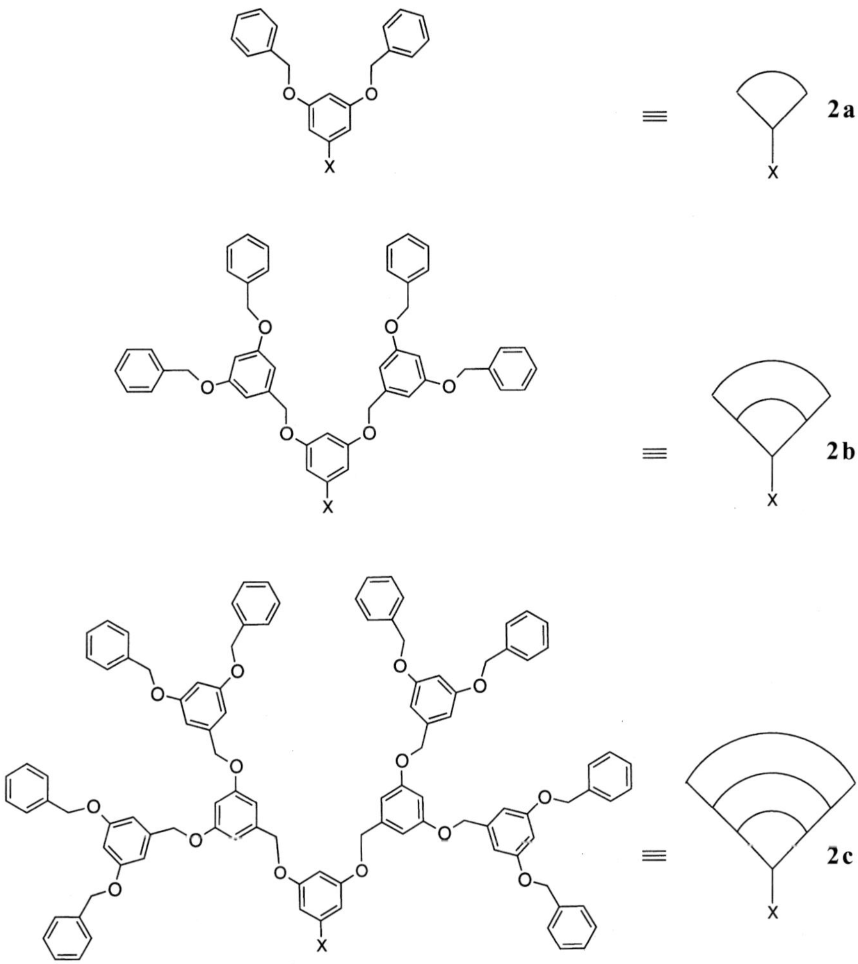

Scheme 3. Fréchet-type dendrons of generations one to three (**2a–c**) with a functional group (X) at the focal point

The convergent syntheses known today all use Fréchet-type dendrons **2a–c** [12] with different functional groups at the focal point (Scheme 3). These dendrons are attached to rigid-rod poly([1.1.1]propellane)s [13] and poly(*para*-phenylene)s [14] with either hydroxy or iodomethyl anchor groups. In the first reported example, polymer **3** is dendronized with a G-1 dendron **2a** to give dendrimer **4** [5c] (Scheme 4a). This potentially interesting reaction did not gain much importance because serious synthetic difficulties were encountered. A critical issue is the poor solubility of polymer **3** and its deprotonated form which is needed to bring about the nucleophilic displacement reaction at the benzylic position of dendron **2a** (X=CH_2Br). With this experience in mind the attachment chemistry and the basic polymer were changed. The hydroxy groups of

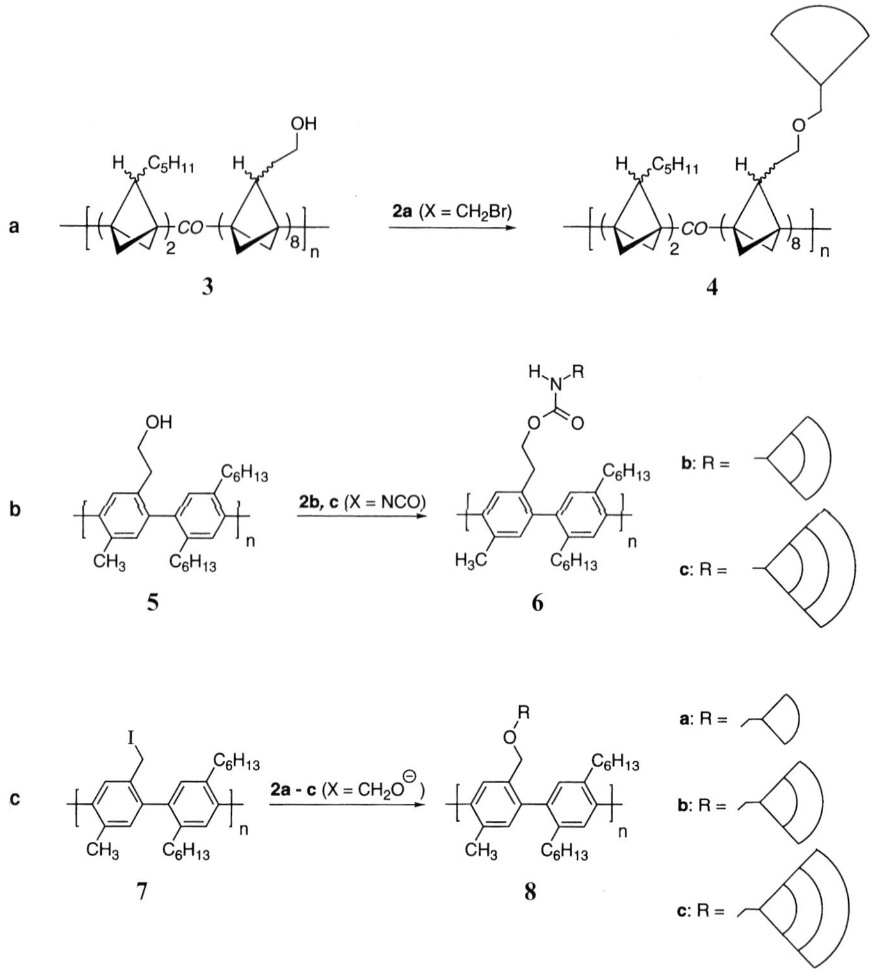

Scheme 4. Examples for the convergently executed route A using poly([1.1.1]propellane) (**a**) and poly(*para*-phenylene) backbones (**b**, **c**)

poly(*para*-phenylene) **5**, were reacted with excess **2b** (X=NCO), a way by which the problematical deprotonation of the polymer can be avoided. Dendrimer **6b** was obtained in a cleanly proceeding reaction [15a]. The degree of dendronization was investigated by 500 MHz NMR spectroscopy and found to be virtually 100%. Figure 1 shows the ^{1}H-NMR spectrum of **6b** used for this purpose to illustrate its high resolution and limited complexity despite the dendrimers' complicated structure. In order to account for differences in relaxation times the spectrum was recorded with a pulse delay.

Whenever a polymer is subjected to a chemical modification (here: dendronization), its degree of polymerization (DP, P_n: number average DP, P_w: weight average DP) has considerable impact. Normally modifications become increas-

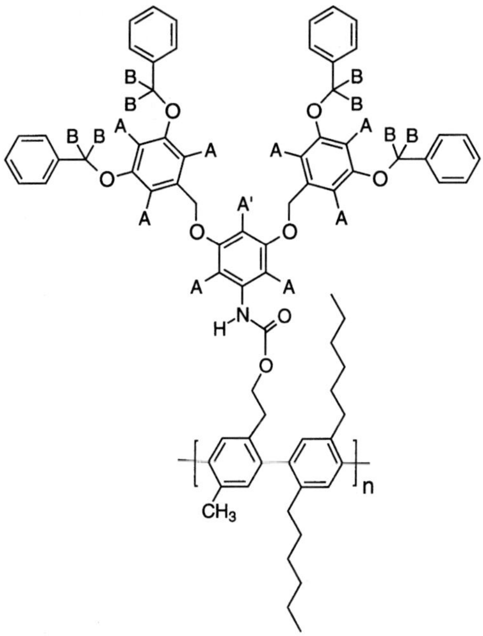

6b

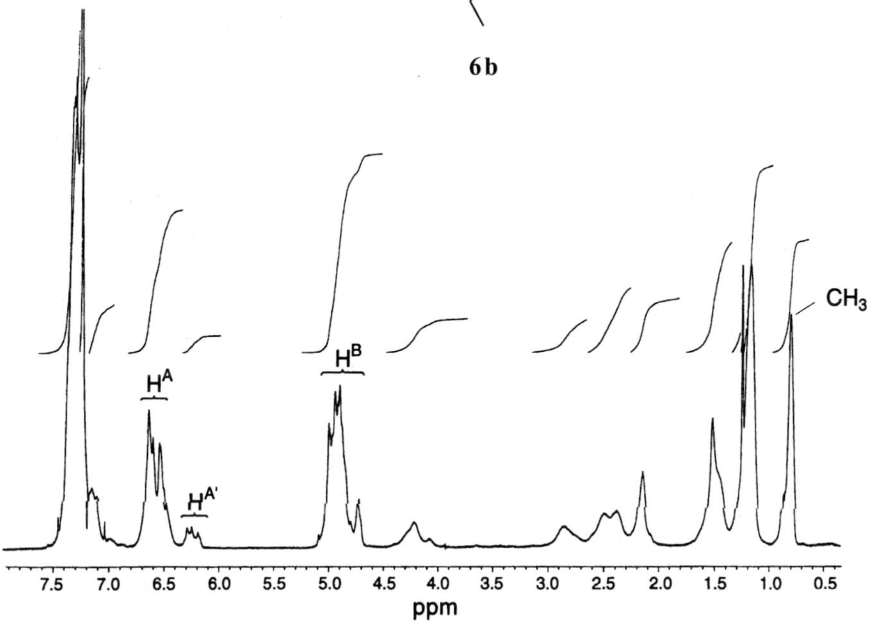

Fig. 1. ^{1}H-NMR spectrum of dendrimer **6b** with some signals assigned and integrals. The degree of dendronization is determined by comparison of the intensities of the signal groups at $\delta = 6.2-6.7$ ppm and $\delta = 4.7-5.0$ ppm with those of the CH_3 signal at $\delta = 0.8$ ppm

ingly more difficult with increasing DP. It is therefore important to note that a sample of polymer **5** was used for these initial experiments which had a relatively low DP (P_n=32, P_w=64; GPC calibrated versus polystyrene standard). It was thus essential for the feasibility of the whole project to prove that this dendronization also works with higher molecular weight core molecules. Fortunately, with a sample of **5** (P_n = 82) the coverage with **2b** (X=NCO) could still be driven to completion [15a]. Obviously each individual coupling step proceeds with very high conversion showing urethane formation to be a powerful tool for dendronization reactions of polyfunctional cores. The limitation of this method, however, becomes visible when the same reaction is carried out with G-3 dendron **2c** (X=NCO). Here the maximum achievable, "magic" coverage is 90–92% [15b]. Though this number would be considered more than sufficient for many chemical reactions, it cannot obscure the fact that the coverage is incomplete. Each individual polymer chain carries 8–10% uncovered anchor groups as integral part of its structure. These defects can neither be healed nor removed by "purification". In addition, the determination of the coverage by NMR integration turns somewhat unreliable because of large intensity differences of signals to be compared.

Limitations with the convergent dendronization of polymers are also found when polymer **7** is reacted with dendrons **2a–c** (X=CH_2O^-) (Scheme 4c). Whereas the first and second generation dendron can be attached completely, the attachment of the third generation dendron cannot be driven beyond a limit of approximately 70% [15b]. In sum, convergent synthesis of polymeric core dendrimers is feasible for low generation dendrons. For larger dendrons incomplete coverage is unavoidable as judged from the presently available examples, even though the dendrimers obtained may still be useful materials.

2.3 Using Macromonomers

Compared with the approach described in the previous section, synthesis of dendronized polymers from macromonomers has been developed more broadly in recent years. A reason for this is the advantage that polymers, if they can be obtained at all by this method, necessarily carry the predetermined number of dendrons at the backbone. The above questions of dendron perfection and coverage degree in divergent and convergent syntheses do not play a role anymore. This only holds true, of course, if the dendrons are compatible with the polymerization conditions, a prerequisite which is fulfilled in many of the cases reported. The polymerization procedures used may be divided into (a) radically initiated and transition-metal catalyzed chain-growth and (b) transition metal-mediated step-growth polymerizations. Scheme 5 gives a structural and chronological overview of the macromonomers reported up to present and

Scheme 5. Structural and chronological overview of the macromonomers used for synthesis of dendronized polymers: chain growth (**a**) and step growth polymerizations (**b**)

a

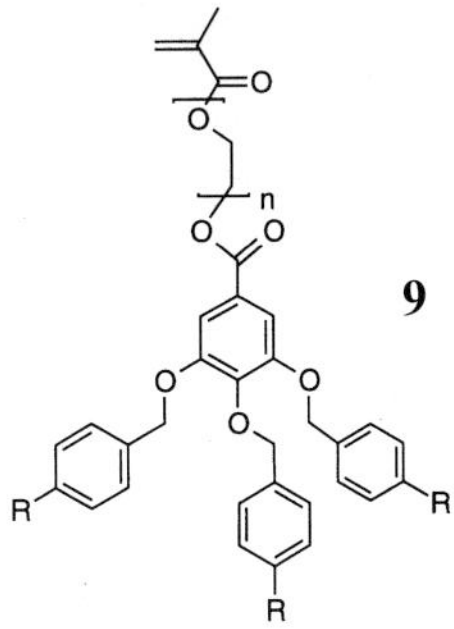

9

G - 1

R = $OC_{12}H_{25}$
n = 1 - 4

Percec 1993

R = $O(CH_2)_m(CF_2)_nF$
m = 4, n = 6

Percec 1995

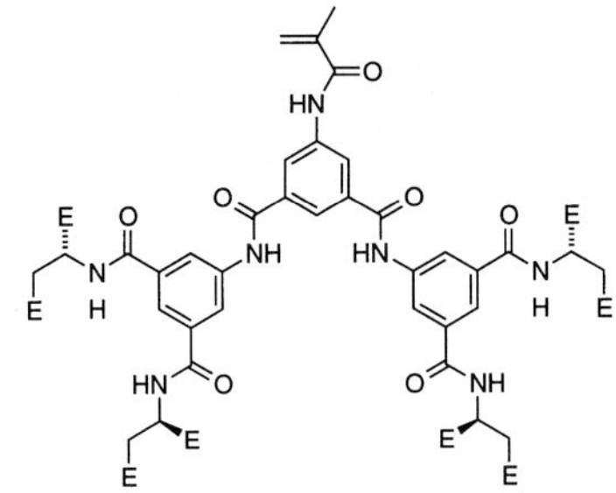

10

G -1 and G - 2

Ritter 1995

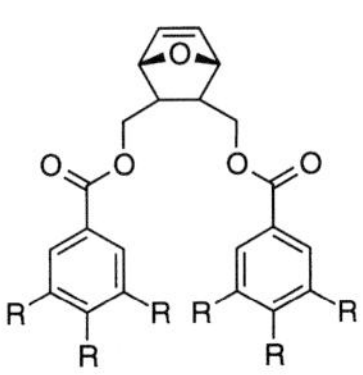

11

R = $OC_{12}H_{25}$

2 x G - 1

Percec 1996

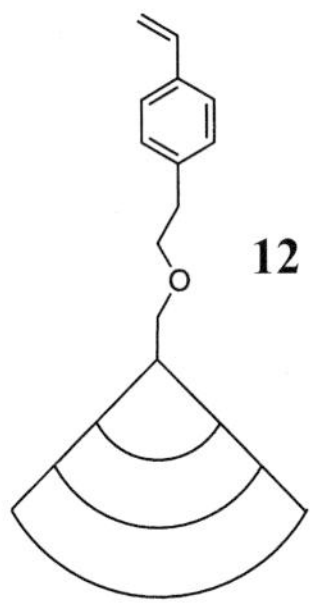

12

G -1, G - 2, and G -3

Schlüter 1996

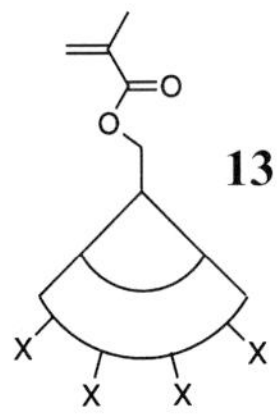

13

X = H, CO_2CH_3

G -1 and G - 2

Xi 1996

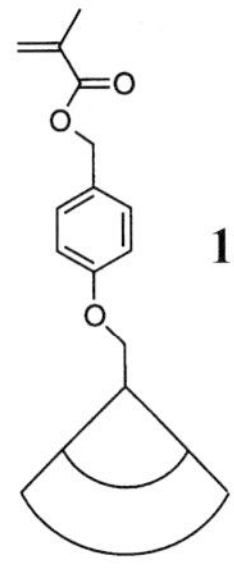

14

G -1 and G - 2

Schlüter 1996

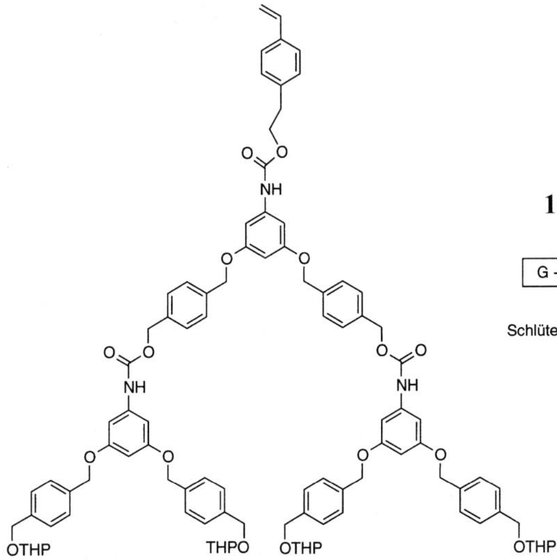

15

G - 2

Schlüter 1996

Scheme 5 (continued)

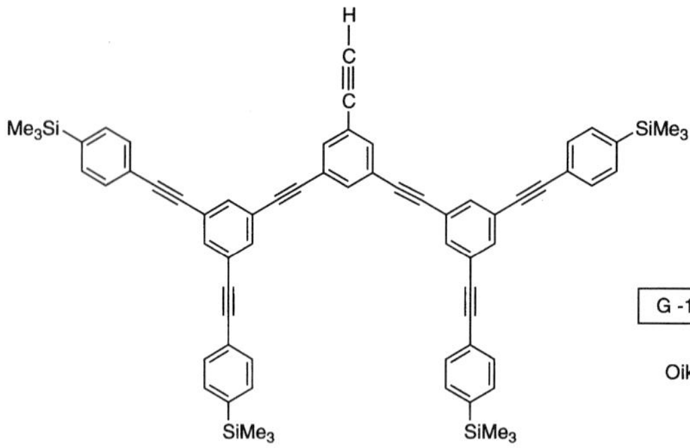

16

G -1 and G - 2

Oikawa 1997

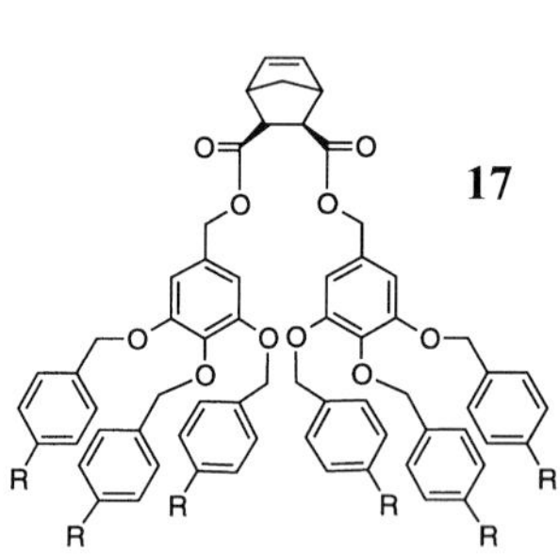

17

$R = OC_{12}H_{25}$

2 x G - 1

Percec 1997

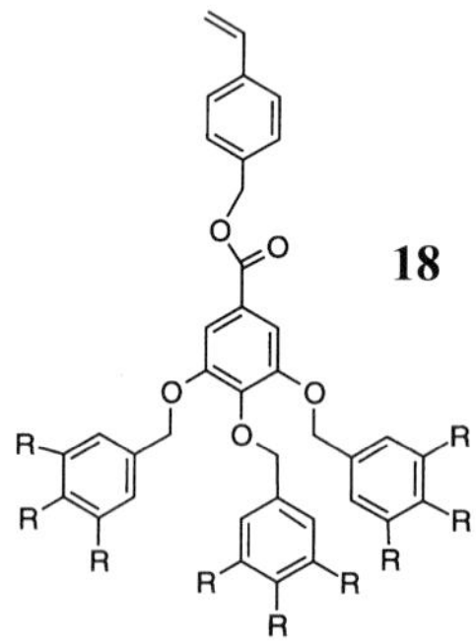

18

$R = C_{12}H_{25}$

G - 1

Percec 1998

b

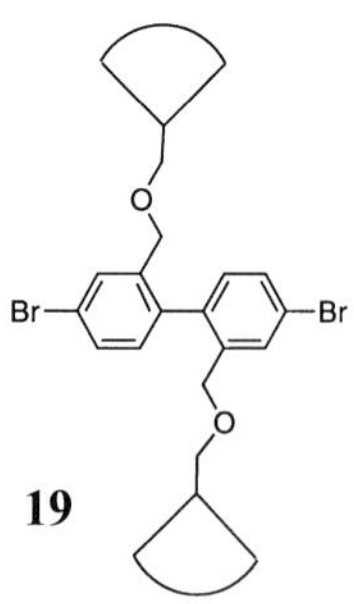

19

2 x G - 1

Schlüter 1995

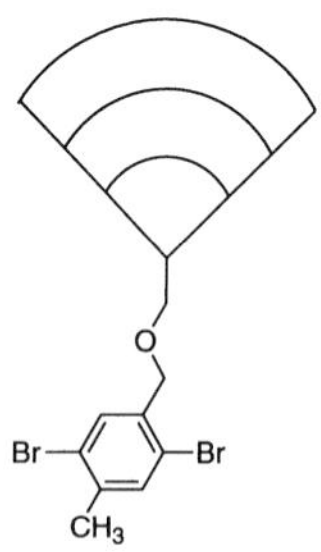

20

G -1, G - 2, and G - 3

Schlüter 1997

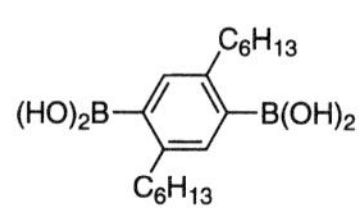

21

grouped into subclasses (a) and (b) according to the polymerization method applied. Monomers are polymerized by radical polymerization (**9** [5a, 16], **10** [17], **12** [18], **13** [19], **14** [20], **15** [20], and **18** [21]), ring-opening metathesis polymerization (ROMP) [22] (**11** [23] and **17** [24], insertion polymerization yielding dendronized polyacetylene (**16** [25]), and Suzuki polycondensation [26] (**19** [27] and **20** [15b]). The structural variety of the dendrons is not very large. Most of them contain ether functions and perhaps one or two esters. Monomer **15** is the only one containing, e.g., carbamate functions and polymerizes nevertheless easily and cleanly. It is also an interesting candidate because it carries tetrahydropyrane (THP) protected functional groups in the periphery which may be used for further chemical modifications. Peripherial modification ought to be also possible for the trimethylsilyl (TMS) substituted **16** by utilizing the known place holder quality of TMS [28]. Except for **12**(G-3) and **20**(G-3) which carry third generation dendrons, all other monomers (**9**–**11** and **13**–**19**) carry first or second generation ones. There is one report in the literature according to which a G-4 styrene derivative cannot be polymerized [29]. From the body of material presently available, it seems, that the third generation is a critical one not only for convergent dendronization but also for the macromonomer approach (see Section 2.2).

Table 1 contains the molecular weights of the dendronized polymers. In practically all cases these values were obtained from GPC versus polystyrene (PS) standard. They therefore cannot be considered correct and directly compared with one another. Nevertheless, some general conclusions can be drawn:

(a) Monomers with more complex structures like **10** or those with the dendron bound too tightly to the polymerizable group like **13** do not give high molecular weight materials (for a discussion on this matter, see Ref. 30),
(b) Rh-catalyzed acetylene polymerization of **16** and Ru-catalyzed ROMP of **11** and **17** give the best results along with a few of the radical polymerizations (**14**(G-1), **18**),
(c) Increasing steric congestion around the polymerizable group reduces its reactivity and requires more drastic conditions. This can go as far as to polymerize under Trommsdorff conditions as was done, e.g., with **12**(G-3), a monomer which does not show any propensity to polymerize in solution or at a temperature below 90 °C.
(d) The yields are not always high which means that these polymerizations do not proceed well or that oligomeric products are formed which are removed during work-up. The latter point relativates the molecular weight values given.

In order to estimate whether the true molar masses of the dendronized polymers are larger or smaller than the "GPC molar masses" reported in Table 1, it is useful to recall that the separation in a GPC experiment is based on the hydrodynamic volume (V_n) of macromolecules. For flexible coils, V_n depends on the chain length (i.e. contour length) $L = M \times b \times M_0^{-1}$ according to

$$V_n \sim \langle R^2 \rangle^{3/2} \sim L^{3\nu};\ 1.5 < 3\nu < 1.8$$

Table 1. Dendronized monomers, polymerization conditions, as well as molecular weights of the polymers obtained. Molecular weights are not comparable (see text). (AIBN = azoisobutyronitrile, DBPO = dibenzoyl peroxide, ^{t}BPB = *tert*-butyl perbenzoate)

Monomer	Initiator or catalyst precursor	$M_n \times 10^{-3}$	M_w/M_n	P_n	Yield [%]
9 ($R=OC_{12}H_{25}$)	AIBN, 60 °C benzene	25–69[a]	1.7–2.5	23–58	44–72
9 [$R=O(CH_2)_m(CF_2)_nF$]	AIBN, 60 °C benzene	23	1.2	12	65
10(G-1)	AIBN, 60 °C DMF	23	3.9	43	90
10(G-2)	AIBN, 60 °C DMF	–	–	≤6[b]	53
11	$RuCl_3 \cdot 3H_2O$, EtOH, 95 °C	217	1.9	147	62
12(G-1)	DBPO, 60 °C	25	1.9	56	60
12(G-2)	DBPO, 70 °C	33	3.9	38	93
12(G-3)	^{t}BPB[c], 90 °C	62	6.7	36	70
13(G-1)	AIBN, 60 °C	9	2.6	24	70–80
13(G-2)	AIBN, 60 °C	6	2.3	7	60
14(G-1)	AIBN, 40 °C	270	2.2	544	75
14(G-2)	DBPO, 60 °C	83	2.2	90	90
15(G-2)	^{t}BPB[c], 90 °C	85	3.3	51	90
16(G-2)	$[Rh(C_7H_8)Cl]_2$ NEt_3	960	1.4	975	36
17	$RuCl_2(CHPh)(PCy_3)_2$ CH_2Cl_2, 22 °C	87	1.3	120	84
18[d]	AIBN, 60 °C	187	2.4	84	–
19 + **21**	$Pd(PPh_3)_4$, 2 N NaOH, toluene	18	3.8	18	80–85
20(G-1) + **21**	$Pd(PPh_3)_4$, 2 N NaOH, toluene	19	3.6	29	90
20(G-2) + **21**	$Pd(PPh_3)_4$, 2 N NaOH, toluene	21	2.8	19	89
20(G-3) + **21**	$Pd(PPh_3)_4$, 2 N NaOH, toluene	52	5.3	27	96

[a] Depending on length of ethylene oxide spacer, [b] MALDI-TOF, [c] in bulk, and [d] a related monomer on acrylate basis gives higher molecular weights.

with b and M_0 the length and molar mass, respectively, of a repeat unit, $\langle R^2 \rangle$ the mean square end-to-end distance of the polymer chain and ν the Flory exponent which varies between 0.5 (poor solvent) and 0.6 (good solvent condition). If the unknown polymer and the calibration standard exhibit identical exponents ν the true and the GPC molar masses differ by a factor $(M_{o,u}/M_{o,s})$ with $M_{o,s}$ and $M_{o,u}$ the mass per unit length of standard and unknown polymer, respectively.

Generally subtle deviations from this simple correction are experimentally observed because the exponent ν is influenced by the solvent quality and the flexibility. The dendronized polymers exhibit a much larger mass per unit length than the calibration standard polystyrene and the true molar mass should be much larger than the GPC molar mass, accordingly. However, the dendronized polymers are expected to exhibit a significantly larger chain stiffness than the calibration standard polystyrene, because the steric repulsion of the voluminous dendritic side chains should stretch the polymer backbone considerably. For the limiting case of a rigid rod $V_n^{rod} \sim L^3$, i.e., $\nu \cong 1$. Thus chain stiffness leads to an increased hydrodynamic volume which causes the GPC molar mass to become much smaller than the true molar mass. It is to be noted that for semiflexible chains the exponent ν gradually changes from 0.6 to 1.0 which makes it impossible to quantify this effect for the dendronized polymers as long as the stiffness is not precisely known.

Two fractions of atactic polystyrene dendrimer derived from monomer **12**(G-2) were investigated by GPC and LS under conditions where no aggregation occurs. Whereas GPC gave M_w=82,000 and M_w=600,000, LS furnished

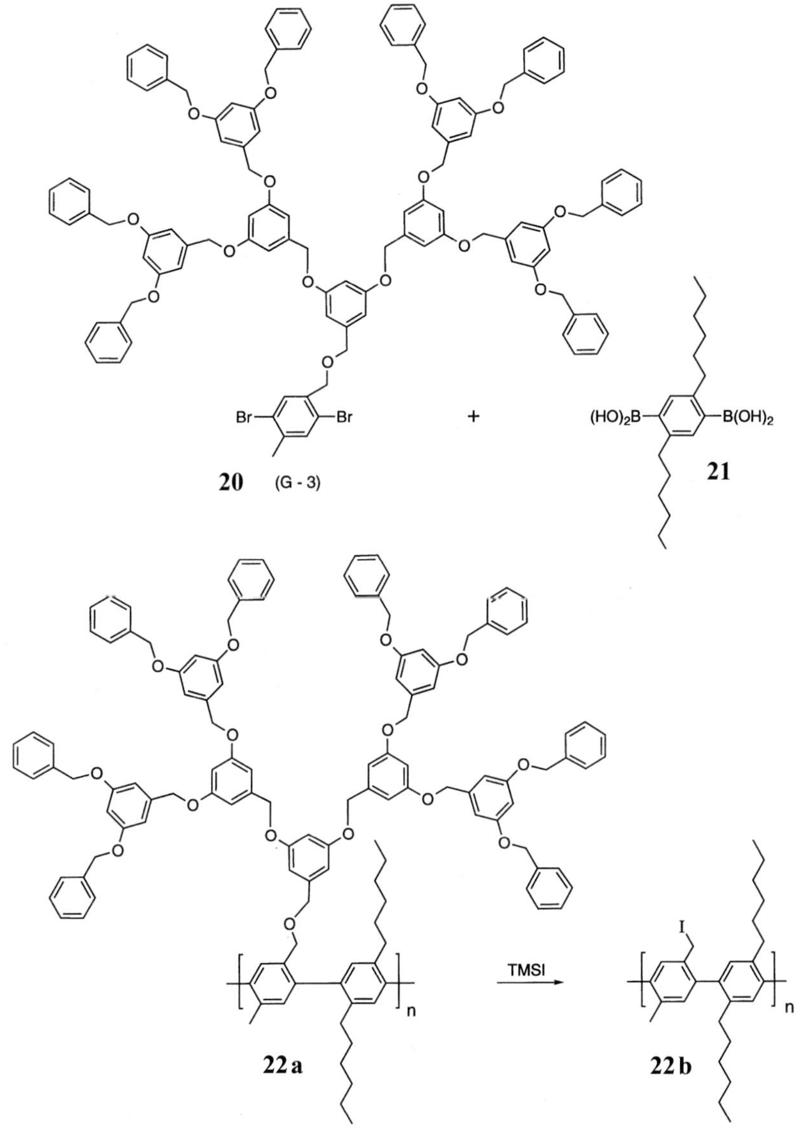

Scheme 6. Synthesis of dendrimer **22a** and its TMSI mediated conversion (dedendronization) to **22b**. **22b** is needed to indirectly determine the molecular weight of **22a**

higher values by a factor of roughly 3 : M_w=213,000 and M_w=2,140,000. A similar experiment was done with a fraction of a dendrimer derived from **12**(G-3). Here GPC gave M_w=101,000 and neutron scattering M_w=176,000 g/mol which amounts to a smaller factor of 1.7.

The following case shows that sometimes one has to go a roundabout way to determine molecular weights of dendronized polymers accurately. According to GPC and LS experiments dendrimer **22a** (Scheme 6) forms aggregates with molecular weights of several millions which cannot be completely deaggregated into the individual constituents regardless of solvent, concentration, and temperature used as well as other factors of sample preparation varied. The DP-value of **22a** could be determined by a chemical trick: De-dendronization of **22a** with a large excess of trimethylsilyl iodide (TMSI), a very potent reagent for ether cleavages [31], afforded polymer **22b** [32]. In contrast to its dendronized precursor and practically all other rod-like polymers, **22b** does not form aggregates and thus was amenable to an in-depth molecular weight determination. Measurements combining state-of-the-art GPC/LS/viscometry gave: M_n=54,000, M_w=139,000, M_w/M_n=2.6. This M_n-value translates into an average P_n=110 of **22b** which, in turn, corresponds to an average length of 90 nm (825 pm/r.u.) for the individual chain. Since the treatment with TMSI should leave the backbone intact, the individual chains of dendrimer **22a** must also have a DP of 110 and, therefore, a contour length of 90 nm. According to Carother's equation [33] this DP results when each coupling step proceeds with a conversion of 99.1 %. From a synthetic point of view this result is truly remarkable. It tells that the Suzuki cross-coupling, despite its considerable mechanistic complexity, is rather insensitive to monomers with high steric demand. Boronic acid function which is relatively prone to undergo protodeboronification remains virtually completely intact. This finding broadens the applicability of the Suzuki polycondensation considerably.

3
Some Properties of Dendronized Polymers

Why are dendronized polymers of interest? How do they differ from the many small core dendrimers known? In the first place it is their unique structural complexity and the enormous degree of control required during synthesis which is a challenge to both synthetic and analytical chemistry. Much is to be learned here for basic research. In the second place they are unique cases to study, e.g., mesophase (LC) behavior, shape, and adsorption on surfaces. Here they complement their small congeners by showing properties which are inherently associated with the polymeric rather than dot-like core. In addition, the question can be raised whether or not they resemble cylindrical nanoobjects. Supposed they can in fact be considered molecular objects, this would open a new path into nanochemistry with all its interesting facets. In the third place, dendronized polymers may in a sense be considered biomimetic compounds and are therefore of interest also under this aspect. The basic structural elements of some of the polymeric dendrimers are reminiscent of naturally occuring macromolecules, like, for example, the Tobacco Mosaic Virus in which proteins coat a nucleic acid attaining a helical conformation much like the way dendrons coat a flexible backbone [34].

The mechanical properties of dendronized polymers depend upon the strength of the interactions between (a) different dendrons of the same molecule, (b) dendrons and backbone of the same molecule, and (c) different dendrimers as a whole. These interactions in turn depend upon factors including the flexiblity of both dendron and backbone, space demand of dendron, distance between anchor groups, hydrophobicity/hydrophilicity of certain parts, and solvent/dendrimer interactions. The dendrimers studied so far differ grossly in regard to these factors and, accordingly, reveal quite different properties. For example, the polyacrylate derived from monomer **9** ($R=OC_{12}H_{25}$) was investigated in regard to its mesophases [5a]. It has a flexible backbone to which flexible and non-compact G-1 dendrons are attached via linkers of variable length. The dendrimer self-assembles into a tubular supramolecular structure exhibiting enantiotropic columnar hexagonal (Θ_h) phases. These phases are characterized by differential scanning calorimetry, wide and small angle X-ray scattering, thermal optical polarized microscopy, and molecular modeling. The structure model proposes the stratum of the column to be formed by the backbone and the linking segments melted and segregated in the center of the column and their melted dendrons radiating towards the column periphery (Figs. 2 and 3). When a second generation dendron is used instead of a first generation one on the same backbone like in the dendrimer derived from **18** an

a

b

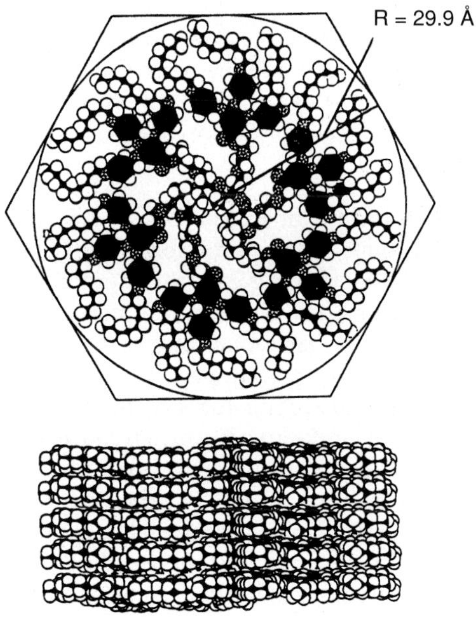

Fig. 2. Schematic representation of the supramolecular cylinders of the dendrimer derived from macromonomer **9** ($R=OC_{12}H_{25}$, $n=3$) in the Qh mesophase: **a** top view of a cylinder containing six repeat units in a stratum with the alkyl tails melted to match the average column radius determined by X-ray scattering experiments; **b** side view of a cylinder containing 30 repeat units of the polymer assembled with melted alkyl tails. Reproduced with permission from references 5a

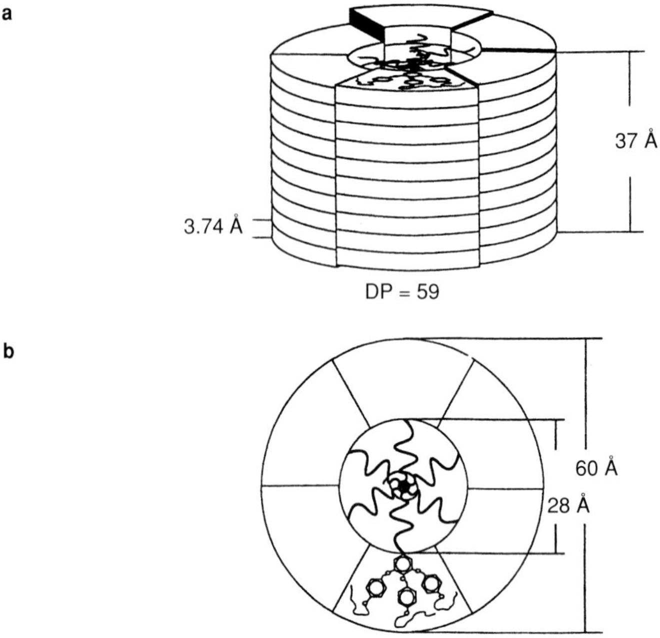

Fig. 3. An idealized representation of the supramolecular cylinder self-assembled from a single chain of the dendrimer derived from monomer **9** (R=$OC_{12}H_{25}$, n=3) in the hexagonal columnar phase, assuming one single backbone per cylinder, drawn to proportion with existing experimental data: **a** tilted side view; **b** top view. Reproduced with permission from references 5a

additionally interesting feature is observed [21]. As the degree of polymerization increases, the dendrimer's shape shifts from a spherical to cylindrical one. This is accompanied by the backbone going from random-coil to extended conformation. This phenomenon provides the unique possibility for molecular engineering of polymer shapes, backbone conformation, and properties. What could show better the interplay and impact of the above-mentioned interactions. Spherical and extended architectures of these dendrimers are basically proved by X-ray diffraction and can also be visualized by a scanning force microscopical (SFM) investigation using bilayers on a mica substrate. The SFM images are resolved to the molecular level. For the extended (higher molecular weight) structures they mostly display parallized but curved features; the molecules do not exhibit two-dimensional crystalline order.

In a complementary approach the backbones of dendrimers are jacketed as tightly as possible with dendrons of the largest possible space demand. Here the dendrons serve as space filling parts to implement sufficient strain to the dendrimers' backbone in order to stretch it to the extreme. A good example for this is dendrimer **22a** with its third generation dendrons [15b]. The molecular dimensions obtained from MD-simulations in vacuo show that (a) it attains an almost stretched conformation, (b) the dendritic layer around the PPP backbone

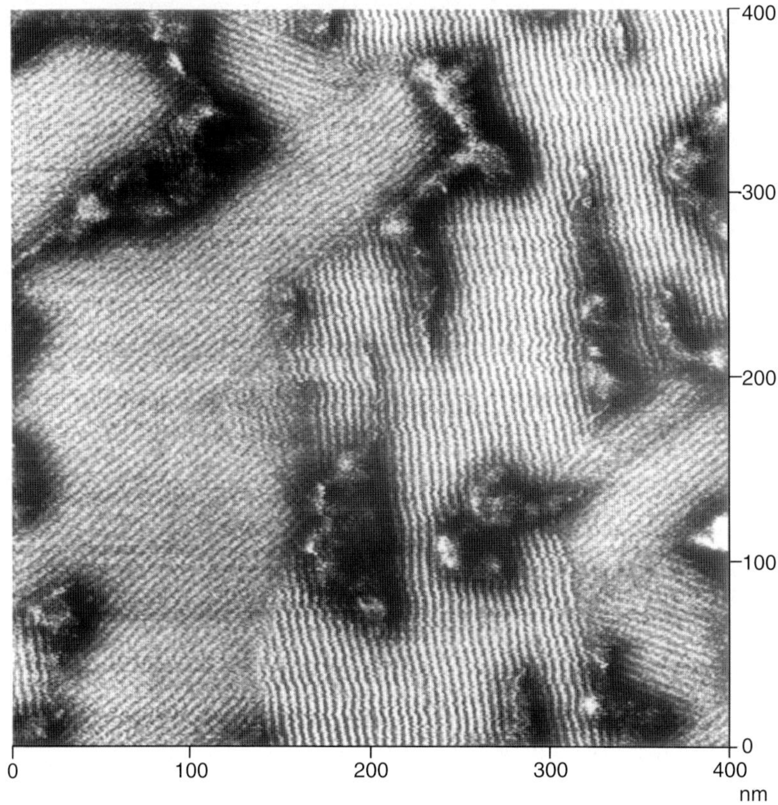

Fig. 4. Phase contrast SFM images of an ultrathin film of dendrimer **22a** on HOPG. The dendrimers are grouped in domains in which they are packed parallel to each other, separated by a periodicity of 4.8 (±0.5) nm. The nature of the dark features is not yet understood

is quite dense but not fully compact and, as a result, (c) the diameter fluctuates between approximately 2 and 4 nm. If the argument with the implemented strain through congestion between adjacent dendrons holds true, one would expect the dendrimer to attain linear, trunk-like shape. SFM of an ultrathin film of **22a** on highly oriented pyrolytic graphite (HOPG) shows a remarkably high degree of order [32] (Fig. 4): Domains are observed whose size varies between 25×25 nm^2 and 100×200 nm^2 and which consist of periodical arrays of linear rows with a periodicity of 4.8 ± 0.5 nm. Since the average length and diameter of a dendrimer is 90 nm (LS) and 2–4 nm (MD), respectively, the SFM features are attributed to the molecular level. The dendrimers are mainly oriented parallel to the surface and are grouped in domains which have different sizes due to the intrinsic polydispersity of **22a**. Within these domains, the dendrimers are packed parallel to each other. A model describes the dendrimers as ellipsoidally flattened "cylinders". Thus, the more dense packing of dendrons about the backbone led the dendrimer to behave on HOPG like a two-dimensional crystal of linear tree-trunks.

Stimulated by this finding it was asked whether steric strain can be increased to the point that the dendrimer stiffens "completely". This would result in loss of all remaining flexibility and the creation of what one could call a molecular object [35]. This object would have a cylindrical shape with defined extensions which (ideally) do not depend on the surrounding medium. Due to its compactness and internal inflexibility the mathematical envelope of such an object can be considered a surface. As mentioned above, **22a** according to MD simulations does not have a compact structure in the sense that all dendrons are squeezed together to van der Waals distance. There are two obvious ways to increase the compactness: Either one uses dendrons with higher generations, which is synthetically not trivial, or the density of the backbones' anchor groups is increased. Scheme 7 gives a rough estimate of the distance between the anchor groups in **22a** and dendrimer **23**, derived from monomer **12**(G-3). In the first case the distance is 8.25 Å, in the latter 2.50 Å.

A molecular model from MD simulation of dendrimer **23** with 50 r.u.'s illustrates the consequences of this reduced distance. The dendrons are packed tightly and a cylindrical envelope is suggested (Fig. 5). The average diameter is $D=4.4(\pm 0.2)$ nm. Molecular scale resolution SFM images of ultrathin films of **23** on HOPG, obtained under dry conditions, show a remarkably high degree of order [36] (Fig. 6). This order is reminiscent of the one observed for **22a** and for Tobacco Mosaic Virus adsorbed on a surface [37]. A packing model assumes a close-packed layer of cylinders (Fig. 7). The dendrimers are cylinders with a defined diameter of approximately 4.8 nm. This model is supported by a small-angle neutron scattering (SANS) investigation of a fraction of dendrimer **23** with $M_n=38{,}000$ ($P_n=22$) and $M_w=101{,}000$ (GPC versus PS) in solution. The scattering curve can only be consistently fitted by assuming the dendrimer exhibits a rod-like structure with cylindrical shape. A contour length $L_z=48.7$ nm and a cylinder diameter of $D=5.1(\pm 0.5)$ nm are obtained. The quality of the scattering data allows one to conclude that the persistence length is on the order of the contour length. Aggregation can be excluded. Thus, the implementation of

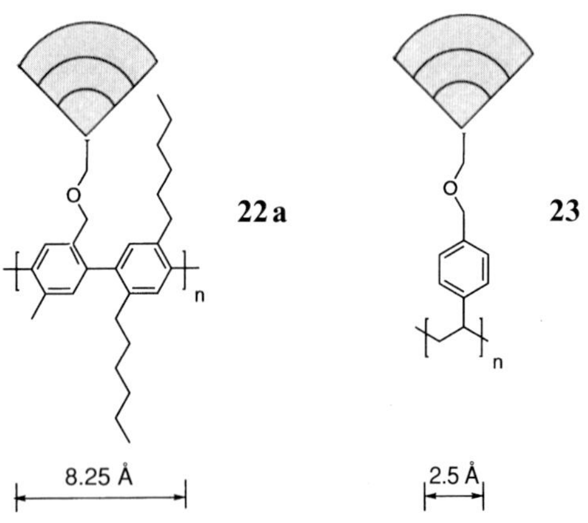

Scheme 7. The lengths of repeat units in dendrimers **22a** (8.25 Å) and **23** (2.5 Å) to qualitatively assess in which of them the average distance between anchor groups is smaller

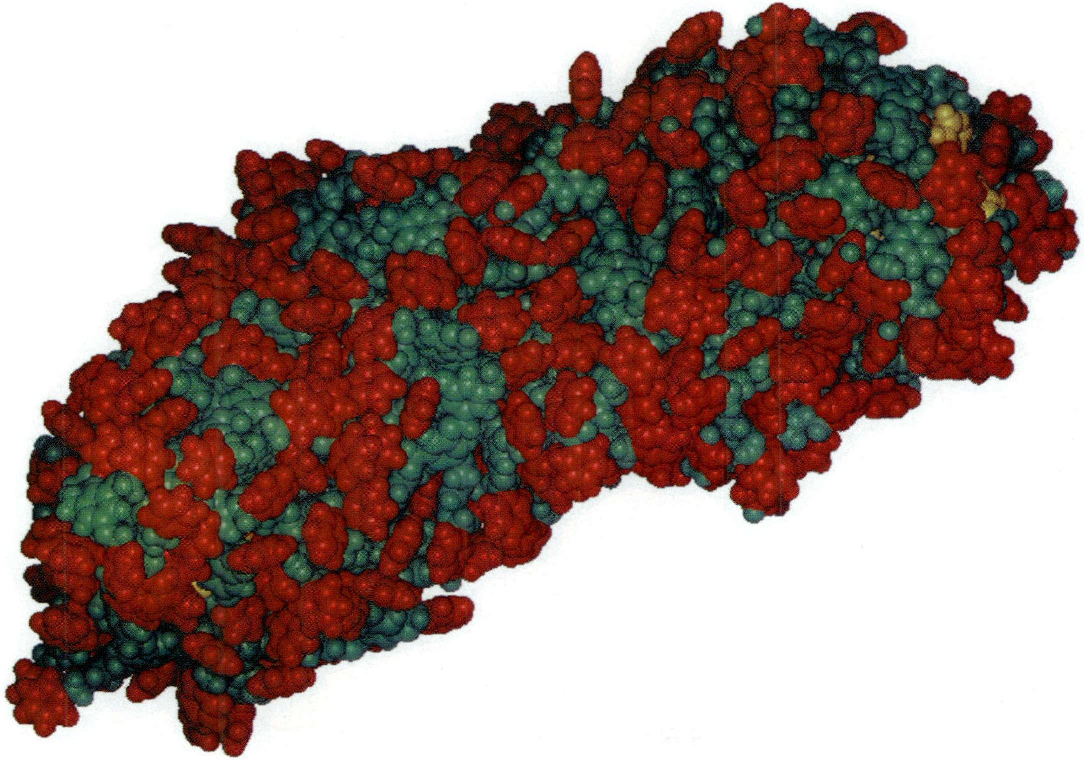

Fig. 5. Molecular dynamics simulations (MD) of dendrimer **23** with 50 repeat units after 300 ps MD. End-to-end distance 9.1 nm, average diameter 4.4 (± 0.2) nm. The backbone atoms are kept in *yellow*, the terminal phenyl rings in *red*, and all other atoms in *green*

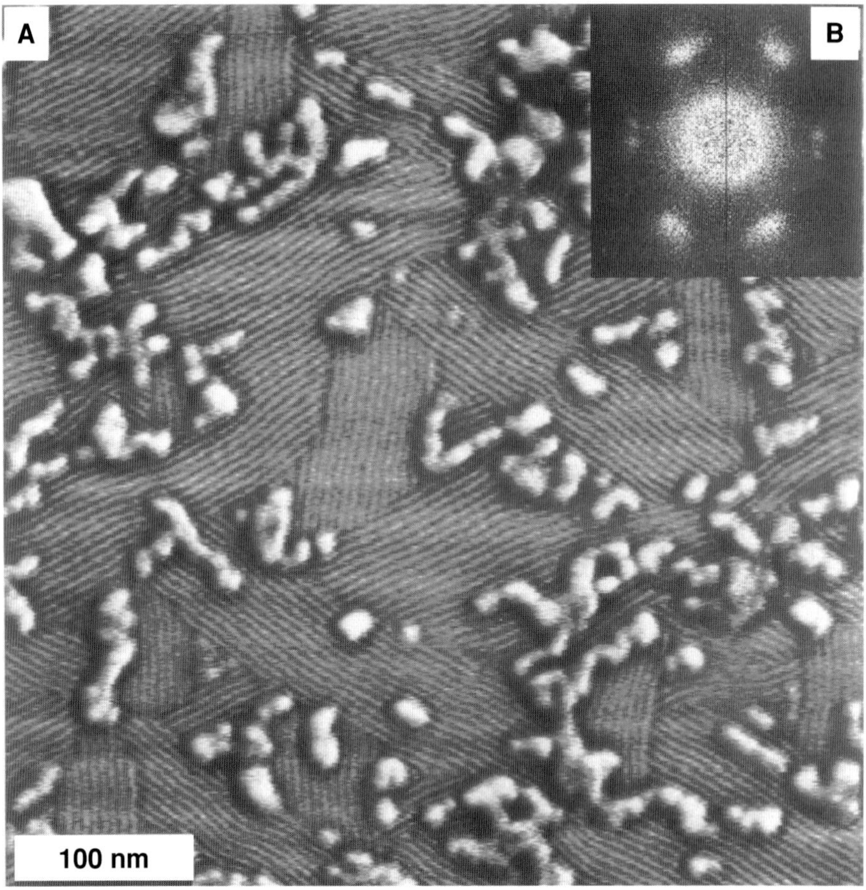

Fig. 6. Scanning force microscopy image of an ultrathin film of dendrimer **23** on HOPG. The cylindrical dendrimers are grouped in domains in which they are packed parallel to each other, with a periodicity of 5.0 (±0.5) nm. Individual chain ends can be detected at grain boundaries. The nature of the unordered features is not yet understood

strain has proven a powerful tool for creating a well-defined molecular object whose shape and dimensions remain practically unchanged regardless of whether this object self-aggregates on a surface or is freely floating in solution.

4 Related Polymers: A Short Comment

Dendronized polymers are a subclass of comb polymers [7]. The synthetic access to comb polymers with dendrons in the side chain is somewhat more difficult and time consuming compared to most other polymers of this class. It is therefore not too far fetched to ask whether the exciting properties described above

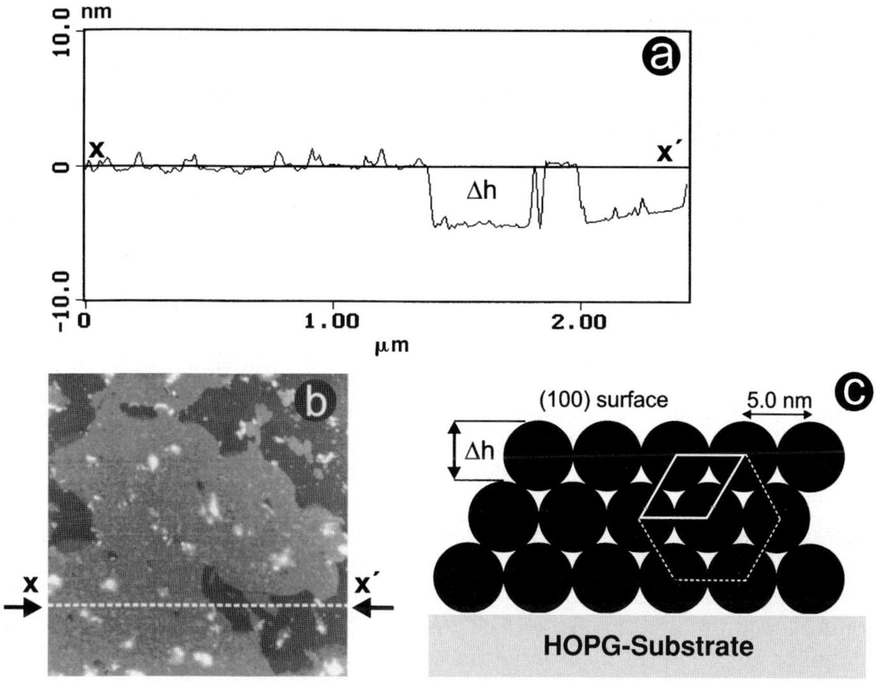

Fig. 7. **a** Cross sectional profile of an ultrathin film of dendrimer **23** on HOPG along x–x′ as indicated in (**b**). The height difference between adjacent terraces has the dimension of a monolayer [$\Delta h = 4.3$ (± 0.2) nm]. Large scale (2.4×2.4 μm^2) SFM image of monomolecular terraces. **c** Schematic model of closely packed molecular cylinders in thin films of **23** on HOPG

and the ones still awaiting to be discovered can also be achieved with other comb polymers. Even though it is certainly too early to arrive at a conclusive answer here, it is worthwhile to take a short look beyond the dendrimer fence and to watch out for related developments with non-dendritic comb polymers.

Comb polymers are generally synthesized by polymerizing a macromonomer resembling what later will become the comb's tooth and a polymerizable group at its terminus. Due to serious synthetic difficulties, the length of the comb polymers obtained seldomly surpassed the tooth's. Only a few years ago this problem was overcome when methacryloyl terminated oligostyrene macromonomers were polymerized to a high molecular weight poly(methacrylate) ($P_n = 1000$) by applying more concentrated solutions than usually used [38]. According to LS this polymer in solution is very rigid with a Kuhn length of 10–20 nm. Its rigidity is like in the case of dendritic side chains attributed to steric repulsion between side chains of considerable molecular weight which are attached in short distances (2.5 Å) to the polymethacrylate backbone. An additional contribution to the rigidity also comes from solvent/side chain interactions. In a good solvent, this kind of comb polymers which are referred to as cylindrical brushes act much like a sponge. Solvent

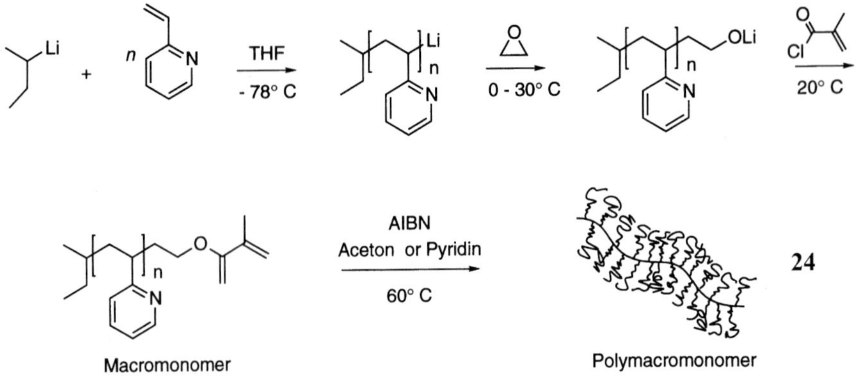

24

Scheme 8. Synthesis of a poly(vinylpyridin)-poly(macromonomer). Reproduced with the permission of Wiley VCH from reference 34

molecules are sucked up leading to a de facto increase of the size of the side chains. It should be mentioned here that this matter is still under debate [39]. In an extension of this work, poly(macromonomer) **24** was synthesized via radical techniques [40] (Scheme 8). According to a SFM visualization this polymer gives rise to disordered features on the molecular level which are approximately 10 nm broad and 2 nm high. The side chains form a soft shell around the backbone whose shape is influenced, e.g., by adhesion forces. The side chains of adjacent polymers do not interdigitate. In the uncharged state (pyridin nitrogen not quaternized), polymer **24** adsorbed on mica shows backfoldings. This is not necessarily an indication of a very low inherent rigidity of the polymer though it clearly tells that the brushes are not to be compared with steel rods or tree trunks. The backfoldings may be caused through the process of film formation whose mechanism is not yet understood. In solution backfoldings are not observed. In regard to its intrinsic (structure inherent) rigidity, polymer **24** resembles to some extent the dendronized polymers with the open, non-compact structures (dendrimers derived from **9** or **18**). If the pyridin nitrogens are quaternized, however, the incorporation of charge associated with this results in a polyelectrolyte which, charge-induced, stretches out as far as possible. Soft shell cylindrical brushes adsorbed on mica are not always deformed by adhesion but may retain shape persistent [41].

The polymers described here are thus similar to the ones described in Sects. 2 and 3. They do however differ in one aspect. Whereas the polymeric dendrimers are monodispers in regard to diameter due to their convergent synthesis, the bottle brushes are polydispers: The teeth have a molecular weight distribution though this can be kept relatively narrow and finite. Another difference relates to the density gradient. On going from the backbone into the outer spheres this gradient decreases faster for the bottle brushes as for the dendrimers because of the increasing degree of branching of the latter. Finally it should be pointed out that the concept of "increased rigidity by implementation of steric strain" is not new. In poly(phthalocyaninato siloxane)s, for example, the phthalocyanine discs are packed so close together that a truly rigid polymer results [42].

5 Outlook

What will be the future directions of this field? One very obvious and important aspect is whether G-3 dendrons are really the maximum. Despite the negative observation in the literature [29] there are massive efforts been presently undertaken to polymerize G-4 macromonomers. Another important issue is to try to incorporate functional groups at the dendrons termini which in the ideal case should furnish functionalized dendrimers with a defined number of functional groups on the surface. These groups would allow one to engineer properties and, thus, further the width of potential applications considerably. A first step in this direction has been done [20]. Additionally it would be very important and synthetically challenging to not only control the diameter of the dendrimers in the mesophase, adsorbed on surfaces, or in solution but also their length. All reported polymerization techniques, though mechanistically quite different, give materials with broad molecular weight distributions. "Living" techniques which are known to give much narrower distributions need to be tried out, regardless of the fact that this will not be easy for a number of reasons. A successful application of these techniques would allow to construct amphiphilic "blockdendrimers" (Scheme 9), intruiging macromolecules of unheard com-

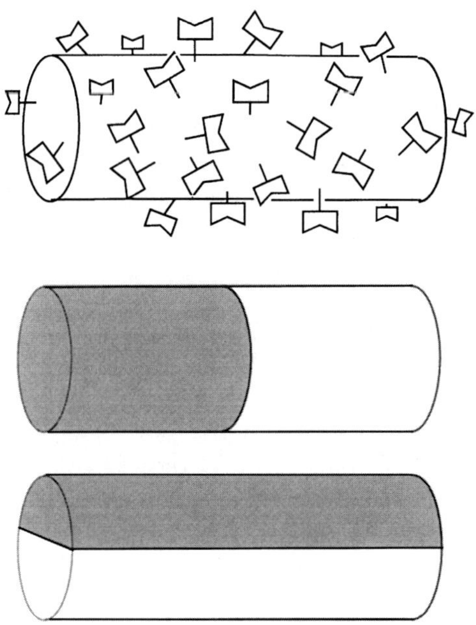

Scheme 9. Cartoons of some dream structures of dendronized polymers and cylindrical objects

plexity. Blockdendrimers could be considered as potential giganto constitutents of new generations of micelles, vesicles, and membranes and would open an exciting avenue into the study of their self-assembly behavior. Interesting future projects lie also in the area of optoelectronical properties. Specifically intriguing would be mechanically stiff cylindrical objects with an electrically active backbone. Due to their rigidity they could be easily moved around on surfaces, e.g., by an SFM tip, and placed between nanoelectrodes. Because of their insulating coat they would be ideal models for the long sought single-stranded electrically conducting wire to study charge transport in nanodevices.

Acknowledgement. I would like to cordially thank the following coworkers and colleagues for a stimulating and fruitful cooperation from which I learned a lot: I. Neubert, B. Karakaya, R. Klopsch, Prof. M. Antonietti (MPIKG, Teltow), Dr. S. Förster (MPIKG, Teltow), Prof. J. P. Rabe (HU, Berlin), Dr. B. Schürmann (HU, Berlin), and last but not least Dr. W. Stocker (HU, Berlin). Helpful discussions with S. Koch and Dr. A. Ingerl are gratefully acknowledged. I also wish to thank Prof. D. A. Tomalia, MMI, Midland, Michigan, for making results available prior to publication (ref. 10) and Prof. M. Schmidt, University of Mainz, for enlightening discussions on how to interpret GPC results of dendronized polymers. This work was supported by the Deutsche Forschungsgemeinschaft (Schl 4–1) and the Fonds der Chemischen Industrie which is gratefully acknowledged.

6
References

1. Newkome GR, Moorefield CN, Vögtle F (1996) Dendritic Molecules – Concepts, Syntheses, Perspectives. VCH, Weinheim
2. Jansen JFGA, de Brabander-van den Berg EMM, Meijer EW (1994) Science 266:1226; Jansen JFGA, Meijer EW, de Brabander-van den Berg EMM (1995) J Am Chem Soc 117:4417
3. Tomalia DA, Baker H, Dewald JR, Hall M, Kallos G, Martin S, Roeck J, Ryder J, Smith P (1985) Polym J (Tokyo) 17:117; Tomalia DA, Baker H, Dewald JR, Hall M, Kallos G, Martin S, Roeck J, Ryder J, Smith P (1986) Macromolecules 19:2466; de Brabander-van den Berg EMM, Meijer EW (1993) Angew Chem Int Ed Engl 32:1308; Wörner C, Mülhaupt R (1993) Angew Chem Int Ed Engl 32:1306
4. Tomalia DA, Kirchhoff PM (1987) US patent 4,694,064
5. (a) Percec V, Heck, J, Tomazos D, Falkenberg F, Blackwell H, Ungar G (1993) J Chem Soc Perk Trans 2799; (b) Schlüter AD (1993) The Polymer Conference. Cambridge, United Kingdom; (c) Freudenberger R, Claussen W, Schlüter AD, Wallmeier H (1994) Polymer 35:4496
6. (a) Gitsov I, Wooley KL, Fréchet JMJ (1992) Angew Chem Int Ed Engl 31:1200; Gitsov I, Fréchet JMJ (1993) Macromolecules 27:7309; (b) Hawker CJ, Fréchet JMJ (1992) Polymer 33:1507; (c) Kawaguchi T, Moore JS (1994) Polym Preprints 35(2):872
7. Elias HG (1997) An Introduction to Polymer Science. VCH, Weinheim, pg 20
8. Yan WW, Kirkland JJ, bly DD (1979) Modern Size-Exclusion Chromatography. Wiley, New York
9. Buhleier E, Wehner W, Vögtle F (1978) Synthesis 155
10. Yin R, Zhu Y, Tomalia DA (1998) J Am Chem Soc, in press
11. Tomalia DA, Naylor AM, Goddard III WA (1990) Angew Chem Int Ed Engl 29:113
12. Fréchet JMJ, Wooley KL, Hawker CJ (1991) J Am Chem Soc 113:4252
13. Schlüter AD, Bothe H, Gosau JM (1991) Makromol Chem 192:2497
14. Schlüter AD (1997) Synthesis of Poly(*para*-phenylene)s. In: Skotheim T, Elsenbaumer R, Reynolds J (eds) Hanbook of Conducting Polymers. Dekker, New York
15. (a) Karakaya B, Claussen W, Schäfer A, Lehmann A, Schlüter AD (1996) Acta Polym 47:79; (b) Karakaya B, Claussen W, Gessler K, Saenger W, Schlüter AD (1997) J Am Chem Soc 119:3296

16. The authors refer to compound **9** as a monomer with tapered side chains and have only recently begun to use dendrimer terminology. Since the side chain meets the criteria for a G-1 dendron, compound **9** is considered here.
17. Draheim G, Ritter H (1995) Macromol Chem Phys 196:2211
18. Neubert I, Amoulong-Kirstein E, Schlüter AD (1996) Macromol Rapid Commun 17:517
19. Chen YM, Chen CF, Liu WH, Xi F (1996) Macromol Rapid Commun 17:401
20. Neubert I, Klopsch R, Claussen W, Schlüter AD (1996) Acta Polym 47:455
21. Percec V, Ahn CH, Ungar G, Yeardley DJP, Möller M, Sheiko SS (1998) Nature 391:161
22. Ivin KJ (1983) Olefin Metathesis. Academic Press, London
23. Percec V, Schlueter D, Ronda JC, Johansson G, Ungar G, Zhou JP (1996) Macromolecules 29:1464
24. Percec V, Schlueter D (1997) Macromolecules 30:5783
25. Kaneko T, Horie T, Asano M, Aoki T, Oikawa E (1997) Macromolecules 30:3118
26. Schlüter AD, Wegner G (1993) Acta Polymer 44:59
27. Claussen W, Schulte N, Schlüter AD (1995) Macromol Rapid Commun 16:89
28. Hensel V, Schlüter AD (1997) Liebigs Ann/Recueil 303
29. Fréchet JMJ, Gitsov I (1995) Macromol Symp 98:441
30. Percec V, Ahn CH, Barboiu B (1997) J Am Chem Soc 119:12978
31. Olah G, Narang SC, Gupta BGB, Malhotra R (1979) J Org Chem 44:1247
32. Stocker W, Karakaya B, Schürmann BL, Rabe JP, Schlüter AD (1998) J Am Chem Soc, in press
33. Cowie JMG (1973) Polymers: Chemistry & Physics of Modern Materials. International Textbook Comp Ltd, Bucks
34. Klug A (1983) Angew Chem Int Ed Engl 22:565
35. Stupp SI, Son S, Li LS, Lin HC, Keser M (1995) J Am Chem Soc 117:5212
36. Stocker W, Schürmann BL, Rabe JP, Förster S, Lindner P, Neubert I, Schlüter AD (1998) Adv Mater, in press
37. Maeda H (1997) Langmuir 13:4150
38. Tsukahara Y, Tsutsumi K, Yamashita Y, Shimada S (1990) Macromolecules 23:5201; Wintermantel M, Gerle M, Fischer K, Schmidt M, Wataoka I, Urakawa H, Kajiwara K, Tsukahara Y (1996) Macromolecules 29:978
39. Wintermantel M, Fischer K, Gerle M, Ries R, Schmidt M, Kajiwara K, Urakawa H, Watoaka I (1995) Angew Chem Int Ed Engl 34:1472
40. Dziezok P, Sheiko SS, Fischer K, Schmidt M, Möller M (1997) Angew Chem Int Ed Engl 36:2894
41. Sheiko SS, Gerle M, Fischer K, Schmidt M, Möller M (1997) Langmuir 13:5368
42. Caseri W, Sauer T, Wegner G (1988) Makromol Chem Rapid Commun 9:651

Electrochemical and Photochemical Properties of Metal-Containing Dendrimers

Margherita Venturi[1] · Scolastica Serroni[2] · Alberto Juris[1] · Sebastiano Campagna[2] · Vincenzo Balzani[1,*]

[1] Dipartimento di Chimica "G. Ciamician", Università di Bologna – I-40126 Bologna, Italy, *E-mail: vbalzani@ciam.unibo.it*

[2] Dipartimento di Chimica Inorganica, Chimica Analitica e Chimica Fisica, Università di Messina – I-98166 Villaggio S. Agata, Messina, Italy

Metal complexes are characterized by a precise molecular geometry related to the characteristic coordination number of the metal ion and also, in some cases, to the rigid structure of the ligands. Furthermore, they can exhibit valuable properties such as absorption of visible light, luminescence, and reduction and oxidation levels at accessible potentials. By using metal complexes to construct a dendrimer it is therefore possible to incorporate in the dendritic structure many "pieces of information". In this paper the available results on the electrochemical and photochemical properties of metal-containing dendrimers are reviewed. It is shown that by a suitable choice of the metal-based building blocks, it is possible to control the number of exchanged electrons at a fixed potential and the pattern of migration of electronic energy. These properties can be exploited for multielectron catalysis and light harvesting.

Keywords. Metal-based dendrimers, electrochemistry, multielectron processes, luminescence, light harvesting.

* Corresponding author.

Topics in Current Chemistry, Vol. 197

List of Symbols and Abbreviations

biq	2,2′-biquinoline
bpm	2,2′-bipyrimidine
2,3-dpp	2,3-bis(2-pyridyl)pyrazine
HAT	1,4,5,8,9,12-hexaazatriphenylene
2,3-Medpp$^+$	2-[2-(1-methylpyridiumyl)]-3-(2-pyridyl)pyrazine
phen	1,10-phenanthroline
ppy	2-phenylpyridine
QP	2,2′:3′,2″:6″,2‴-quaterpyridine
tpphz	tetrapyrido[3,2-a:2′,3′-c:3″,2″-h:2″,3″j]phenazine
tpy	2,2′:6′,2″-terpyridine
β-CD	β-cyclodextrin

1 Introduction

Chemical synthesis has long been considered a realm of organic chemistry. In the last decade, however, outstanding progress has been made in the field of the synthesis of metal complexes, and it can now be stated that as far as the preparation of new (non-natural) compounds is concerned, metal coordination chemistry competes in imagination and achievement with organic chemistry. This is also the case for dendrimers. This new class of well-defined macromolecules exhibiting tree-like structure, first developed by organic chemists [1], has considerably extended its aesthetic appeal and range of potential applications in the hands of coordination chemists [2].

Metal complexes are characterized by a precise molecular geometry related to the characteristic coordination number of the metal ion and also, in some cases, to the rigid structure of the ligands. Furthermore, they can exhibit valuable properties such as absorption of visible light, luminescence, and reduction and oxidation levels at accessible potentials. By using metal complexes to construct a dendrimer it is therefore possible to incorporate in the dendritic structure many "pieces of information" that, when placed in suitable sites of the array, can be used to perform valuable functions such as light harvesting and catalysis. The aim of this chapter is to illustrate the most significant properties of photo- and electroactive metal-containing dendrimers.

From a structural viewpoint, most of the metal-containing dendrimers can be classified according to the following four categories:

1) *Dendrimers built around a metal complex as a core.* These compounds can be considered metal complexes of ligands carrying dendritic substituents (Fig. 1a). The most commonly used metal complex cores are porphyrin complexes, polypyridine complexes, and ferrocene-type compounds.
2) *Dendrimers containing metal complexes as peripheral units.* These dendrimers (Fig. 1b) belong to the class of dendrimers functionalized on the surface. Dendrimers coated with up to 48 $Ru(Cp)(CO)_2R$ [3], 64 $Fe(Cp)_2$ [4], and 3072 AuCl [5] units have been reported.

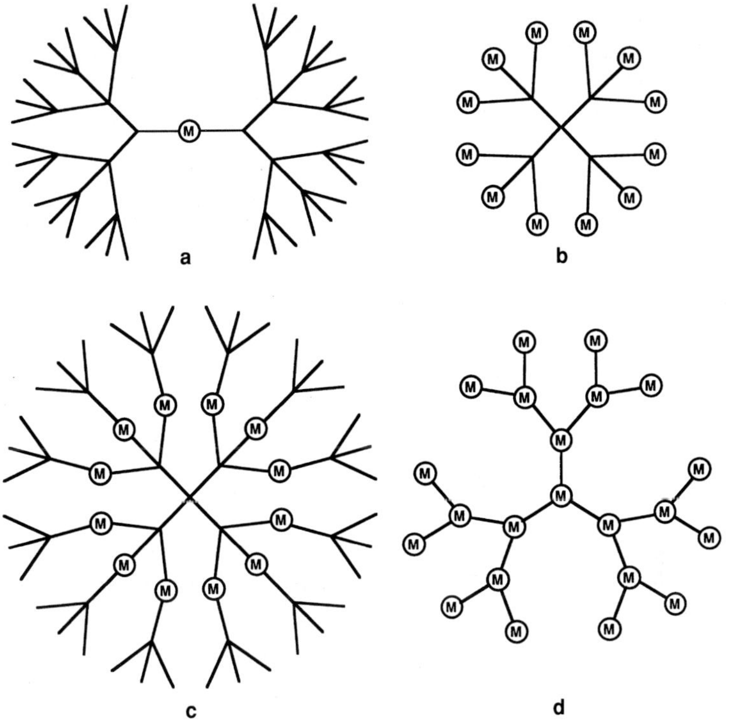

Fig. 1a–d. Different kinds of metal-containing dendrimers: **a** a metal complex as a core; **b** metal complexes only as peripheral units; **c** metal complexes only in the branches; **d** metals as branching centers

3) *Dendrimers containing metal complexes in the branches.* In these compounds (Fig. 1c), metal complexes may play the role of connectors along the branches of a dendritic structure as in the case of $(tpy)Ru(tpy)^{2+}$ (tpy=2,2′:6′,2″-terpyridine) [6], or may be attached to specific sites as in the case of cobalt carbonyls [7].
4) *Dendrimers based on metals as branching centers.* These compounds represent the extension of the old concept of polynuclear complex [8]. They are based on ligands capable to coordinate, and therefore to bridge, two (or more) metal ions (Fig. 1d) [2b, e]. Besides the metal ions and the bridging ligands, they contain normal (i.e., non bridging) ligands at the periphery of the structure. Compounds made of 22 Ru polypyridine moieties [9], and 28 Pt [10] and 47 Pd [11] cyclometalated units have been synthesized.

There are, of course, metal-containing dendrimers that belong to more than one of the above-mentioned categories. Examples are the heptametallic dendrimer made of a central $Fe(Cp)(C_6Me_6)^+$ core and coated with 6 ferrocene moieties [12], and the heterometallic dendrimers made of an organic core, containing up to 6 Pt(IV)-based organometallic species in the branches, and coated with up to 12 ferrocene units [13].

2
Electrochemical Properties

2.1
Introduction

The electroactive units in the dendrimers that we are going to discuss are the metal-based moieties. An important requirement for any kind of application is the chemical redox reversibility of such moieties. The most common metal complexes able to exhibit a chemically reversible redox behavior are ferrocene and its derivatives and the iron, ruthenium and osmium complexes of polypyridine ligands. Therefore it is not surprising that most of the investigated dendrimers contain such metal-based moieties. In the electrochemical window accessible in the usual solvents (around +2/–2 V) ferrocene-type complexes undergo only one redox process, whereas iron, ruthenium and osmium polypyridine complexes undergo a metal-based oxidation process and at least three ligand-based reduction processes.

When the only metal complex of a dendrimer is that constituting the core of the structure (Fig. 1a), the most interesting problem is whether and, if so, how much the electrochemical properties (potential value, kinetic reversibility) of the metal-based core are modified by the surrounding branches.

When the metal complexes constitute the peripheral units (Fig. 1b) and/or belong to the branches (Fig. 1c) of a dendrimer, a number of equivalent metal-based centers are present since dendrimers are usually highly symmetric species by their own nature. The metal-based centers may or may not interact, depending on distance and nature of the connector units. Multielectron redox processes can therefore be observed, whose specific patterns are related to the degree of interaction among the various units.

In dendrimers based on metals as branching centers (Fig. 1d), the electrochemical behavior is even more complex since (i) each unit of the dendrimer is electroactive, (ii) the chemical nature of the metal-based units constituting the dendrimer may be different, (iii) chemically equivalent units can be different from the topological viewpoint, and (iv) the degree of interaction among the moieties depends on their chemical nature and distance.

For space reasons, we will deal mainly with the electrochemical behavior of large dendritic compounds. Therefore, the electrochemical properties of a number of borderline compounds [14] between metal complexes and dendrimers have not been included in this review.

2.2
Dendrimers Containing Ferrocene-Type Units

The electrochemical behavior of ferrocene-based polymers [15] and of complexes containing a large, well defined number of ferrocene-type units [16] had already been reported when several groups became interested in dendrimer research. In the past few years several dendrimers of different chemical nature and structure carrying ferrocene-type units in the periphery have been synthesized and their electrochemical behavior has been reviewed [2d]. In principle, these compounds can be used as (i) systems for homogeneous multielectron catalysis [17], (ii) sensors for anions [2a,c,d], and (iii) materials to modify electrode surfaces [1d, 2a,c,d].

In **1** [18a], the nine peripheral $Fe(Cp)(MeC_6H_5)^+$ moieties are reduced simultaneously, indicating that such units are equivalent and independent. Since this reduction process is both chemically and electrochemically reversible, **1** can be considered as a "reservoir" of nine electrons [18]. In **2** [19], **3** [20], and **4** [21] only one oxidation process is observed, with a number of exchanged electrons equal

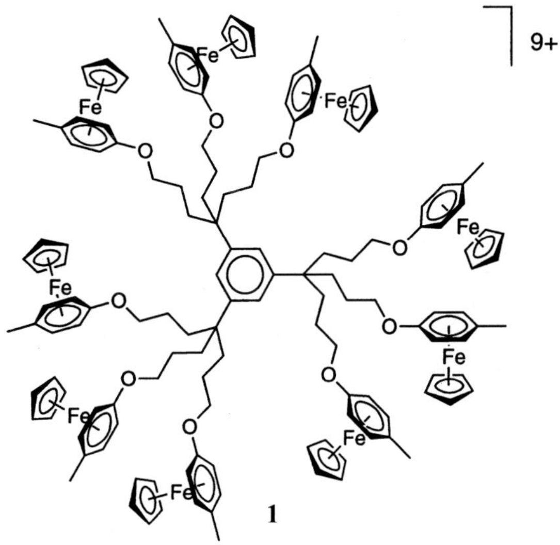

1

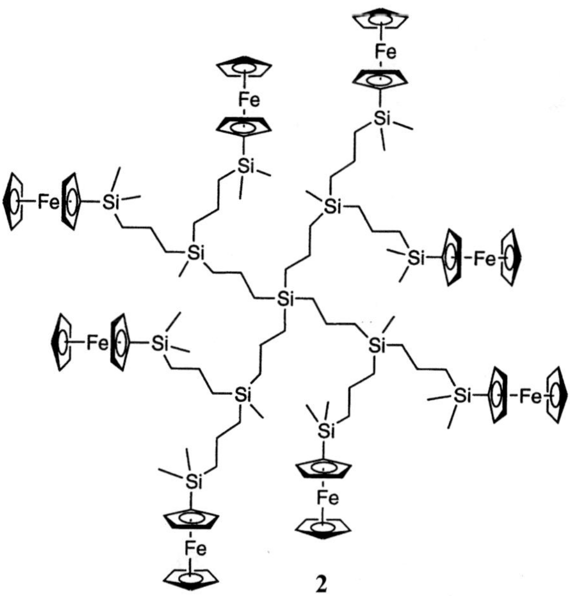

2

to the number of ferrocene units contained in the periphery of the dendrimer. This result indicates that the ferrocene units behave independently from one another. For the dendrimers of the aryl ether-based structure like **4** it has also been demonstrated that the $E_{1/2}$ value for the ferrocene redox couple remains essentially constant in going from the first to the second and third (**4**) generation [21]. For **5** and its first and second generation analogs [22], two well separated oxidation processes of equal intensity can be observed, each one corresponding to half the ferrocene units contained in the dendrimer. This behavior is consistent with the existence of significant interaction between the two ferrocene units linked to the same Si atom.

Compound **6** contains seven iron-based units [12], of which the six peripheral ones are chemically and topologically equivalent, whereas that constituting the core ($(Fe(Cp)(C_6Me_6)^+)$) has a different chemical nature. Accordingly, two redox processes are observed, i.e., oxidation of the peripheral ferrocene moieties and reduction of the core, whose cyclic voltammetric waves have current intensities in the 6:1 ratio. Clearly, the one-electron process of the core is a convenient internal standard to calibrate the number of electron exchanged in the multi-electron process. In the absence of an internal standard, the number of exchanged electrons has to be obtained by coulometry measurements, or by comparison with the intensity of the wave of an external standard after correction for the different diffusion coefficients [15].

Decafunctionalization of ferrocene with allyl units has been used to construct a dendrimer (**7**) having a ferrocene-type core and ten ferrocene-type

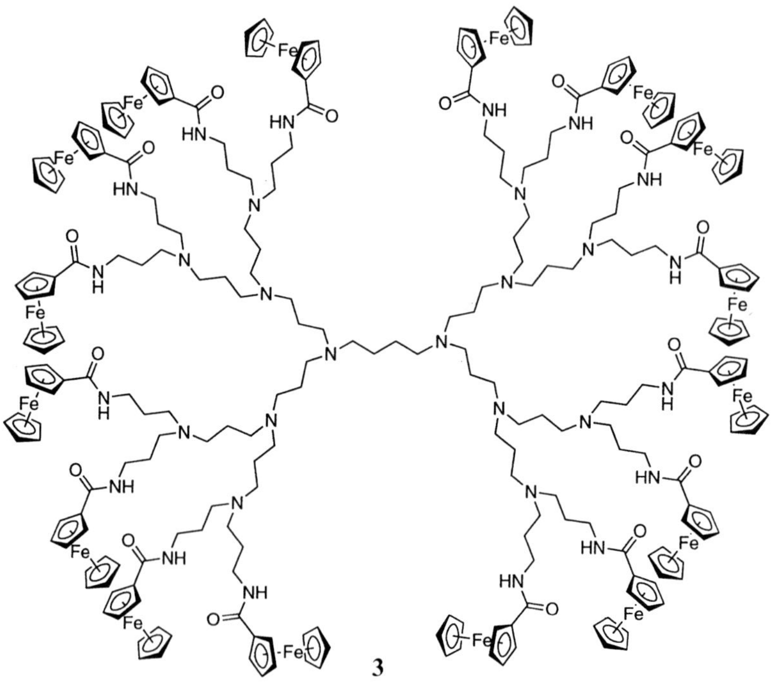

3

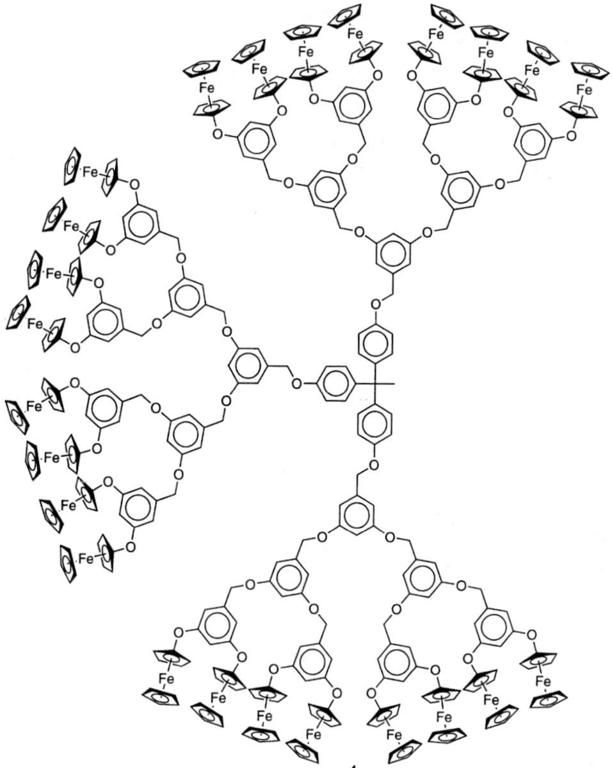

4

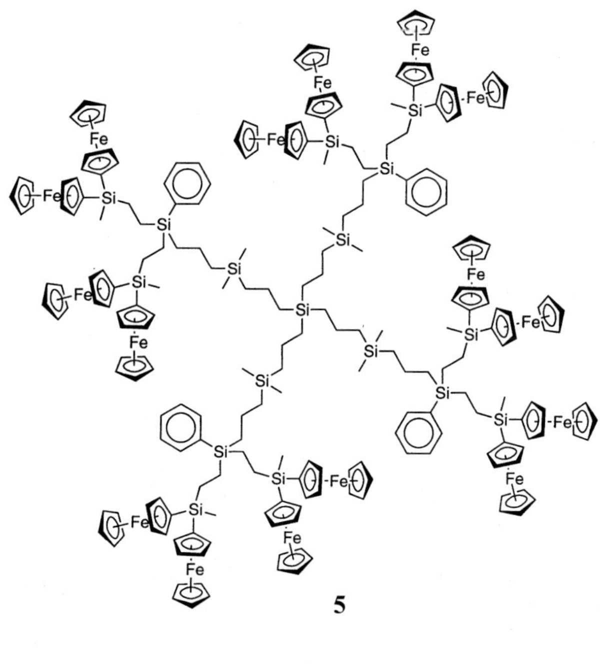

5

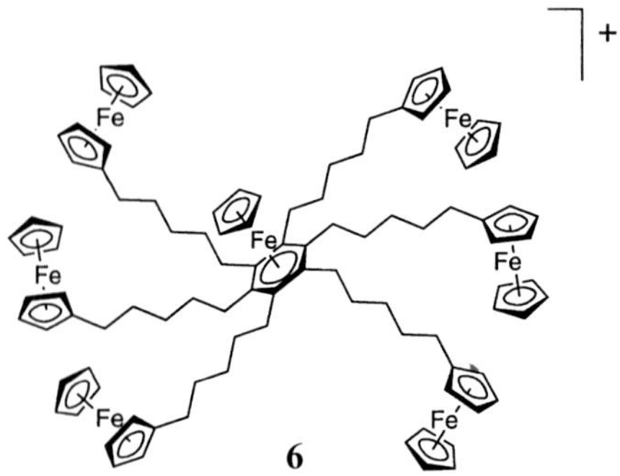

6

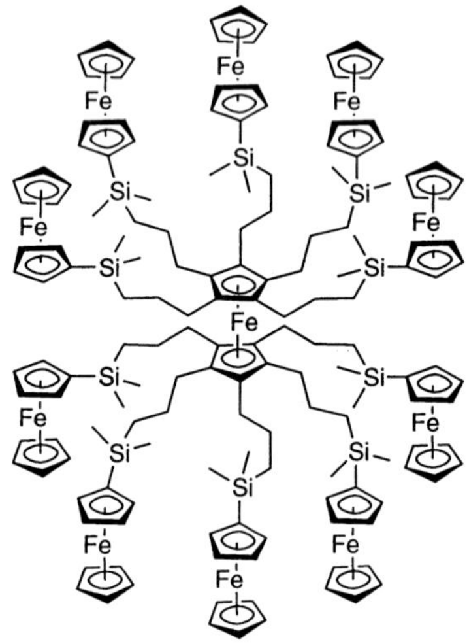

7

peripheral moieties [23]. Cyclic voltammetry experiments have shown that the peripheral ferrocene-type moieties are oxidized at the same potential. At a slightly negative potential value, a much weaker cyclic voltammetric wave, assigned to the core unit, is also observed.

Mixed-metal dendrimers containing up to 6 Pt(IV)-based organometallic species in the branches and 12 peripheral ferrocene units (**8**) have recently been synthesized and their electrochemical behavior investigated [13]. As in the previously discussed examples, multi-electron reversible oxidation processes, assigned to the equivalent, non-interacting ferrocene units, have been observed. The authors point out that cyclic voltammetry is a powerful tool to support the structure of the dendrimers containing ferrocene units.

The electrochemical behavior of dendrimers based on the $Fe(Cp)(C_6Me_6)^+$ core with attached six branches bearing in the periphery $Ru(tpy)_2^{2+}$ or $Ru(bpy)_3^{2+}$ units [24] will be discussed in the next section.

The electrochemical properties of redox-active moieties of a dendrimer can be influenced by other species added to the solution. This may open the way to

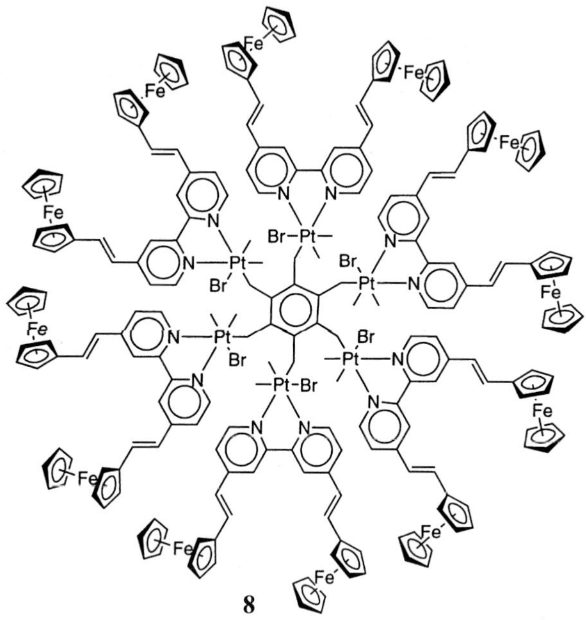

8

molecular recognition and to its interesting applications. Compound **3**, as well as its lower generation analogs, associate with β-CDs (β-CD=β-cyclodextrin) to give compounds in which the ferrocene peripheral units are enclosed in the cyclodextrin cavity [20]. Figure 2 shows the adduct **9a** obtained by capping the second generation dendrimer **9**. Such an inclusion process increases the solubility of the dendrimer in water and displaces the oxidation process of the ferrocene units towards more positive potentials. In the first and second generation dendrimers only one positively shifted voltammetric wave is observed, indicating that all the peripheral ferrocene units can be included in the β-CD hosts. For the third generation dendrimer **3** two waves are observed, one corresponding to enclosed and one to free ferrocene units. This behavior shows that steric congestion prevents inclusion of some of the 16 ferrocene units. From a conceptual viewpoint, such dendrimers play the role of tri-dimensional templates to organize supramolecular structures.

Compounds containing several ferrocene units linked to an organic core can play the role of anion sensors. In **10**, which contains 18 ferrocene peripheral units [25], only one voltammetric wave is observed, indicating that all the units are independent and equivalent. Addition of $H_2PO_4^-$ causes the formation of another wave at less positive potentials and the disappearance of the initial wave. Other anions (HSO_4^-, Cl^-, and NO_3^-) only displace the initial wave to less positive potentials. Such effects decrease in passing from **10** to similar compounds containing a smaller number (9 and 3) of ferrocene units. This shows that there is a "dendritic effect", defined as the ability for a dendrimer to achieve a better sensing and recognition of anions, related to a synergic effect between the interaction of anion with oxidized ferrocene and amide hydrogen [26].

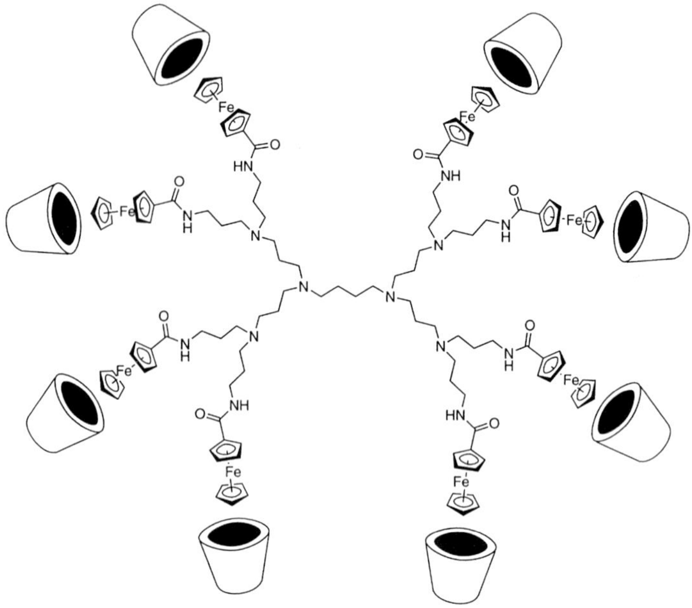

Fig. 2. A supramolecular array (**9a**) obtained enclosing the eight ferrocene units of dendrimer **9** into β-CD molecules

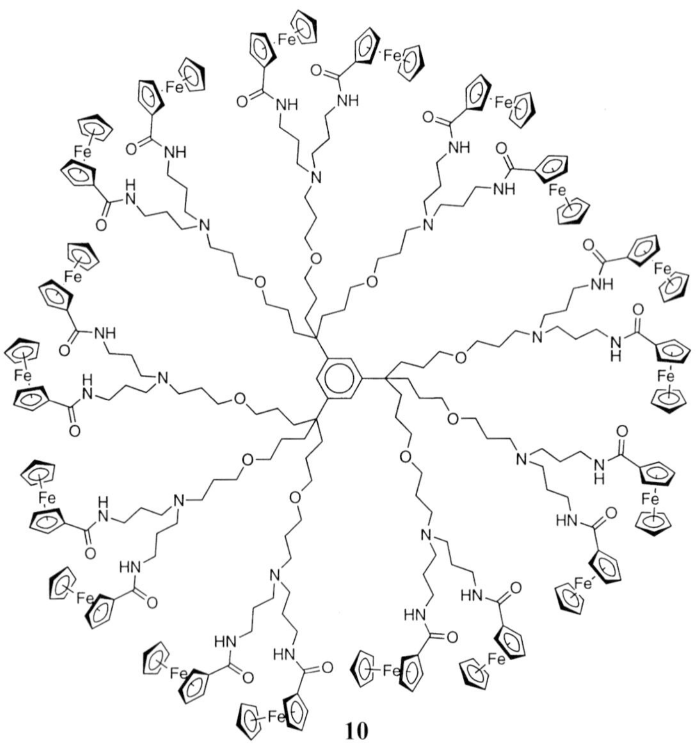

10

2.3 Dendrimers Containing Polypyridine-Type Metal Complexes

The first attempt to construct a dendrimer with an electroactive Ru-polypyridine core was based on the reaction of $Ru(bpy)_2Cl_2$ with a branched polyether-substituted phenanthroline ligand (**11**) [27]. In the potential window +2/−2 V, this compound shows a one-electron oxidation process and three distinct one-electron reduction processes that, by comparison with the behavior of the

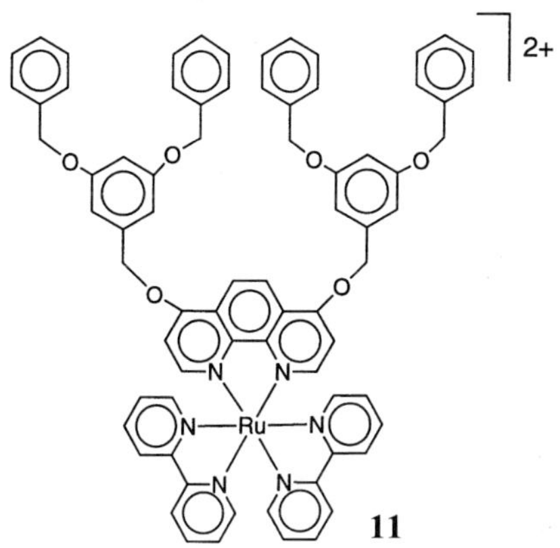

11

parent $[Ru(bpy)_2(phen)]^{2+}$ complex (phen = 1,10-phenanthroline), can be straightforwardly assigned as metal- and ligand-centered processes, respectively [28]. The relatively small appended branches exhibit an effect essentially equivalent to that of benzyloxy electron-donating groups.

Larger dendrimers based on a $Ru(bpy)_3^{2+}$ core and containing up to 54 peripheral methylester units (**12**) have recently been obtained [29a]. Both the metal-centered oxidation and ligand-centered reduction processes become less reversible on increasing dendrimer size [29b].

In dendrimers containing a $Fe(tpy)_2^{2+}$ core with two arborols appended as 4,4′-substituents of the tpy ligands (**13**), the reversibility of the metal-based oxidation and ligand-based reduction decreases with increasing generation of the dendrimer [30]. This effect has been ascribed to encapsulation of the redox-active units into the two insulating polyether dendritic sectors. As we will see when discussing porphyrin-based dendrimers, such a behavior is reminiscent of that exhibited by protein cytochrome c. A similar trend has been observed for dendrimers made of a $Ru(tpy)_2^{2+}$ core with appended two hydrocarbon long chains bearing dendritic branches, one of which increases in size along the series of examined compounds [31]. In the high generation compounds of this series, such as **14**, the electrochemical irreversibility is accompanied by a chemical irreversibility assigned to a redox driven locking/unlocking mechanism.

$Ru(tpy)_2^{2+}$ units have also been placed in the branches of dendrimers based on tris- and hexa-substituted benzene cores [32]. Apparently, the electrochemical investigation of these compounds is made difficult by adsorption processes

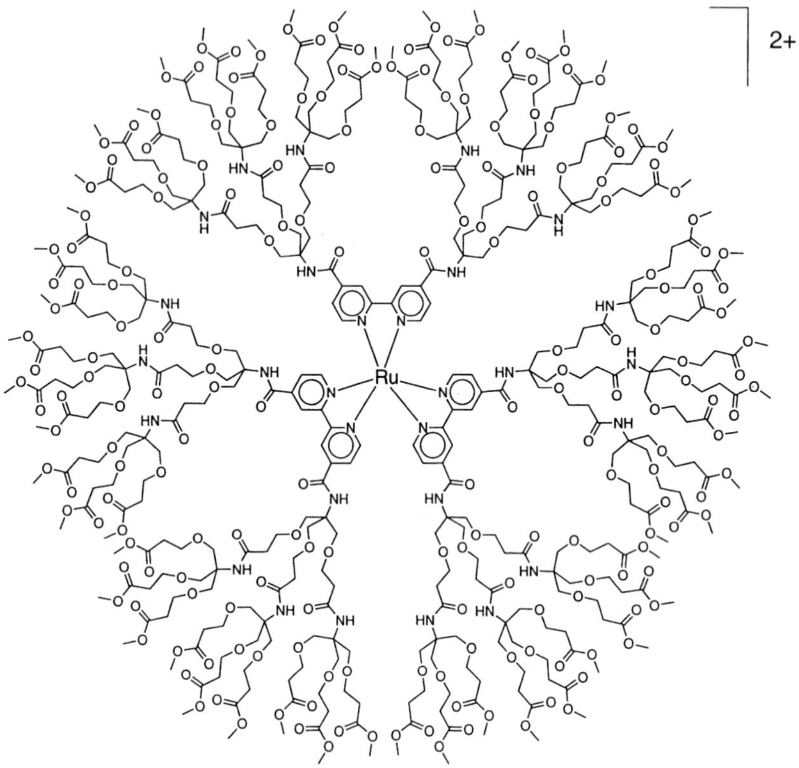

12

to the electrode surface. The dendrimer carrying six branches (**15**) shows only an oxidation wave that was claimed to indicate that the 18 Ru-based centers are equivalent. No further detail has been given.

Four $Ru(tpy)_2^{2+}$ units with or without carborane-type substituents have been linked to a pentaerythritol core. The resulting compounds exhibit only one cyclic voltammetric wave assigned to the simultaneous oxidation of the four metal-based centers [33].

The electrochemical behavior of dendrimers based on a C_6Me_6 or $Fe(Cp)(C_6Me_6)^+$ cores with attached six branches bearing in the periphery $Ru(tpy)_2^{2+}$ (**16a** and **16b**) or $Ru(bpy)_3^{2+}$ (**17a** and **17b**) units has been investigated [24]. Cyclic voltammetric experiments have evidenced that in each case the six peripheral Ru-based units are reversibly oxidized at the same potential, showing that they are equivalent and non-interacting. In compounds containing the $Fe(Cp)(C_6Me_6)^+$ core, the Fe^{II}/Fe^{I} process is also expected to occur, which would be useful as an internal reference for the exchange of one electron. Unfortunately this process is partially masked by the bpy- and tpy-based reduction processes.

$M(bpy)_3^{2+}$ (M=Fe or Co) cores have been functionalized with six $Ru(tpy)_2^{2+}$ moieties. Only the simultaneous oxidation of the peripheral $Ru(tpy)_2^{2+}$ moieties has been observed [34].

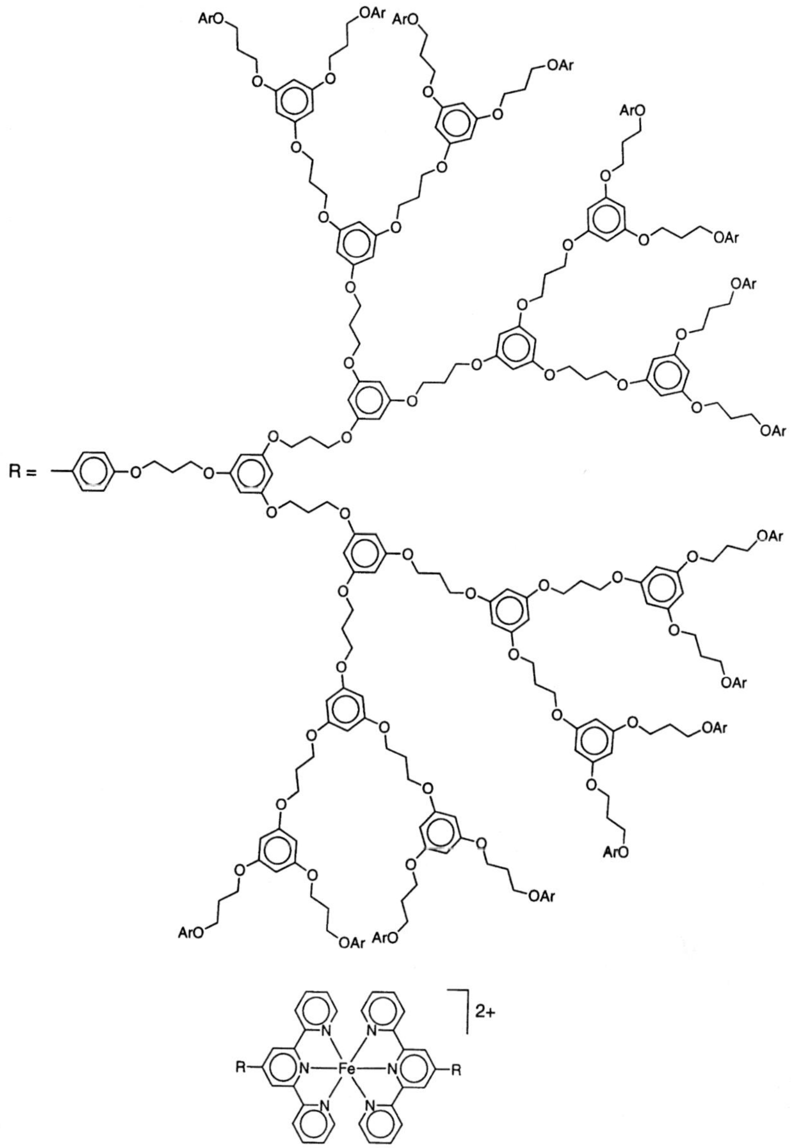

13

Starting from investigation on oligonuclear metal complexes [8], a great number of dendrimers based on metals as branching centers have been synthesized in the past few years [1c, 2a, b, c, e]. In these species, such as, for example, compounds **18–20**, the metal centers are linked by bridging ligands made of two, closely connected bis-chelating units such as 2,3-dpp (2,3-dpp=2,3-bis(2-pyridyl)pyrazine) [35, 36]. In the dendritic species, each mononuclear component brings its own redox properties, more or less affected by intercomponent

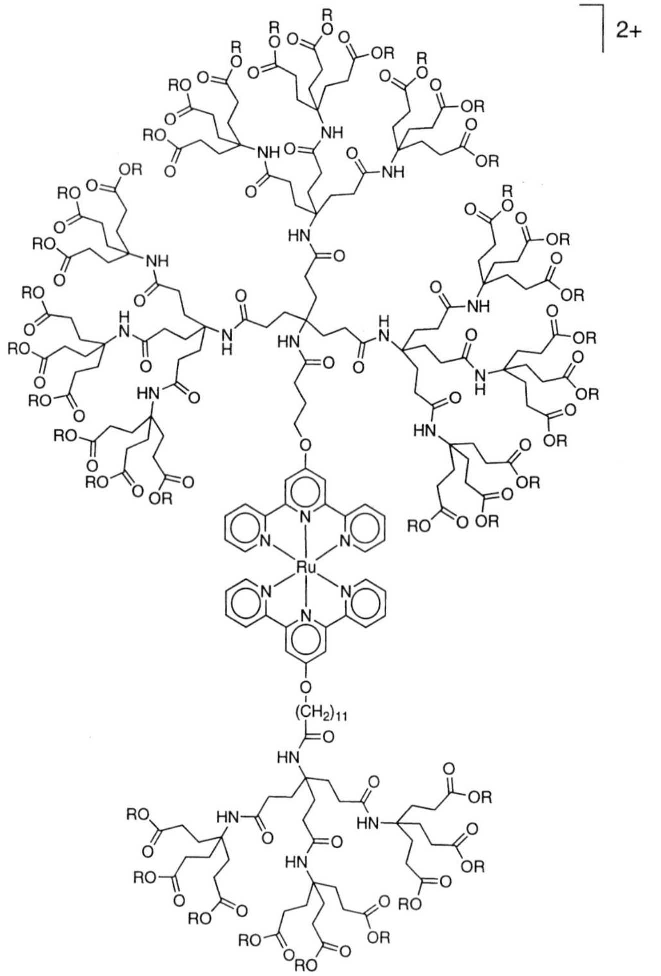

14

interactions. Metal-metal and ligand-ligand interactions are noticeable for metals coordinated to the same bridging ligand and for ligands coordinated to the same metal, whereas they are negligibly small for metals or ligands that are sufficiently far apart. Since non-interacting units undergo electrochemical processes at the same potential, the number of electrons lost or gained at a certain potential can be controlled by placing in the dendrimer the desired number of suitable, equivalent and non-interacting units.

On oxidation, interesting results have been obtained for decanuclear dendrimers of general formula $[M_1\{(\mu\text{-}2,3\text{-dpp})M_2[(\mu\text{-}2,3\text{-dpp})M_3(bpy)_2]_2\}_3]^{20+}$, where $M_1=M_2=M_3=$ Ru (**18**), $M_1=$Os and $M_2=M_3=$Ru (**19**), and $M_1=M_3=$Os and $M_2=$Ru (**20**) [36, 37]. Since Os(II) is easier to oxidize than Ru(II) and the bpy ligands are better electron-donors than the bridging 2,3-dpp ligands, different oxidation patterns are expected.

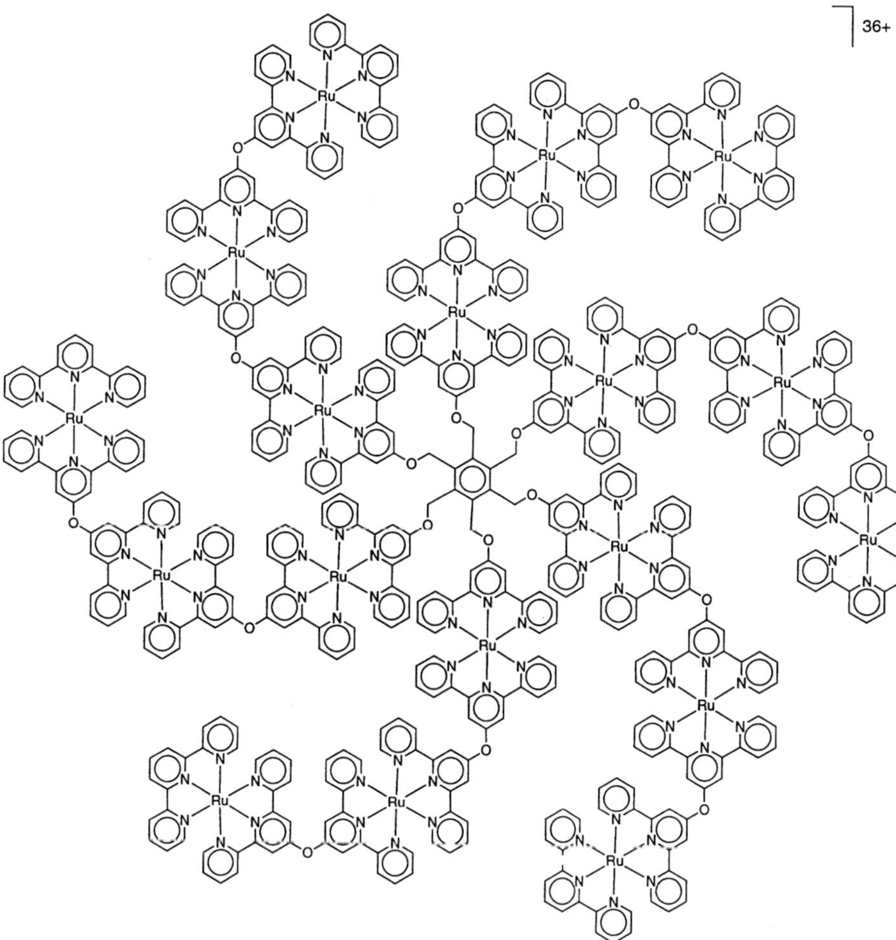

15

Compound **20** is made of an Os(II)-based core, three Ru(II)-based units in the intermediate positions, and six Os(II)-based units in the peripheral positions. Oxidation should thus first involve the six peripheral Os(II)-based units (which contain the stronger electron-donor bpy ligand in their coordination sphere), followed by the Os(II)-based core, and by the intermediate Ru(II)-based units, according to a 6-1-3 electron-exchange pattern. This is consistent with the 6-1 observed pattern (Fig. 3a), considering that the oxidation of the intermediate Ru(II)-based units is shifted to inaccessible potentials because of their interaction with the already oxidized core and peripheral units.

In **19**, which is made of an Os(II)-based core and nine Ru(II)-based units, a 1-6-3 oxidation pattern is expected resulting from oxidation of the Os-based core, followed by the six peripheral Ru-based units (which are coordinated to the stronger electron-donor bpy ligand), and by the three intermediate Ru-based

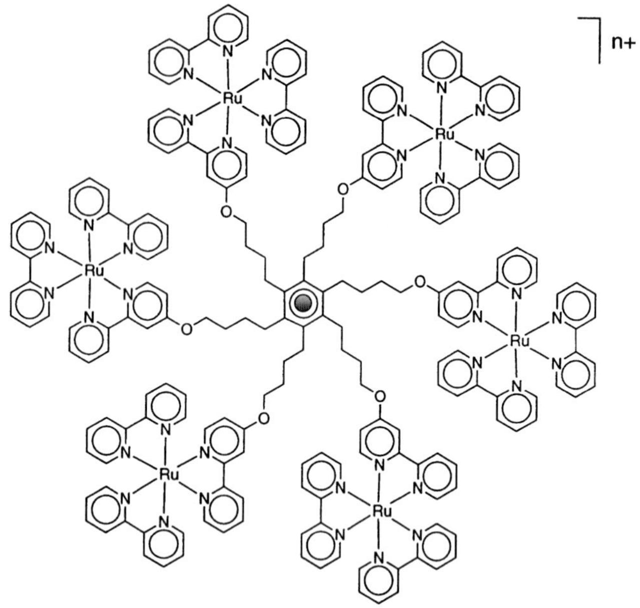

16a ● = nothing, n = 12
16b ● = $Fe(\eta_5\text{-}C_5H_5)^+$, n = 13

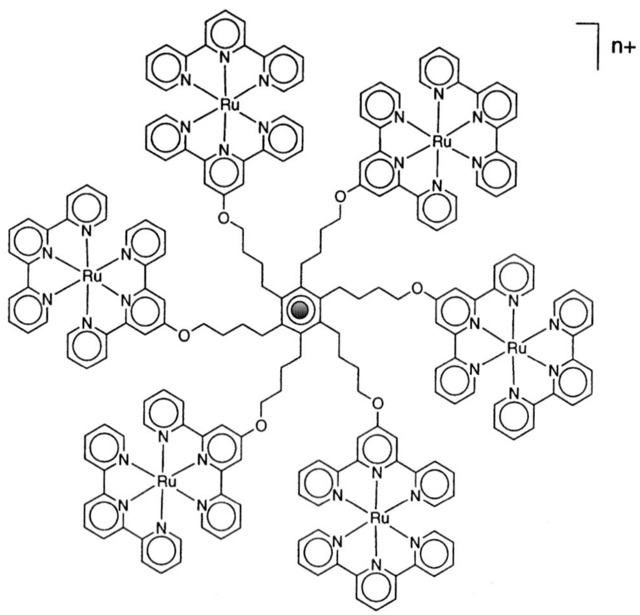

17a ● = nothing, n = 12
17b ● = $Fe(\eta_5\text{-}C_5H_5)^+$, n = 13

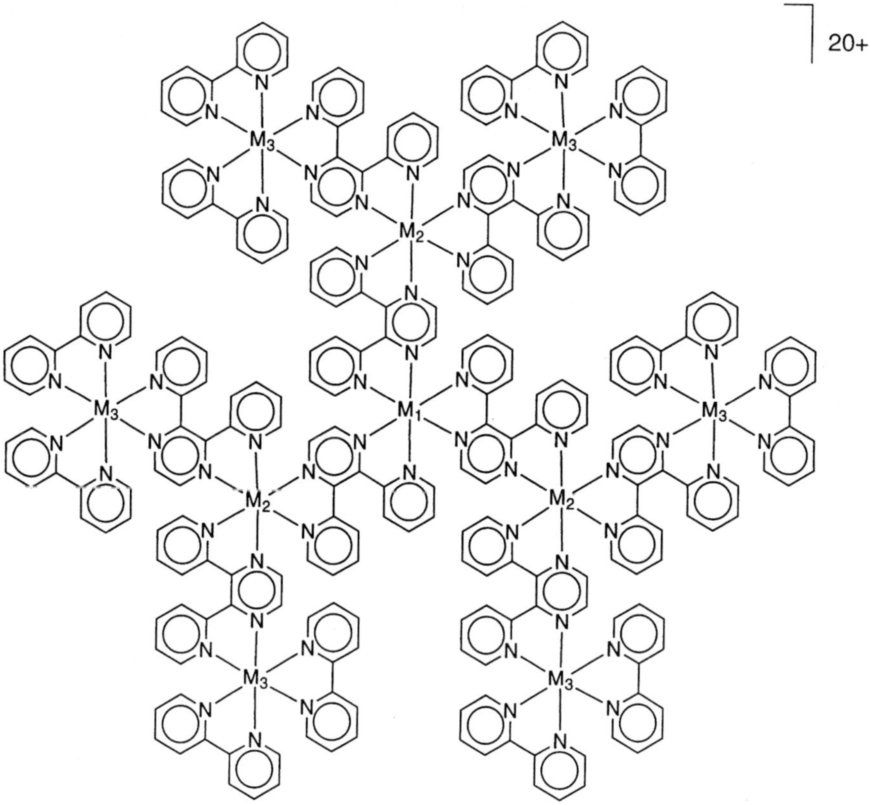

18 $M_1 = M_2 = M_3 = Ru$
19 $M_1 = Os, M_2 = M_3 = Ru$
20 $M_1 = M_3 = Os, M_2 = Ru$

units. The observed 1-6 pattern (Fig. 3b) is consistent with these expectations considering again that oxidation of the intermediate Ru(II)-based units is shifted to inaccessible potentials.

In **18**, on the basis of the different electron donor power of the ligands and the metal-metal interaction, the expected pattern is 6-1-3. In practice, only a six-electron process is observed (Fig. 3c) in the accessible potential window.

In larger structures, the number of equivalent units becomes huge. In the docosanuclear dendrimer made of an Os(II)-based core and 21 Ru(II)-based units, a one-electron oxidation process, assigned to the Os(II)-based unit, is followed by a 12-electron process, due to the simultaneous oxidation of the 12 equivalent and non-interacting peripheral Ru(II)-based units [36].

Because of the presence of many polypyridine ligands, each capable of undergoing several reduction processes [38], the electrochemical reduction of this type of dendritic compound produces very complex electron exchange patterns.

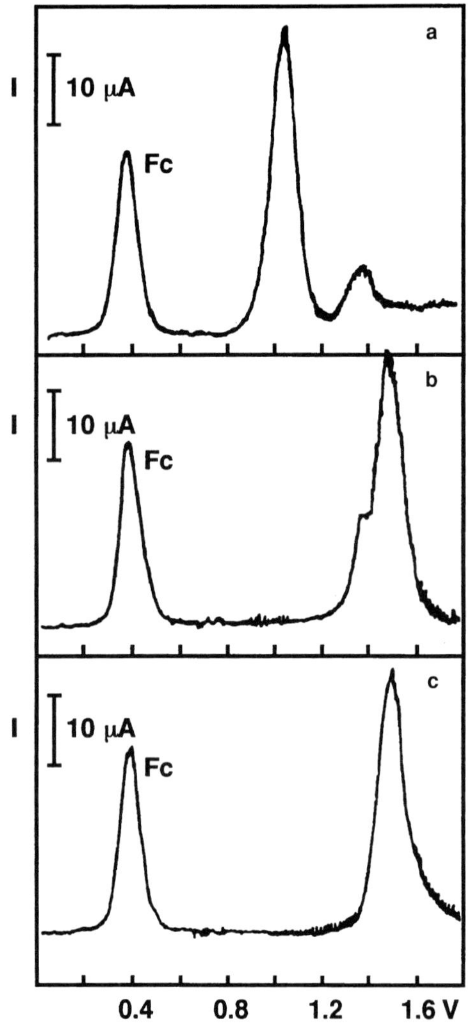

Fig. 3 a–c. Differential pulse voltammetric patterns for oxidation of decanuclear compounds in acetonitrile solution: **a 20**; **b 19**; **c 18**. Fc indicates the oxidation peak of ferrocene, used as an internal standard

For example, **18** shows a differential pulse voltammogram with two, broad peaks, followed by several other overlapping peaks [35]. The first peak, which corresponds to the exchange of six electrons, is due to the one-electron reduction of the six outer equivalent bridging ligands. The width of the peak, compared to that observed on oxidation, suggests a non-negligible interaction between the two ligands coordinated to the same metal, resulting in two closely lying three-electron processes. The second broad peak, which involves three electrons, is assigned to the one-electron reduction of the three inner bridging

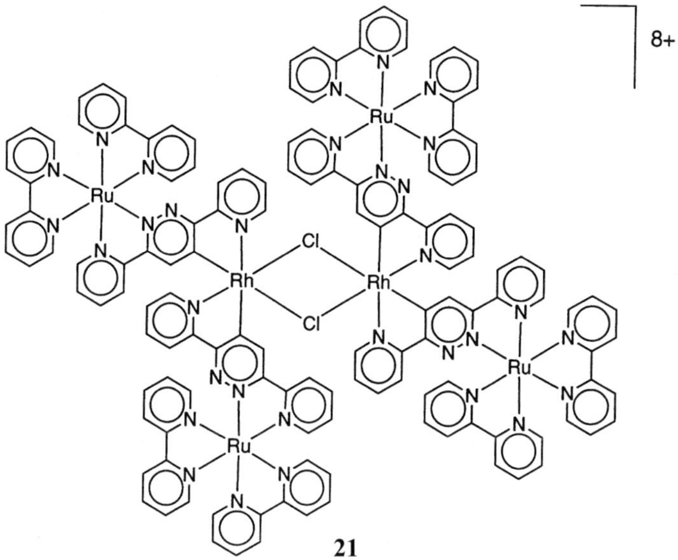

21

ligands, occuring at close potential values. Since the interaction between reduced ligands depends strongly on the nature of the metal, the electron exchange pattern can be modified by replacing Ru(II) with Os(II) [36].

In conclusion, the electrochemical data offer a fingerprint of the chemical and topological structure of these dendrimers. Furthermore, the knowledge of the electrochemical properties of the mononuclear components and the synthetic control of the supramolecular structure allow the design of dendrimers with predetermined redox patterns.

Different metals and/or ligands can be incorporated in this type of dendrimers. As a step along this direction, four tetranuclear bimetallic Ru(II)-Rh(III)$_3$ Ru(II)-Ir(III)$_3$, Os(II)-Rh(III)$_3$, and Os(II)-Ir(III)$_3$ complexes [39] and a hexanuclear Ru(II)$_4$-Rh(III)$_2$ (**21**) complex [40], containing cyclometalating terminal and bridging ligands respectively, have been synthesized and their electrochemical behavior investigated. In the hexanuclear complex, the four (weakly interacting) Ru-based units are oxidized at the same potential. No other oxidation process can be observed. On reduction, a complex pattern of ligand-centered processes is observed, starting with a two electron process assigned to reduction of the *N-N* chelating moieties of two remote bridging ligands [40].

The dendrimer-type tetranuclear Ru(II)-Os(II)$_3$ complex (**22**, protonated form) shows an interesting electrochemical behavior due to the presence of free basic sites in its bridging ligands [41]. The protonated form shows a 3-1 oxidation pattern due to the simultaneous oxidation of the three Os-based units, followed by the one-electron oxidation of the Ru-based unit. On addition of base, the six chelating moieties (three on the Ru center and one on each Os center) undergo deprotonation. This causes changes in the oxidation potential of the metal ions, with a consequent switching from 3-1 to 1-3 in the oxidation pattern.

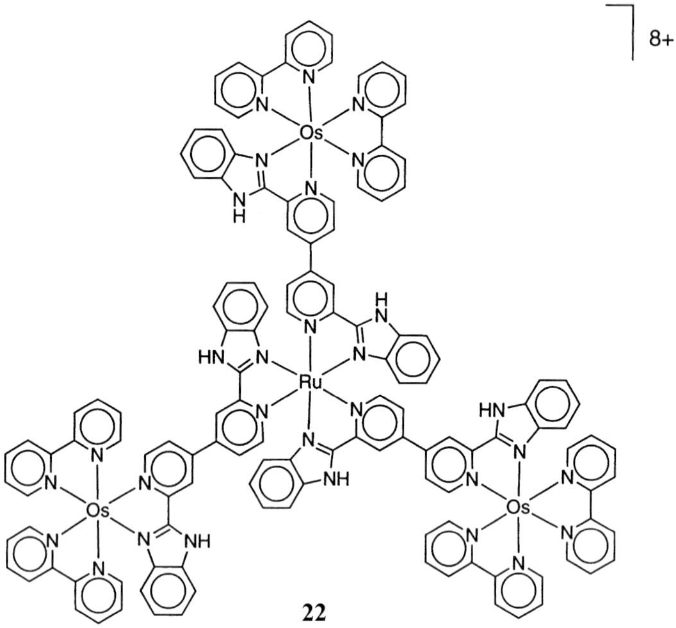

22

2.4
Dendrimers with a Porphyrin Metal Complex as a Core

Porphyrin complexes are particularly suitable cores to construct dendrimers and to investigate how the behavior of an electroactive species is modified when surrounded by dendritic branches. In particular, dendritic porphyrins can be regarded as models for electron-transfer proteins like cytochrome c [42, 43]. Electrochemical investigation on Zn-porphyrins bearing polyether-amide branches has shown that the first reduction and oxidation processes are affected by the electron-rich microenvironment created by the dendritic branches [42]. Furthermore, for the third generation compound all the observed processes become irreversible.

Two dendrimers based on Fe-porphyrin core carrying peptide-like branches of different sizes have been synthesized in order to have more open and a more densely packed (**23**) structures [43]. The electrochemical behavior has been examined in CH_2Cl_2 and in aqueous solution. In the less polar solvent, the two dendrimers show similar potentials for the Fe^{III}/Fe^{II} couple, suggesting that the iron porphyrins in both the more open and the more densely packed dendrimers experience similar microenvironments. On the contrary, in water the behavior of the two dendrimers is very different since the reduction from Fe^{III} to Fe^{II} is much easier for the densely packed dendrimer. This result can be explained considering that in the dendrimer with the relatively open structure the aqueous solvation of the iron porphyrin is still possible, whereas in the densely packed one the contact between the heme and the external solvent is signifi-

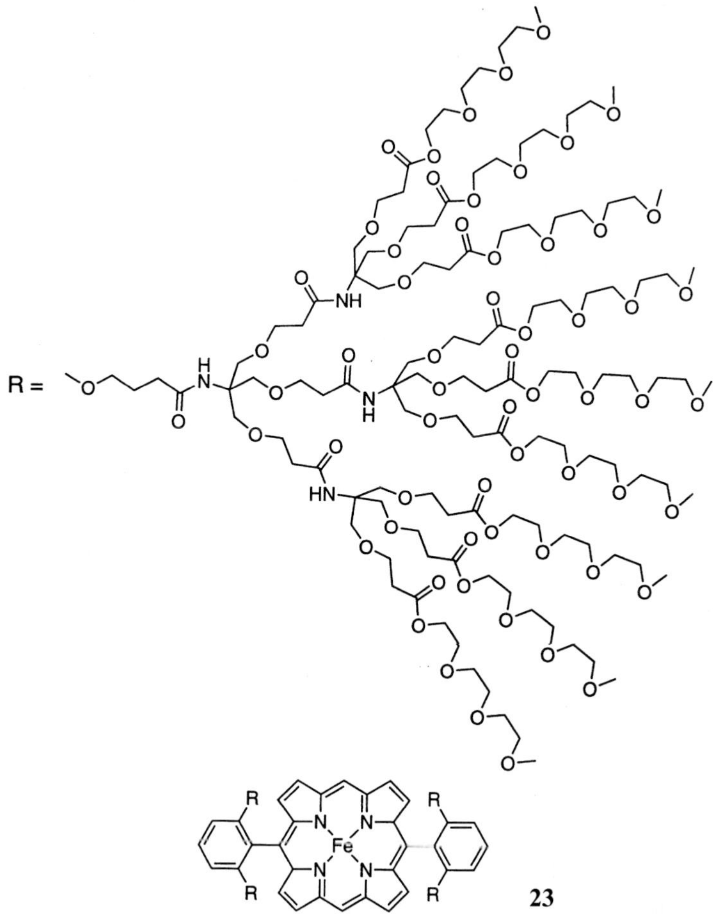

23

cantly reduced, destabilizing the oxidized, more charged state and favouring the reduction of the iron porphyrin core. This study can help to clarify the reasons for the large changes in redox potential of the Fe^{III}/Fe^{II} couple observed in the biological systems. The difference in the reduction potential exhibited by these two purely synthetic dendrimers is, in fact, very similar to that found for the cytochrome c and a more solvent-exposed cytochrome c heme octapeptide [43].

2.5 Other Metal-Containing Dendrimers

Several other metal-containing dendrimers have been synthesized, but in most cases the electrochemical behavior has not been investigated. An interesting family of compounds has been obtained by appending dendrons according to a tetrahedral geometry to an electroactive iron-sulfur core [44]. On increasing generation of the dendrons from 1 to 4 (**24**), the core becomes more difficult to

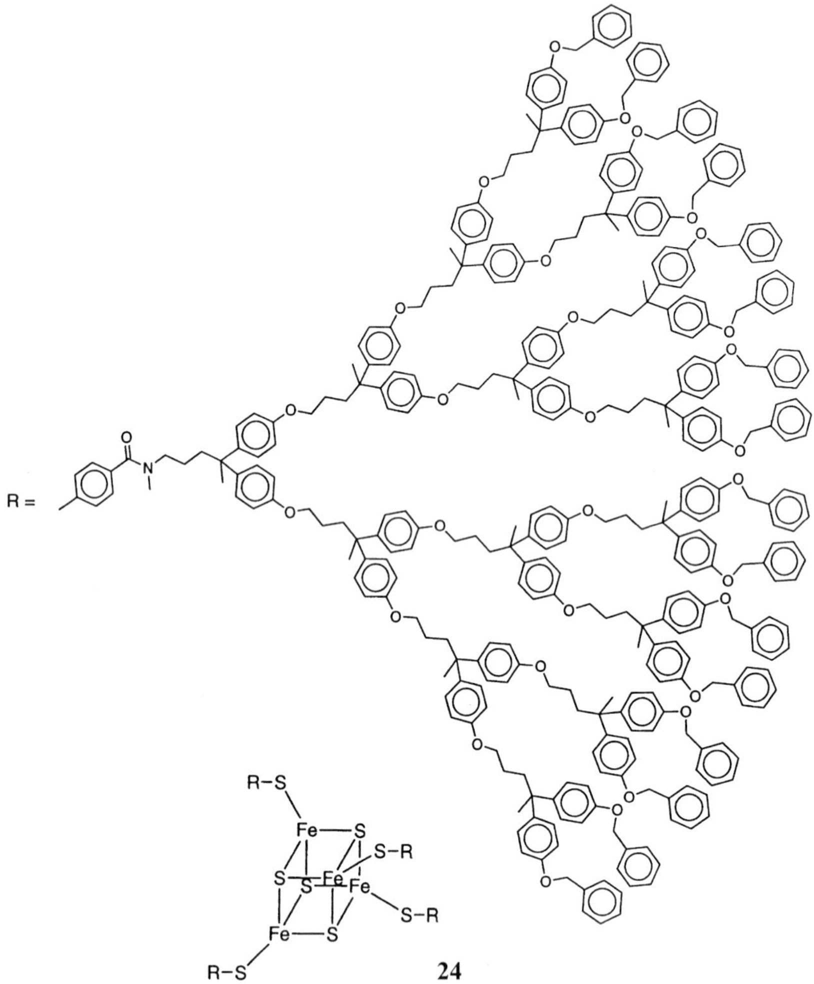

24

reduce both kinetically and thermodynamically because of the increased insulating effect towards electrode and solvent.

2.6 Conclusions

This review of the electroactive metal-containing dendrimers, not exhaustive for space reasons, clearly indicates that several interesting dendrimers have been synthesized, but in most cases their electrochemical properties have not been investigated in detail. A few carefully performed studies have shown that electrochemistry is a powerful technique (i) to elucidate the structure (and purity) of the dendrimers, a task not at all easy in the case of highly charged compounds,

(ii) to evaluate the degree of electronic interaction of the various, chemically and/or topologically equivalent or non-equivalent moieties of a dendrimer, and (iii) to study the endo- and exoreceptor capabilities of the dendrimers. Detailed electrochemical studies on dendrimers with an electroactive core promise to be quite useful in modelling natural electroactive species and in investigating the problem of insulation of such a unit from the electrode surface.

3 Photochemical and Photophysical Properties

3.1 Introduction

In all of the photoactive metal-containing dendrimers investigated so far, the excited states properties can be discussed within the frame of the localized molecular orbital approximation and are essentially controlled by the characteristics of the metal-based units [8]. In most cases interesting photochemical and photophysical properties are associated with luminescent, long-lived excited states at suitable energies. The processes most commonly observed are energy and electron transfer either among the units of the same dendrimer or involving external species.

Well-known classes of luminescent species are Ru(II), Os(II), and Re(I) polypyridine and Rh(III) and Ir(III) cyclometalated complexes [8, 45, 46]. Therefore the choice of these compounds as metal-based components to construct photoactive dendrimers has been straightforward. All the dendrimers discussed in this review, with few exception, are based on such building blocks. Generally speaking in these compounds the excited states involved in the photochemical and photophysical processes are metal-to-ligand charge-transfer (MLCT) in nature [8, 45, 46]. To a first approximation, the energy of such states depends on the oxidation potential of the metal and the reduction potential of the ligand [8].

As far as structure is concerned, most of the photochemical and photophysical studies concern dendrimers of type *d* shown in Fig. 1. Only a few examples of type *a* and none of types *b* and *c* have been investigated from the photochemical and photophysical viewpoint. For dendrimers containing metal-based units only in the core, an interesting problem is whether the excited state properties of such units are affected by the surrounding branches. For dendrimers based on metals as branching centers, the most interesting problems are the mutual perturbation between the chromophores and the occurrence of (intradendrimer) photoinduced electron- and/or energy-transfer processes. In most cases, such processes are highly efficient between units sharing a bridging ligand, whereas they can be much less efficient between metal-based units which are far apart [8]. Moreover, due to the presence of many metal-based chromophores, each characterized by its own visible MLCT absorption bands, *d*-type dendrimers exhibit noticeable light harvesting capabilities. This property, together with the possibility to control the energy migration patterns, justifies why these compounds are often considered artificial antennae [37, 47].

3.2 Dendrimers Built Around a Metal Complex as a Core

The first photoactive dendrimer in which purely organic branches have been convergently attached to a metal-based core is **11** [27]. The photophysical properties of this compound are very similar to those of its $[Ru(bpy)_2(phen)]^{2+}$ core. The slight perturbation caused by the dendritic branches cannot be ascribed to their structure but only to the electron-donating character of the benzyloxy groups. The expected energy transfer from the benzyloxy groups to the central core could not be investigated because the absorption bands of the organic chromophores of the branches are hidden by the more intense bands of the polypyridine ligands.

More extended organic branches of the benzyloxy type have been grafted on a Zn-porphyrin core to obtain a family of photoactive dendrimers [48]. Compound **25** is a high generation member of this family. Photophysical investigation showed that, in the dendrimer of higher generation, long-range photoinduced electron transfer *through* the dendrimer framework occurs between the Zn-porphyrin core and methylviologen molecules electrostatically interacting with the peripheral units of the dendrimers.

Compound **12** is the largest photoactive dendrimer in which organic branches have been attached around a polypyridine metal core [29]. This dendrimer was obtained by reacting the metal precursor with a modified bpy carrying highly-

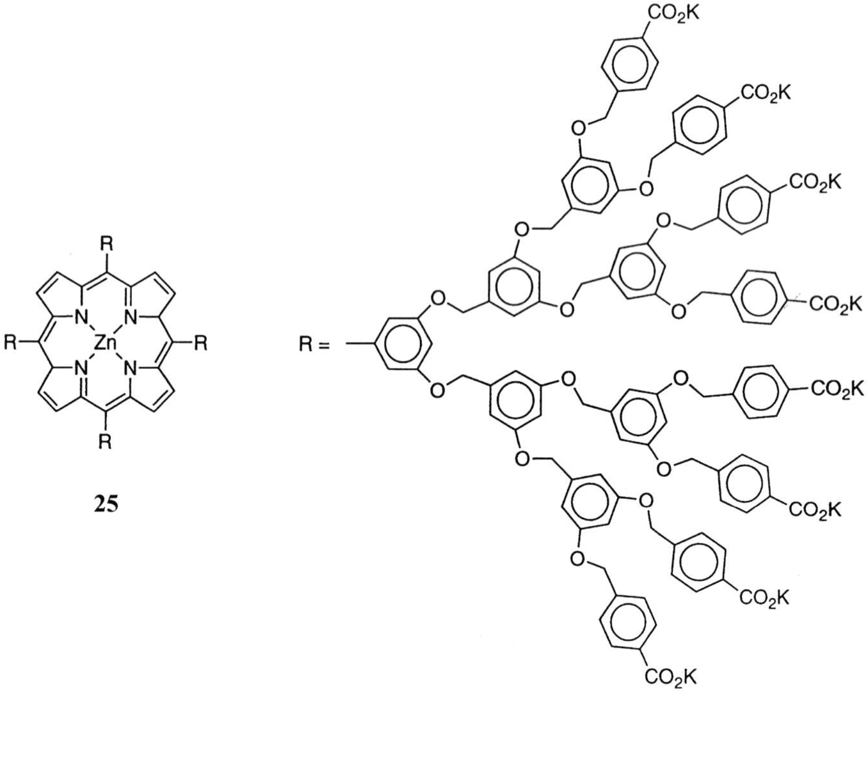

25

branched methylester substituents. The photophysical properties of the dendrimer recall those of the metal-based core. Interestingly, molecular oxygen is less efficient in quenching the luminescent MLCT state on passing from the parent $[Ru(bpy)_3]^{2+}$ complex to the dendrimer, suggesting that the organic branches are able to shield the metal-based core.

3.3 Dendrimers Based on Metals as Branching Centers

Within this family of compounds (Fig. 1d) a distinction can be made between species in which the core is (i) a polychelating ligand, or (ii) a metal complex. Even if for most of these species only the first generation has been obtained so far, they are taken into consideration as reference compounds for larger arrays that will most likely be prepared.

A typical ligand capable of generating a dendritic structure is 1,4,5,8,9,12-hexaazatriphenylene (HAT). Photophysical studies of trinuclear species based on HAT have been reported [14a, 49]. Representative example of complexes of this type are **26**, **27**, and **28**. For some of these complexes, the luminescence, originating from MLCT levels involving the central HAT ligand, was found to decay with multiexponential kinetics. Furthermore, the vibrational modes responsible for the nonradiative decay of the luminescent MLCT states are different in the polynuclear dendritic edifices with respect to the mononuclear $[M(L)_2(HAT)]^{2+}$ compounds [14a].

Other first-generation dendrimers built around multichelating ligands are the trinuclear complexes **29**–**38** [50–52] and several hexanuclear complexes (for representative examples, see **39** and **40**) [53]. In all of these compounds, the photophysical properties are equivalent to those of their building blocks. In the mixed-metal complexes **31** and **34** [50b] the Ru(II)- and Os(II)-based chromophores are only weakly-coupled and a dual luminescence is observed.

The ligand more extensively used to build up dendrimers having a metal core is 2,3-dpp. By using this bridging ligand, tetra- [35, 36, 39, 54], hepta- [55], deca- [36, 37, 56] (see **18**–**20**), trideca- [57], and docosanuclear [9, 35, 36] dendrimers have been obtained and their photophysical properties studied. Several photoinduced energy migration patterns have been obtained changing the metals (Ru(II), Os(II), Ir(III), and Rh(III)) and/or the peripheral ligands (bpy, biq, ppy, 2,3-Medpp$^+$; biq=2,2′-biquinoline, ppy=2-phenylpyridine, 2,3-Medpp$^+$=2-[2-(1-methylpyridiumyl)]-3-(2-pyridyl)pyrazine). In the dendrimers having an Os(II)-based core and Ru(II)-based units in the branches, the efficiency of energy transfer from the (higher energy) Ru(II) units to the (lower energy) Os(II) core is unitary for the tetranuclear species [54b]. On increasing generation, intermediate Ru(II)-based units, whose MLCT excited states lie at higher energy than those of the Ru(II)-based peripheral ones, are also present. Such units represent a barrier to energy transfer from the periphery to the center of the dendrimer, so that the energy transfer efficiency decreases [36, 37].

First generation dendrimers (tetranuclear complexes) of the same family with branches containing different metals have also been synthesized and energy migration patterns leading to one or two peripheral units have been obtained [58].

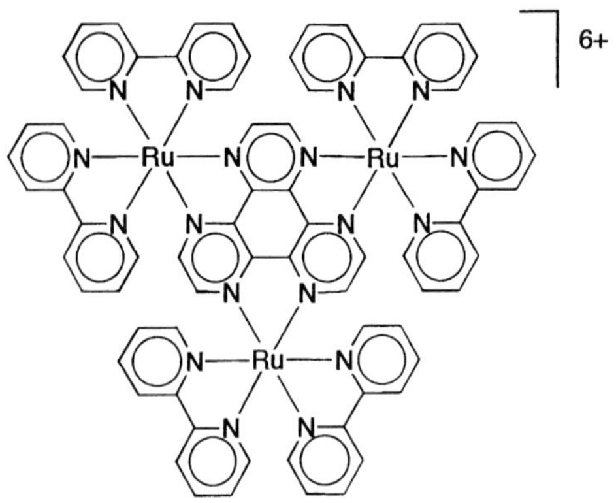

26

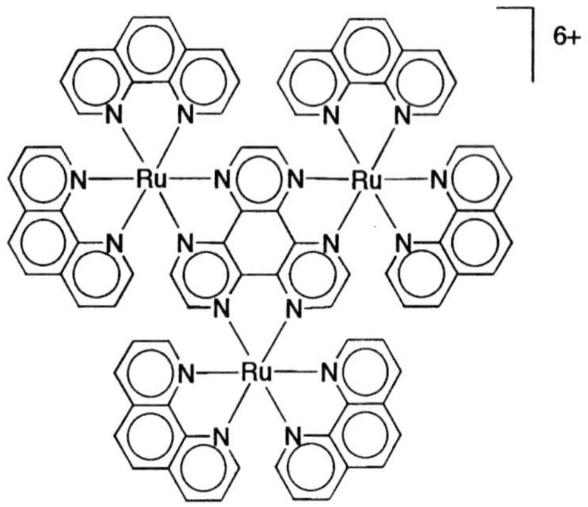

27

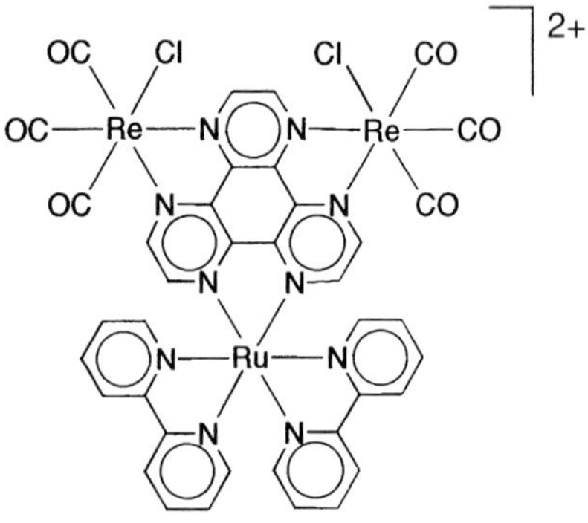

28

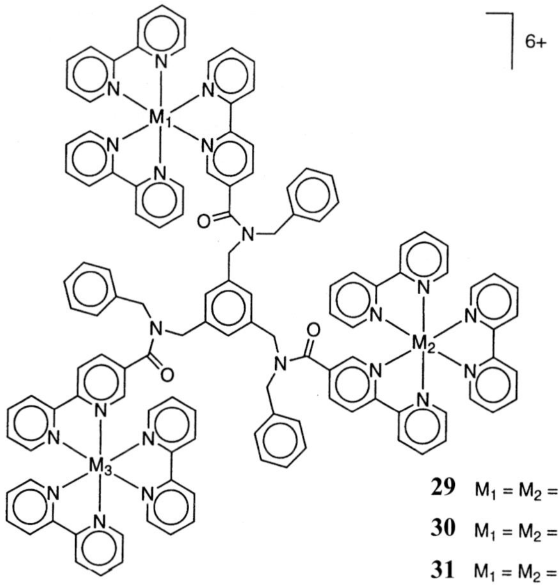

29 $M_1 = M_2 = M_3 = Ru$

30 $M_1 = M_2 = M_3 = Os$

31 $M_1 = M_2 = Ru, M_3 = Os$

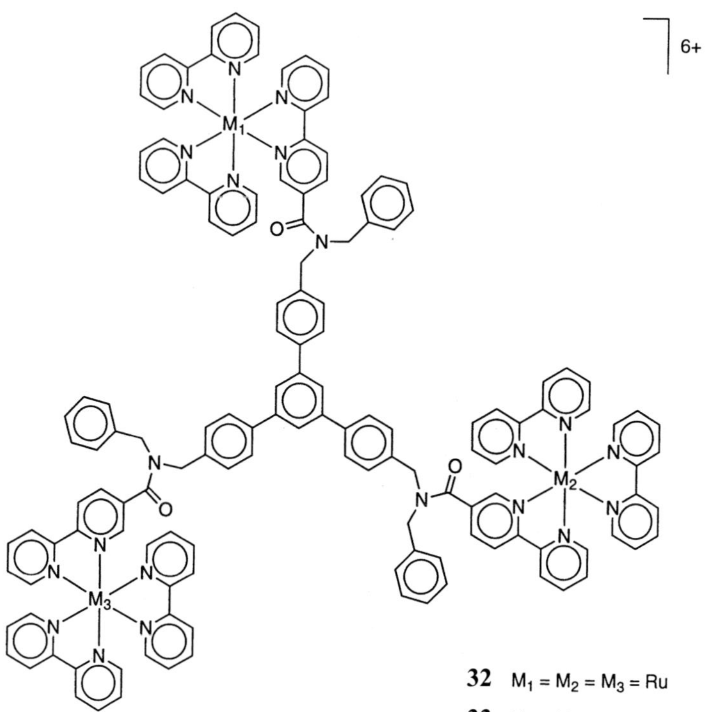

32 $M_1 = M_2 = M_3 = Ru$

33 $M_1 = M_2 = M_3 = Os$

34 $M_1 = M_2 = Ru, M_3 = Os$

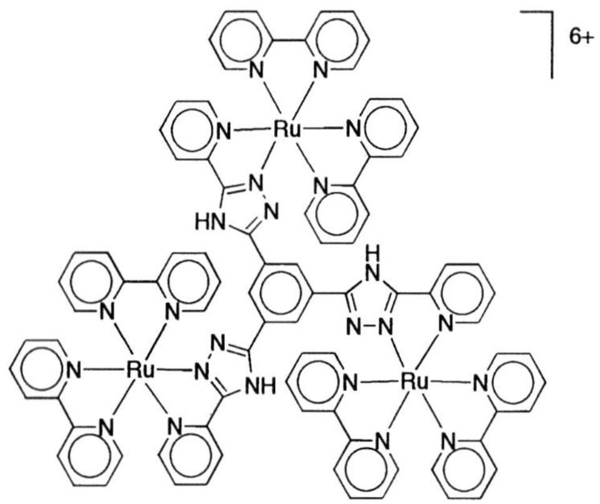

35

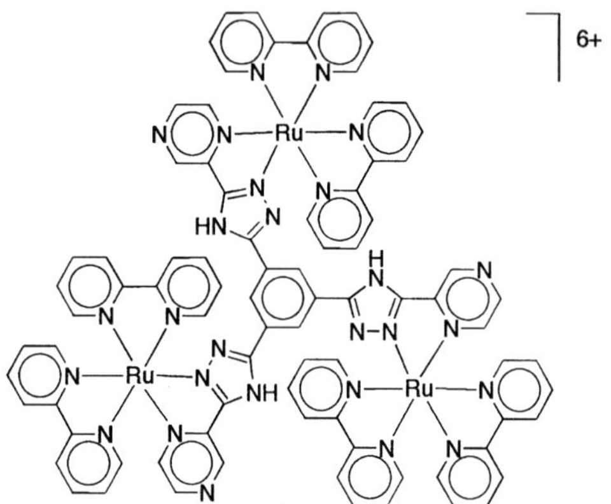

36

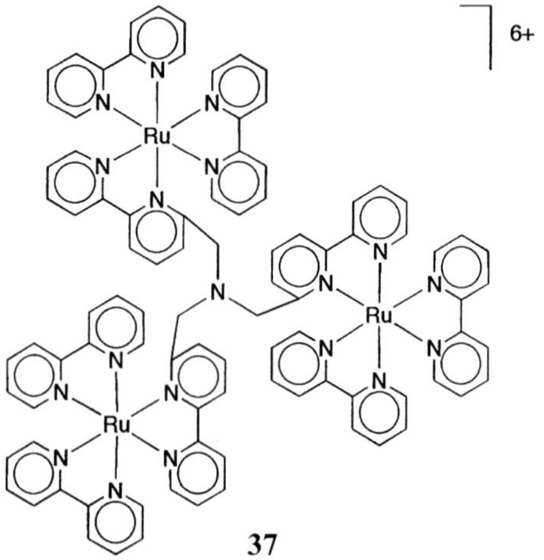

37

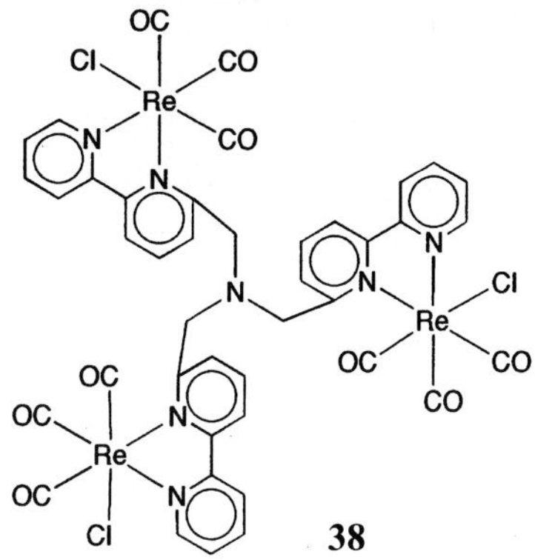

38

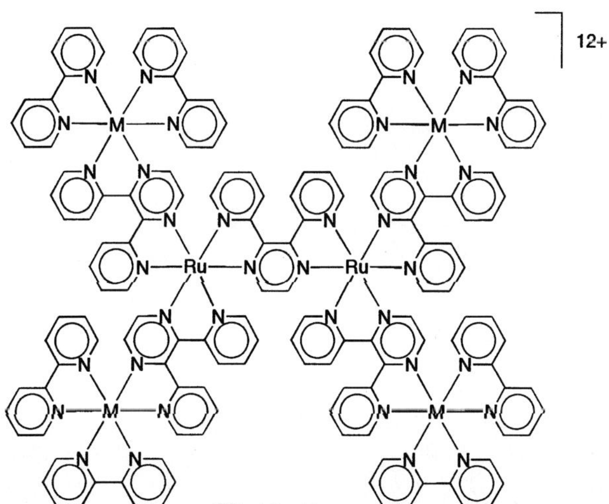

39 M = Ru

40 M = Os

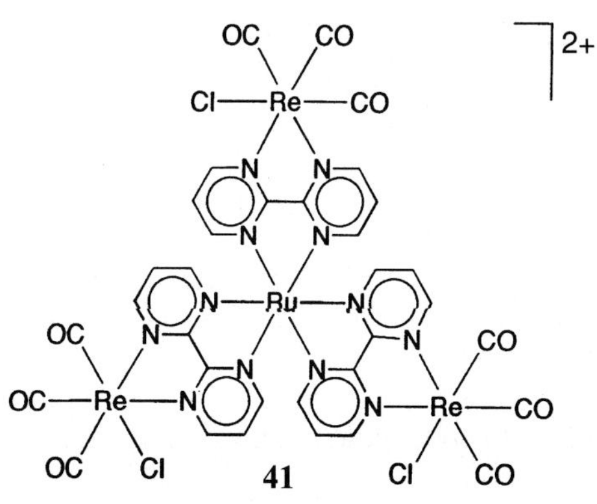

41

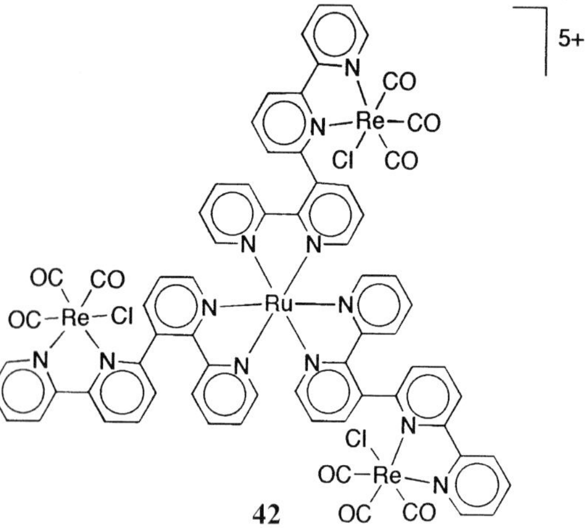

42

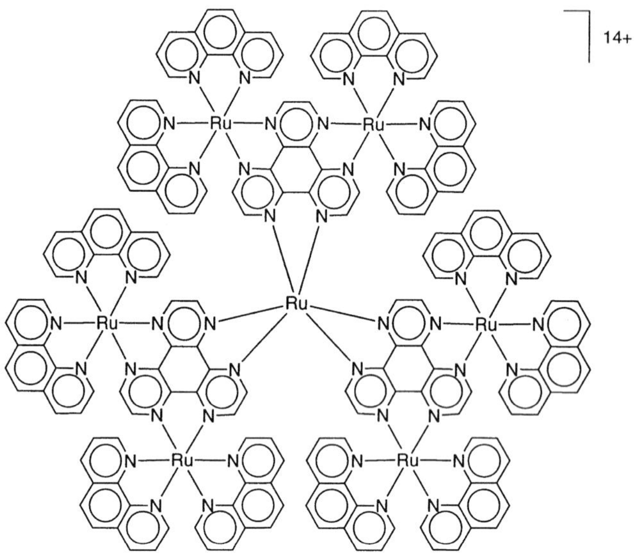

43

Photophysical studies have been performed on dendrimers **41** [49], built around a $[Ru(bpm)_3]^{2+}$ core (bpm=2,2′-bipyrimidine), and **42** [59], built around a $[Ru(QP)_3]^{2+}$ core (QP=2,2′:3′,2″:6″,2‴-quaterpyridine). In both compounds energy transfer from the peripheral Re(I)-based chromophores to the central Ru(II)-based unit occurs with unitary efficiency.

The $[Ru(HAT)_3]^{2+}$ complex has also been used as a metal core for the dendritic heptanuclear species **43** [60] which exhibits luminescence in the infrared region.

3.4
Stereochemically Pure Metal-Based Dendrimers

Polynuclear complexes based on octahedral building blocks may be structurally not well defined because of stereogenic problems [8]. However, clever synthetic strategies have recently been devised to obtain chirally pure species [61–65]. Synthesis and, of course, photophysical and photochemical studies of stereochemically pure metal-based dendrimers are still in their infancy.

Preliminary studies have been performed on the different diasteroisomers of compound **44** [63b], which show typical MLCT luminescence. However, no comparison between the different isomers has been made.

Photophysical investigations performed on chirally resolved forms of compounds **26** and **27** [63b, 66] showed that the diastereoisomeric forms do not exhibit any significant difference at room temperature. However, in a glass at low temperature, the luminescence lifetimes of the heterochiral diastereoisomers were slightly shorter than those of the homochiral forms [66].

The first photophysical investigation performed on stereochemically pure metal-based dendrimers having a metal complex as the core is that concerning the tetranuclear species based on a $[Ru(tpphz)_3]^{2+}$ core (tpphz = tetrapyrido[3,2-a:2′,3′-c:3″,2″-h:2″,3″j]phenazine) [67]. Dendrimer **45** is an example of this family. In this compound, two different types of MLCT excited states, coupled by a medium- and temperature-dependent photoinduced electron transfer, are responsible for the luminescence behavior. However, the properties of all the optical isomers of this family of compounds are very similar. This finding is also in

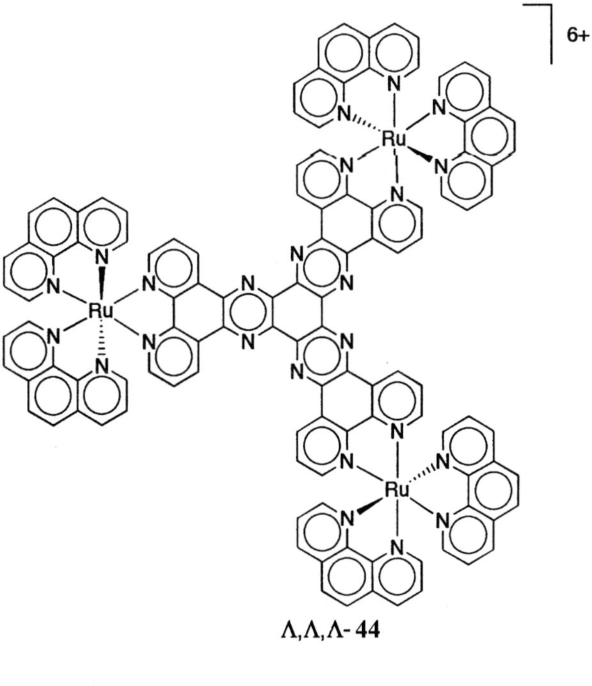

Λ,Λ,Λ-**44**

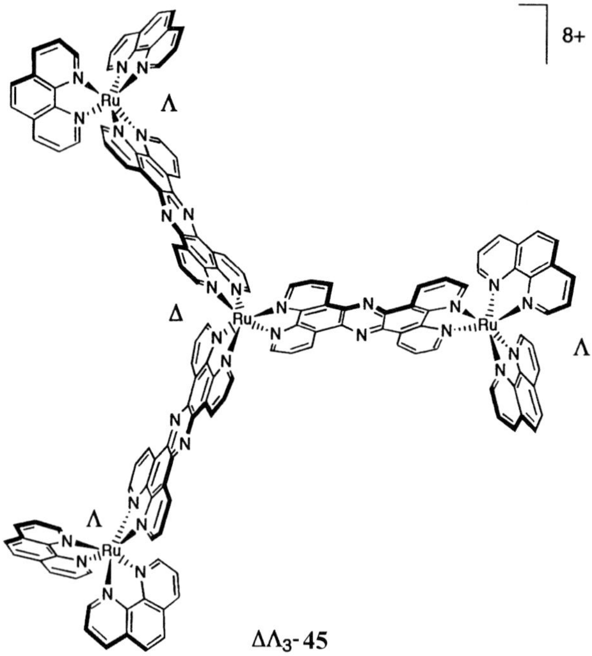

ΔΛ$_3$-**45**

agreement with the photophysical properties of stereochemically-resolved trinuclear polypyridine Ru(II) complexes [65].

3.5 Other Systems

In the pentaporphyrin array **46**, light absorption by the four peripheral Zn-porphyrins is followed by efficient energy transfer to the central free base porphyrin [68], as it happens in the polypyridine dendrimers discussed above [53b].

The hexanuclear compound **21** is a particular case of dendrimer in that a $[Rh(\mu\text{-}Cl)_2Rh]^{4+}$ moiety can be identified as the core. The photophysical properties of this compound [40] indicate that the bimetallic core plays the role of "insulator" with respect to the peripheral luminescent Ru(II)-based units.

Many other metal-containing dendrimers which are potentially interesting from a photochemical viewpoint [43, 69] have been synthesized, but their photochemical and photophysical properties have not yet been investigated.

3.6 Conclusions

Metal dendrimers containing photoactive units exhibit very interesting properties, both from a fundamental and applicative viewpoint. For example, well-designed photoactive dendrimers can play the role of artificial antennae in supramolecular structures devoted to solar energy conversion [37, 47]. Photophysical

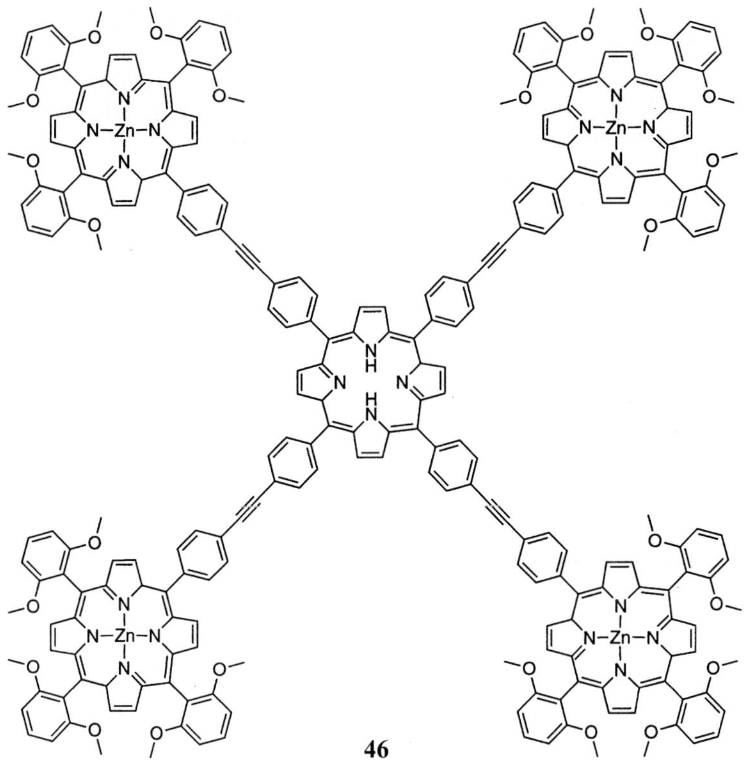

46

studies of metal-containing dendrimers can also give information on the accessibility of the inner shell or core to external reagents (e.g., molecular oxygen) and on interchromophoric interactions within the dendritic structures.

4 Perspectives

Dendrimers have a number of applications related to their endo- and exoreceptor properties deriving from their three-dimensional branching structures [1]. Dendrimers based on metal complexes can exhibit other peculiar properties due to the characteristics of the component metal complexes, namely the presence of unsaturated coordination centers, the possibility to undergo a number of reversible redox processes and to show low energy, long lived charge-transfer excited states. Because of such properties, one can foresee a noticeable expansion in this research area. The use of dendrimers based on metal complexes for catalytic processes, already exploited in a few, very interesting cases [70–72], is a very promising field particularly because the (dendrimer) homogeneous catalyst can be physically separated at the end of the process [72]. Furthermore, when stereochemically resolved species are obtained, dendrimers based on metal complexes will take advantage from the presence of many chiral centers.

It is also worth recalling that many metal-based dendrimers, because of their absorption in the visible region and the possibility to exhibit predetermined energy migration patterns, can play the role of antennas for light harvesting. Therefore it is particularly interesting to couple in a dendrimer the light harvesting function with the capability to exchange several electrons with the aim of exploiting successive one-photon/one-electron events to perform multielectron transfer processes. This would open the way to solar energy conversion by artificial photochemical systems performing water splitting and CO_2 reduction [73–76].

Acknowledgement. This work was supported by the University of Bologna (Funds for Selected Research Topics) and by the EU (TMR grant FMRX-CT96-0031).

5 References

1. (a) Tomalia DA, Durst HD (1993) Top Curr Chem 165:193; (b) Fréchet JMJ (1994) Science 263:1710; (c) Newkome GR (ed) (1994–1996) Advances in dendritic macromolecules, vols 1–3. JAI Press, London; (d) Newkome GR, Moorefield CN, Vögtle F (eds) (1996) Dendritic macromolecules: concepts, syntheses, perspectives. VCH, Weinheim; (e) Zeng F, Zimmerman SC (1997) Chem Rev 97:1681
2. (a) Ardoin N, Astruc D (1995) Bull Soc Chim Fr 132:875; (b) Serroni S, Campagna S, Denti G, Juris A, Venturi M, Balzani V (1996) Dendrimers based on metal complexes. In: Newkome GR (ed) Advances in dendritic macromolecules, vol 3. JAI Press, London;, p 61; (c) Bryce MR, Devonport W (1996) Redox-active dendrimers, related building blocks, and oligomers. In: Newkome GR (ed) Advances in dendritic macromolecules, vol 3. JAI Press, London, p 115; (d) Cuadrado I, Morán M, Losada J, Casado CM, Pascual C, Alonso B, Lobete F (1996) Organometallic dendritic macromolecules: organosilicon and organometallic entities as core or building blocks. In: Newkome GR (ed) Advances in dendritic macromolecules, vol 3. JAI Press, London, p 151; (e) Constable EC (1997) J Chem Soc Chem Commun 1073
3. Liao Y-H, Moss JR (1996) Organometallics 15:4307
4. Cuadrado I, Morán M, Casado CM, Alonso B, Lobete F, García B, Ibisate M, Losada J (1996) Organometallics 15:5278
5. Slany M, Bardají M, Casanove M-J, Caminade A-M, Majoral J-P, Chaudret B (1995) J Am Chem Soc 117:9764
6. Newkome GR, Cardullo F, Constable EC, Moorefield CN, Cargill Thompson AMW (1993) J Chem Soc Chem Commun 925
7. Newkome GR, Moorefield CN, Baker GR, Johnson AL, Behera RK (1991) Angew Chem Int Ed Engl 30:1176
8. Balzani V, Juris A, Venturi M, Campagna S, Serroni S (1996) Chem Rev 96:759
9. Serroni S, Denti G, Campagna S, Juris A, Ciano M, Balzani V (1992) Angew Chem Int Ed Engl 31:1493
10. (a) Achar S, Puddephatt RJ (1994) J Chem Soc Chem Commun 1895; (b) Achar S, Vittal JJ, Puddephatt RJ (1996) Organometallics 15:43
11. Huck WTS, Vanveggel FGJM, Reinhoudt DN (1996) Angew Chem Int Ed Engl 35:1213
12. Fillaut J-L, Linares J, Astruc D (1994) Angew Chem Int Ed Engl 33:2460
13. Achar S, Immoos CE, Hill MG, Catalano VJ (1997) Inorg Chem 36:2314
14. (a) Jacquet L, Kirsch-De Mesmaeker A (1992) J Chem Soc Faraday Trans 88:2471; (b) Steiger B, Walder L (1992) Helv Chim Acta 75:90; (c) Sessler JL, Capuano VL, Burrell AK (1993) Inorg Chim Acta 204:93; (d) Toma HE, Chavez-Gil TE (1997) Inorg Chim Acta 257:197

15. Flanagan JB, Margel S, Bard AJ, Anson FC (1978) J Am Chem Soc 100:4248
16. Beer PD, Tite EL (1988) Tetrahedron Lett 29:2349
17. Bard AJ (1995) Nature 374:13
18. (a) Moulines F, Djakovitch L, Boese R, Gloaguen B, Thiel W, Fillaut J-L, Delville M-H, Astruc D (1993) Angew Chem Int Ed Engl 32:1075; (b) Astruc D, Valério C, Fillaut J-L, Ruiz J, Hamon J-R, Varret F (1996) Electron-reservoir sandwich complexes: from mono- and bimetallic complexes to molecular trees. In: Kahn (ed) Magnetism: a supramolecular function. Kluwer Academic Publishers, Netherlands, p 107
19. Alonso B, Morán M, Casado CM, Lobete F, Losada J, Cuadrado I (1995) Chem Mater 7:1441
20. Castro R, Cuadrado I, Alonso B, Casado CM, Morán M, Kaifer AE (1997) J Am Chem Soc 119:5760
21. Shu C-F, Shen H-M (1997) J Mater Chem 7:47
22. Cuadrado I, Casado CM, Alonso B, Morán M, Losada J, Belsky V (1997) J Am Chem Soc 119:7613
23. Jutzi P, Batz C, Neumann B, Stammler H-G (1996) Angew Chem Int Ed Engl 35:2118
24. (a) Marvaud V, Astruc D (1997) Chem Commun 773; (b) Marvaud V, Astruc D, Leize E, Dorsselaer AV, Guittard J, Blais J-C New J Chem (in press)
25. Valério C, Fillaut J-L, Ruiz J, Guittard J, Blais J-C, Astruc D (1997) J Am Chem Soc 119:2588
26. (a) Beer P (1992) Adv Inorg Chem 39:79; (b) Beer P, Graydon AR, Johnson AOM, Smith DK (1997) Inorg Chem 36:2112
27. Serroni S, Campagna S, Juris A, Venturi M, Balzani V, Denti G (1994) Gazz Chim Ital 124:423
28. Vlcek AA (1982) Coord Chem Rev 43:39
29. (a) Issberner J, Vögtle F, De Cola L, Balzani V (1997) Chem Eur J 3:706; (b) De Cola L, Credi A (unpublished results)
30. Chow H-F, Chan IY-K, Chan DTW, Kwok RWM (1996) Chem Eur J 2:1085
31. Newkome GR, Güther R, Moorefield CN, Cardullo F, Echegoyen L, Pérez-Cordero E, Luftmann H (1995) Angew Chem Int Ed Engl 34:2023
32. (a) Constable EC, Harverson P (1996) Inorg Chim Acta 252:9; (b) Constable EC, Harverson P (1996) Chem Commun 33
33. Armspach D, Cattalini M, Constable EC, Housecroft CE, Phillips D (1996) Chem Commun 1823
34. Constable EC, Harverson P, Oberholzer M (1996) Chem Commun 1821
35. Campagna S, Denti G, Serroni S, Juris A, Venturi M, Ricevuto V, Balzani V (1995) Chem Eur J 1:211
36. Serroni S, Juris A, Venturi M, Campagna S, Resino IR, Denti G, Credi A, Balzani V (1997) J Mater Chem 7:1227
37. Balzani V, Campagna S, Denti G, Juris A, Serroni S, Venturi M Acc Chem Res (in press)
38. Roffia S, Marcaccio M, Paradisi C, Paolucci F, Balzani V, Denti G, Serroni S, Campagna S (1993) Inorg Chem 32:3003
39. Serroni S, Juris A, Campagna S, Venturi M, Denti G, Balzani V (1994) J Am Chem Soc 116:9086
40. Campagna S, Serroni S, Juris A, Venturi M, Balzani V (1996) New J Chem 20:773
41. Haga M, Ali MM, Arakawa R (1996) Angew Chem Int Ed Engl 35:76
42. Dandliker PJ, Diederich F, Gross M, Knobler CB, Louati A, Sanford EM (1994) Angew Chem Int Ed Engl 33:1739
43. Dandliker PJ, Diederich F, Gisselbrecht J-P, Louati A, Gross M (1995) Angew Chem Int Ed Engl 34:2725
44. Gorman CB, Parkhurst BL, Su WY, Chen K-Y (1997) J Am Chem Soc 119:1141
45. (a) Meyer TJ (1986) Pure & Appl Chem 58:1193; (b) Juris A, Balzani V, Barigelletti F, Campagna S, Belser P, von Zelewsky A (1988) Coord Chem Rev 84:85; (c) Kalyanasundaram K (1992) Photochemistry of polypyridine and porphyrin complexes. Academic Press, London
46. Maestri M, Balzani V, Deuschel-Cornioley C, von Zelewsky A (1992) Adv Photochem 17:1
47. Bar-Haim A, Klafter J, Kopelman R (1997) J Am Chem Soc 119:6197

48. Sadamoto R, Tomioka N, Aida T (1996) J Am Chem Soc 118:3978
49. Sahai R, Rillema DP, Shaver R, Van Wallendael S, Jackman DC, Boldaji M (1989) Inorg Chem 28:1022
50. (a) De Cola L, Barigelletti F, Balzani V, Belser P, von Zelewsky A, Seel C, Frank M, Vögtle F (1991) Coord Chem Rev 111:255; (b) Belser P, von Zelewsky A, Frank M, Seel C, Vögtle F, De Cola L, Barigelletti F, Balzani V (1993) J Am Chem Soc 115:4076
51. Lempers HEB, Haasnoot JG, Reedijk J, Hage R, Weldon FN, Vos JG (1994) Inorg Chim Acta 225:67
52. Lehn JM, Ziessel R (1987) J Chem Soc Chem Commun 1292
53. (a) Campagna S, Denti G, Serroni S, Ciano M, Balzani V (1991) Inorg Chem 30:3728; (b) Denti G, Serroni S, Campagna S, Ricevuto V, Juris A, Ciano M, Balzani V (1992) Inorg Chim Acta 198–200:507; (c) Juris A, Balzani V, Campagna S, Denti G, Serroni S, Frei G, Güdel HU (1994) Inorg Chem 33:1491
54. (a) Murphy WR Jr, Brewer KJ, Gettliffe G, Petersen JD (1989) 28:81; (b) Campagna S, Denti G, Sabatino L, Serroni S, Ciano M, Balzani V (1989) J Chem Soc Chem Commun 1500; (c) Denti G, Campagna S, Sabatino L, Serroni S, Ciano M, Balzani V (1990) Inorg Chem 29:4750
55. Denti G, Campagna S, Sabatino L, Serroni S, Ciano M, Balzani V (1990) Inorg Chim Acta 176:175
56. (a) Serroni S, Denti G, Campagna S, Ciano M, Balzani V (1991) J Chem Soc Chem Commun 944; (b) Denti G, Campagna S, Serroni S, Ciano M, Balzani V (1992) J Am Chem Soc 144:2944
57. Campagna S, Denti G, Serroni S, Ciano M, Juris A, Balzani V (1992) Inorg Chem 31:2982
58. (a) Denti G, Serroni S, Campagna S, Ricevuto V, Balzani V (1991) Inorg Chim Acta 182:127; (b) Denti G, Serroni S, Campagna S, Ricevuto V, Balzani V (1991) Coord Chem Rev 11:227; (c) Serroni S, Campagna S, Denti G, Keyes TE, Vos JG (1996) Inorg Chem 35:4513
59. Cleary RL, Byrom KJ, Bardwell DA, Jeffery JC, Ward MD, Calogero G, Armaroli N, Flamigni L, Barigelletti F (1997) Inorg Chem 36:2601
60. Moucheron C, Kirsch-De Mesmaeker A, Dupont-Gervais A, Leize E, Van Dorsselaer A (1996) J Am Chem Soc 118:12,834
61. (a) Hua X, von Zelewsky A (1991) Inorg Chem 30:3796; (b) Hayoz P, von Zelewsky A, Stoeckli-Evans H (1993) J Am Chem Soc 115:5111; (c) Hua X, von Zelewsky A (1995) Inorg Chem 34:5791
62. (a) Rutherford TJ, Quagliotto MG, Keene FR (1995) Inorg Chem 34:3857; (b) Rutherford TJ, Keene FR (1997) Inorg Chem 36:3580
63. (a)Warnmark K, Thomas JA, Heyke O, Lehn JM (1996) Chem Commun 701; (b) Warnmark K, Heyke O, Thomas JA, Lehn JM (1996) Chem Commun 2603
64. (a) MacDonnell FM, Bodige S (1996) Inorg Chem 35:2601; (b) Bodige S, Torres AS, Maloney DJ, Tate D, Kinsel G, Walker A, MacDonnell FM J Am Chem Soc (in press)
65. Fletcher NC, Keene FR, Viebrock H, von Zelewsky A (1997) Inorg Chem 36:1113
66. Rutherford TJ, Van Gijte O, Kirsch-De Mesmaeker A, Keene FR (1997) Inorg Chem 36:4465
67. MacDonnell FM, Campagna S (manuscript in preparation)
68. Prathapan S, Johnson TE, Lindsey JS (1993) J Am Chem Soc 115:7519
69. Constable EC (1997) Chem Commun 1073
70. Astruc D (1986) Acc Chem Res 19:377
71. Miedaner A, Curtis CJ, Barkley RM, DuBois DL (1994) Inorg Chem 33:5482
72. Knapen JWJ, van der Made AW, de Wilde JC, van Leeuwen PWNM, Wijkens P, Grove DM, van Koten G (1994) Nature 372:659
73. Kalyanasundaram K, Graetzel M (eds) (1993) Photosensitization and photocatalysis using inorganic and organometallic compounds. Kluwer Academic Publishers, Netherlands
74. Amouyal E (1995) Sol Energy Mat Sol Cells 38:249
75. Balzani V, Scandola F (1996) Photochemical and photophysical devices. In: Reinhoudt DN (ed) Comprehensive supramolecular chemistry, vol 10. Elsevier Science, Oxford, p 687
76. Balzani V, Credi A, Venturi M (1997) Current Opinion Chem Biol 1:506

Author Index Volumes 151–197

Author Index Vols. 26–50 see Vol. 50
Author Index Vols. 51–100 see Vol. 100
Author Index Vols. 101–150 see Vol. 150

The volume numbers are printed in italics